WELL LOGGING FOR EARTH SCIENTISTS

WELL LOGGING FOR EARTH SCIENTISTS

DARWIN V. ELLIS

ELSEVIER
NEW YORK • AMSTERDAM • LONDON

Elsevier Science Publishing Co., Inc.
52 Vanderbilt Ave., New York, New York 10017

Sole distributors outside the United States and Canada:
Elsevier Science Publishers B. V.
P.O. Box 211, 1000 AE Amsterdam, The Netherlands

© 1987 by Elsevier Science Publishing Co., Inc.

This book has been registered with the Copyright Clearance Center, Inc. For further information, please contact the Copyright Clearance Center, Salem, Massachusetts.

Library of Congress Catalog Number 87-81289

ISBN 0-444-01180-3

Current printing (last digit):

10 9 8 7 6 5 4 3 2

Manufactured in the United States of America

Dedicated to Ellen, Sam, and Minou for encouragement and patience.

CONTENTS

PREFACE	xiii
1 AN OVERVIEW OF WELL LOGGING	**1**
Introduction	1
What Is Logging?	2
Properties of Reservoir Rocks	5
Well Logging: The Narrow View	6
Measurement Techniques	8
How is Logging Viewed by Others?	9
References	13
2 INTRODUCTION TO WELL LOG INTERPRETATION: FINDING THE HYDROCARBON	**15**
Introduction	15
Rudimentary Well Site Interpretation	15
The Borehole Environment	18
Qualitative Interpretation	20
Reading a Log	22
Examples of Curve Behavior and Log Display	27
A Sample Rapid Interpretation	32
References	34
Problems	34
3 BASIC RESISTIVITY AND SPONTANEOUS POTENTIAL	**37**
Introduction	37
The Concept of Bulk Resistivity	38
Electrical Properties of Rocks and Brines	42
Spontaneous Potential	46
Log Example of the SP	51
References	52
Problems	52
4 EMPIRICISM: THE CORNERSTONE OF RESISTIVITY INTERPRETATION	**57**
Introduction	57
Early Electric Log Interpretation	58
Empirical Approaches to Quantifying R_t	61
Archie's Formation Factor	61
Archie's Synthesis	62

A Review of Electrostatics	65
A Thought Experiment for a Logging Application	66
References	70
Problems	70

5 RESISTIVITY: ELECTRODE DEVICES AND HOW THEY EVOLVED — 73

Introduction	73
Unfocused Devices	73
The Short Normal	73
Estimating the Borehole Size Effect	75
Focused Devices	84
Laterolog Principle	84
Spherical Focusing	87
Pseudogeometric Factor	91
Dual Laterolog Example	94
References	97
Problems	97

6 INDUCTION DEVICES — 99

Introduction	99
Review of Magnetostatics and Induction	100
The Two-Coil Induction Device	105
Geometric Factor for the Two-coil Sonde	107
Focusing the Two-coil Sonde	111
Skin Effect	114
Induction or Electrode?	118
Induction Log Example	119
References	121
Problems	122

7 ELECTRICAL DEVICES FOR MEASUREMENTS OTHER THAN R_t — 123

Introduction	123
Microelectrode Devices	124
Uses for R_{xo}	128
A Note on Reality	138
Characterizing Dielectrics	142
Tools to Measure Formation Dielectric Properties	146
Measurement Technique	147
Log Example	149
Appendix	
Propagation of E–M Waves in Conductive Dielectric Materials	152
References	155
Problems	157

8 BASIC NUCLEAR PHYSICS FOR LOGGING APPLICATIONS: GAMMA RAYS   161
Introduction   161
Nuclear Radiation   162
Radioactive Decay and Statistics   163
Radiation Interactions   166
Fundamentals of Gamma Ray Interactions   167
Attenuation of Gamma Rays   172
Gamma Ray Detectors   173
References   179
Problems   179

9 GAMMA RAY DEVICES   181
Introduction   181
Sources of Natural Radioactivity   183
Gamma Ray Devices   186
Uses of the Gamma Ray Measurement   188
Spectral Gamma Ray Logging   190
    Spectral Stripping   195
A Note on Depth of Investigation   197
References   198
Problems   199

10 GAMMA RAY SCATTERING AND ABSORPTION MEASUREMENTS   201
Introduction   201
Density and Gamma Ray Attenuation   202
Density Measurement Technique   204
Lithology Logging   212
Estimating Porosity from Density Measurements   222
References   223
Problems   224

11 BASIC NEUTRON PHYSICS FOR LOGGING APPLICATIONS   227
Introduction   227
Fundamentals of Neutron Interactions   228
Nuclear Reactions and Neutron Sources   233
Useful Bulk Parameters   234
Neutron Detectors   240
References   241
Problems   242

12 NEUTRON POROSITY DEVICES   243
Introduction   243

Basis of Measurement	244
Measurement Technique	249
The Neutron Porosity Device: Response Characteristics	254
Shale Effect	254
Matrix Effect	256
Gas Effect	259
Depth of Investigation	264
Log Examples	266
Appendix	
Depth of Investigation of Nuclear Logging Tools	269
References	278
Problems	279

13 PULSED NEUTRON DEVICES   281

Introduction	281
Thermal Neutron Die-away Logging	282
Thermal Neutron Capture	282
Measurement Technique	285
Instrumentation and Interpretation	288
Pulsed Neutron Spectroscopy	293
Appendix	
A Solution of the Time-dependent Diffusion Equation	299
References	301
Problems	302

14 NUCLEAR MAGNETIC LOGGING   305

Introduction	305
Nuclear Resonance Magnetometers	306
Why Nuclear Magnetic Logging?	307
A Look at Magnetic Gyroscopes	308
The Precession of Atomic Magnets	309
Paramagnetism of Bulk Materials	310
Some Details of Nuclear Induction	313
Operation of a Conventional Nuclear Magnetic Logging Tool	321
Interpretation	326
Factors Affecting T_1 and T_2	326
Surface Interactions	328
Applications	331
Free Fluid Index	331
Permeability from Surface/Volume Information	331
Measurement of Residual Oil	333
A New Approach	334
References	336

| 15 | INTRODUCTION TO ACOUSTIC LOGGING | 339 |

Introduction — 339
A Short history of Acoustic Logging in Boreholes — 340
Application of Borehole Acoustic Logging — 342
Review of Elastic Properties — 342
Wave Propagation — 349
Rudimentary Acoustic Logging — 353
Rudimentary Interpretation — 355
References — 357
Problems — 358

| 16 | ACOUSTIC WAVES IN ROCKS | 359 |

Introduction — 359
A Review of Laboratory Measurements — 360
Acoustic Models of Rocks — 366
Acoustic Waves in Boreholes — 371
References — 377
Problems — 378

| 17 | ACOUSTIC LOGGING METHODS AND APPLICATIONS | 381 |

Introduction — 381
Transducers — 382
Conventional Sonic Logging — 382
  Some Typical Problems — 390
  Newer Devices — 391
Acoustic Logging Applications — 397
  Lithology and Pore Fluid Identification — 400
  Formation Fluid Pressure — 402
  Mechanical Properties and Fractures — 403
  Permeability — 406
Ultrasonic Devices — 408
References — 412
Problems — 413

| 18 | LITHOLOGY IDENTIFICATION FROM POROSITY LOGS | 417 |

Introduction — 417
Graphical Approach for Binary Mixtures — 418
Combining Three Porosity Logs — 425
Lithology Logging: Incorporating P_e — 430
Numerical Approaches to Lithology Determination — 432
References — 436
Problems — 436

19 CLAY TYPING AND QUANTIFICATION FROM LOGS — 439
Introduction — 439
What is Clay/Shale? — 440
   Distribution of Clay — 444
   Influence on Logging Measurements — 446
Some Traditional Indicators — 448
Some New Methods of Clay Quantification — 451
   Interpretation of P_e in Shaly Sands — 451
   Clay Mineral Parameters and Neutron Porosity Response — 454
   Response of Σ to Clay Minerals — 457
   Aluminum Activation — 458
   Clay Typing (Geochemical Logging) — 459
References — 467
Problems — 468

20 SATURATION ESTIMATION — 471
Introduction — 471
Clean Formations — 472
Shaly Formations — 478
   V_{sh} Models — 480
   Effect of Clay Minerals on Resistivity — 481
   Double Layer Models — 483
   Saturation Equations — 484
References — 486
Problems — 487

21 EXTENDING MEASUREMENTS AWAY FROM THE BOREHOLE — 489
Introduction — 489
Depth of Investigation and Resolution of Logging Measurements — 490
   Nuclear — 490
   Electrical — 491
   Acoustic — 492
Surface Seismics — 494
Borehole Seismics — 498
The VSP — 500
   Rudimentary VSP Processing — 504
   Information from the VSP — 508
   Increasing the Lateral Investigation: Offset VSP — 510
Appendix
   Acoustic Impedance and Reflection Coefficients — 517
References — 519

INDEX — 521

PREFACE

A poll of casual users of the geophysical information derived from well logging might indicate that the two words most frequently associated with well logging are "black magic." An equally disheartening survey conducted among students might produce a single associative word: "boring." It was in this climate that I was invited to accept the challenge to produce a course on well logging to combat these negative associations. This text is my attempt to demystify well logging and to communicate the excitement of the research which underlies these frequently derided geophysical measurements. The reader will have to be the judge of the extent to which I have succeeded in this effort. The text is directed at students heading toward a career in the petroleum industry but should be useful to practitioners who want to know more about the logging measurements than can be extracted from conventional sources. It should also appeal to a growing number of earth scientists who have discovered that sub-surface "field" work can be performed with logging measurements.

This book is the outgrowth of a course taught at Stanford University's School of Earth Sciences during the winter quarters of 1985 and 1986. The students enrolled were primarily graduate students from the Departments of Petroleum Engineering, Geophysics, and Geology. A further critical review of the course was provided by Dr. J. Jensen of Heriot-Watt University whose students in the Master's of Engineering program used a draft copy of the present text for the well logging sequence of their petroleum engineering course.

Until recent years, relatively few books on well logging have appeared, and none were particularly suited for classroom use. The material that was available tended to provide only a superficial discussion of the measurements with little discussion of the underlying physics. My bias in emphasizing the importance of the physics of well logging is clearly reflected throughout the text. However, I have attempted to balance this by including discussions and problems which explain the mechanics of obtaining the usual petrophysical parameters from well log data.

The approach taken is to begin with a survey of the major categories of measurement: electrical, nuclear, and acoustic. The measured physical parameters are related to desired petrophysical parameters. As a first step in the interpretation process the measurements are examined in terms of limitations imposed by practical engineering considerations and the borehole environment in which the logging tools must operate. Later chapters provide discussion and examples of combinations of measurements and the types of interpretations which can be made from them. Throughout, large sections of the text have been set off in italics which may be skipped by the casual reader. These detailed sections are included for those who would like to know more about the particular measurement techniques exploited and may be of more interest to researchers.

Logging can be considered as an experiment. One which is conducted, not in a pristine laboratory environment, but in an uncontrolled and hostile environment with unknown conditions. The goal is to infer petrophysical information from rather indirect means. The extraction of this information sometimes comes from rather unscientific rules of thumb, and other times from straightforward application of physical principles. The elementary extraction of the petrophysical parameters usually desired from logging measurements, porosity, and water saturation, are treated from a historical perspective to show the origin of the rules of thumb and indicate their weaknesses. Directions of current research to refine the estimate of the desired petrophysical properties from logging measurements are also discussed. Problem sets at the end of the chapters are designed either to familiarize the student with the mechanics of obtaining useful answers from the logging measurements or to illustrate non-conventional utilization of logging responses to questions which will hopefully appeal to earth scientists outside the petroleum industry. The message I have tried to communicate is that, if you are interested in the subsurface of the earth, well logging has something to offer.

For the opportunity to develop this course I am indebted to Schlumberger for their permission to devote preparation and teaching time away from the laboratory, to the School of Earth Sciences at Stanford University for welcoming me to their faculty for two consecutive years, and to Prof. G. Stewart for arranging the use of the draft text and for inviting me to participate in the lectures at Heriot-Watt.

For indispensable help in the refinement of this text I would like to thank Jay Tittman for his more than careful reading and rereading of major portions of the manuscript. Additional thanks for reviewing and comments go to J. Robinson of Shell, and Schlumberger colleagues H. Scott, J. Singer, and J. Tabanou.

For relief from the first year's teaching load I would like to thank guest lecturers, P. Day, W. Kenyon, M. Herron, T. Plona, and M. Oristaglio. For efforts in suggesting, locating, and providing figures, references, and generally helpful suggestions I am indebted to: L. Jacobsen, P. Day, J. Howard, M. Herron, J. Jensen, J. Banavar, R. Kleinberg, R. Plumb, S. Chang, T. Timur, J. Kent, T. Obenchain, J. Schweitzer, and C. Straley.

For assistance with grueling details of the physical production, I would like to thank Mary Savoca, C. Green and T. Killpack. And finally, for their hard work and patience in providing the graphics for the text I wish to credit and thank Allison Fazio, R. Giuliano and M. Dalton.

WELL LOGGING FOR EARTH SCIENTISTS

1
AN OVERVIEW OF WELL LOGGING

INTRODUCTION

The French translation of the term *well logging* is *carottage électrique**, literally "electrical coring," a fairly exact description of this geophysical prospecting technique when it was invented in 1927.[1,2] A less literal translation might be "a record of characteristics of rock formations traversed by a measurement device in the well bore." However, well logging means different things to different people. For a geologist, it is primarily a mapping technique for exploring the subsurface. For a petrophysicist, it is a means to evaluate the hydrocarbon production potential of a reservoir. For a geophysicist, it is a source of complementary data for surface seismic analysis. For a reservoir engineer, it may simply supply values for use in a simulator.

The initial uses of well logging were for correlating similar patterns of electrical conductivity from one well to another, sometimes over large distances. As the measuring techniques improved and multiplied, applications began to be directed to the quantitative evaluation of hydrocarbon-bearing formations. Much of the following text is directed toward the understanding of the measurement devices and interpretation techniques developed for this type of formation evaluation.

---

* The French definition was chosen for two reasons: as an acknowledgment of the national origin of well logging and as one of the rare cases in which Anglo-Saxon compactness is outdone by the French.

Although well logging grew from the specific need of the petroleum industry to evaluate hydrocarbon accumulations, it is relevant to a number of other areas of interest to earth scientists. New measurements useful for subsurface mapping have evolved which have applications for structural mapping, reservoir description, and sedimentological identification. The measurements can be used to identify fractures or provide the formation mineralogy. A detailed analysis of the measurement principles precedes the discussion of these applications. In this process, well logging is seen to require the synthesis of a number of diverse physical sciences: physics, chemistry, electrochemistry, geochemistry, acoustics, and geology.

The goal of this first chapter is to discuss well logging in terms of its traditional application to formation hydrocarbon evaluation and to describe the wide variety of physical measurements which address the relevant petrophysical parameters. We begin with a description of the logging process, to provide an idea of the experimental environment in which the measurements must be made.

WHAT IS LOGGING?

The process of logging involves a number of elements, which are schematically illustrated in Fig. 1-1. Our primary interest is the measurement device, or *sonde*. Currently, over fifty different types of these logging tools exist in order to meet various information needs and functions. Some of them are passive measurement devices; others exert some influence on the formation being traversed. Their measurements are transmitted to the surface by means of a special armored cable, referred to as a wire line.

Much of what follows in succeeding chapters is devoted to the basic principles exploited by the measurement sondes, without much regard to details of the actual devices. It is worthwhile to mention a few general points regarding the construction of the measurement sondes. Superficially, they all resemble one another. They are generally cylindrical devices with an outside diameter on the order of 4" or less; this is to accommodate operation in boreholes as small as 6" in diameter. Their length varies depending on the sensor array used and the complexity of associated electronics required. It is possible to connect a number of devices concurrently, forming tool strings as long as 100'.

Some sondes are designed to be operated in a centralized position in the borehole. This operation is achieved by the use of bow-springs attached to the exterior or more sophisticated hydraulically actuated "arms." Some measurements require that the sensor package (in this case called a pad) be in intimate contact with the formation. This is also achieved by the use of a hydraulically actuated back-up arm. Fig. 1-2 illustrates the measurement portion of four different sondes. On the left is an example of a centralized device which uses four actuated arms. There is a measurement pad at the extremity of each arm. Second from the left is an example of a tool which is

Figure 1–1. The elements of well logging: a measurement sonde in a borehole, the wireline, and a mobile laboratory. Courtesy of Schlumberger.

generally kept centered in the borehole by external bow-springs, which are not visible in the photo. The third from the left is a more sophisticated pad device, showing the actuated back-up arm in its fully extended position. The tool on the right is similar to the first device but has an additional sensor pad which is kept in close contact with the formation being measured.

These specially designed instruments, which are sensitive to one or more formation parameters of interest, are lowered into a borehole by a surface instrumentation truck. This mobile laboratory provides the down-hole power to the instrument package. It provides the cable and winch for the lowering and raising of the sonde, and is equipped with computers for data processing, interpretation of measurements, and permanent storage of the data.

Most of the measurements which will be discussed in succeeding chapters are continuous measurements. They are made as the tool is slowly raised toward the surface. The actual logging speeds vary depending on the nature

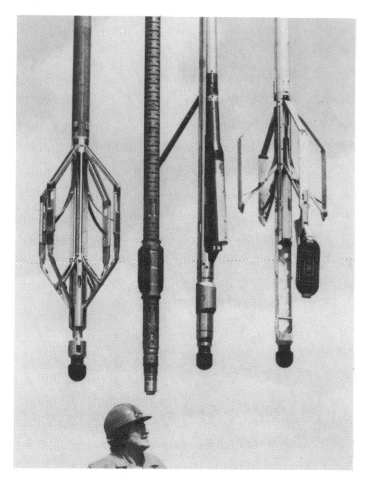

Figure 1-2. Examples of four logging tools. The dipmeter, on the left, has sensors on four actuated arms, which are shown in their fully extended position. It is followed by a sonic logging tool and a density device with its hydraulically activated back-up arm fully extended. The second dipmeter tool has an additional electrode array attached to the bottom of one of its four arms. Courtesy of Schlumberger.

of the device. Measurements which are subject to statistical precision errors or require mechanical contact between sensor and formation tend to be run more slowly, between 600' to 1800' per hour. Some acoustic and electrical devices can be withdrawn from the well, while recording their measurements, at much greater speeds. The traditional sampling provides one averaged measurement for every 6" of tool travel. For some devices that have good vertical resolution, the sampling interval is 1.2". There are special devices with geological applications (such as the determination of depositional environment) which have a much smaller vertical resolution; their data are sampled so as to resolve details on the scale of millimeters.

In the narrowest sense, logging is an alternate or supplement to the analysis of cores, side-wall samples, and cuttings. Although often preferred because of the possibility of continuous analysis of the rock formation over a given interval, economic and technical problems limit the use of cores. Side-wall cores obtained from another phase of wireline operations give the possibility of obtaining samples at discrete depths after drilling has been completed. Side-wall cores have the disadvantage of returning small sample sizes, as well as the problem of discontinuous sampling. Cuttings, extracted from the drilling mud return, are one of the largest sources of subsurface sampling. However, the reconstitution of the lithological sequence from cuttings is imprecise due to the problem of associating a depth with any given sample.

Although well logging techniques (with the exception of side-wall sampling) do not give direct access to the physical rock specimens, they do, through indirect means, supplement the knowledge gained from the three preceding techniques. Well logs provide continuous, in-situ measurements of parameters related to porosity, lithology, presence of hydrocarbons, and other rock properties of interest.

PROPERTIES OF RESERVOIR ROCKS

Before discussing the logging measurements which are used to extract information concerning the rock formations encountered in the borehole, let us briefly consider some of the properties of reservoir rocks, in order to identify parameters of interest and to gain some insight into the reasons for the indirect logging measurements which will be described later. The following description could be modified or augmented depending on the application.[3,4] It would be different when prepared by a geologist, a reservoir engineer, a geophysicist, or a log analyst.

The intergranular nature of the porous medium which constitutes the reservoir rock is fundamental. Above all, the rock must be porous. A measurement of its porosity is of primary consideration.

The rock may be clean or it may contain clays. The clean rock is of a given lithological type which in itself is an important parameter. The presence of clays can affect log readings as well as have a very important impact on the permeability which is a measure of the ease of extraction of fluids from the pore space.

The rock may be consolidated or unconsolidated. This mechanical property will influence the acoustic measurements made and have an impact on the stability of the borehole walls as well as on the ability of the formation to produce flowing fluids.

The formation may be homogeneous or fractured. The existence of fractures, natural or induced, alter the permeability significantly. Thus the detection of fractures and the prediction of the possibility of fracturing is of some importance.

The internal surface area of the reservoir rock is used to evaluate the possibilities of producing fluids from the pore space. It is related to the granular nature, which can be described by the grain size and distribution.

Although we have concentrated, so far, on the properties of the rock, it is usually the contained fluid which is of commercial interest. It is crucial to distinguish between hydrocarbons and brine which normally occupy the pore space. A term frequently used to describe the partitioning of the hydrocarbon and the brine is the "saturation"; the water saturation is the percentage of the porosity occupied by brine rather than hydrocarbons.

In the case of hydrocarbons, it important to distinguish between liquid and gas. This can be of considerable importance not only for the ultimate production procedure but also for the interpretation of seismic measurements, since gas-filled formations often produce distinct reflections.

Although the nature of the fluid is generally inferred from indirect logging measurements, there are wireline devices which are specifically designed to take samples of the formation fluids and measure the fluid pressure at interesting zones. The pressure and temperature of the contained fluids are important for both the drilling and production phases. Overpressured regions must be identified and taken into account to avoid blow-outs. Temperature may have a large effect on the fluid viscosity and thus limit the producibility.

The contained fluids are closely linked to the structural shape of the rock body. It is of importance to know whether the rock body corresponds, for example, to a small river bar of a minor meandering stream or a vast limestone plain. This will have an important impact on the estimates of reserves and the subsequent drilling for production.

WELL LOGGING: THE NARROW VIEW

Well logging plays a central role in the successful development of a hydrocarbon reservoir. Its measurements occupy a position of central importance in the life of a well, between two milestones: the surface seismic survey, which has influenced the decision for the well location, and the production testing. The traditional role of wireline logging has been limited to participation primarily in two general domains: formation evaluation and completion evaluation.

The goals of formation evaluation can be summarized by a statement of four questions of primary interest in the production of hydrocarbons:

- Are there any hydrocarbons?

    First, it is necessary to identify or infer the presence of hydrocarbons in formations traversed by the well bore.

- Where are the hydrocarbons?

    The depth of formations which contain accumulations of hydrocarbons must be identified.

- How much hydrocarbon is contained in the formation?

    An initial approach is to quantify the fractional volume available for hydrocarbon in the formation. This quantity, porosity, is of utmost importance. A second aspect is to quantify the hydrocarbon fraction of the fluids within the rock matrix. The third concerns the areal extent of the bed, or geological body, which contains the hydrocarbon. This last item falls largely beyond the range of traditional well logging but is accessible using recently developed borehole seismic measurements.

- Are the hydrocarbons producible?

    In fact, all the questions really come down to just this one practical concern. Unfortunately, it is the most difficult to answer from inferred formation properties. An approach to this problem is a determination of permeability. Many empirical methods are used to extract this parameter from log measurements with varying degrees of success.

One important characteristic of formation evaluation, as it has evolved, is that it is essentially performed on a well-by-well basis. A number of measurement devices and interpretation techniques have been developed. They provide, principally, values of porosity and hydrocarbon saturation, as a function of depth, generally without much recourse to additional geological knowledge. Because of the wide variety of subsurface geological formations, many different logging tools are needed to give the best possible combination of measurements for the rock type anticipated. Despite the availability of this rather large number of devices, each providing complementary information, the final answers derived are mainly two: the location of hydrocarbon-bearing formations and an estimate of the quantity of hydrocarbon in place in the reservoir.

The second domain of traditional wireline logging is completion evaluation. This area is comprised of a diverse group of measurements concerning cement quality, pipe and tubing corrosion, and pressure measurements, as well as a whole range of production logging services. Although completion evaluation is not the primary focus of this book, some of the measurement techniques used for this purpose, such as clay mineral identification and estimation of rock mechanical properties, are discussed.

MEASUREMENT TECHNIQUES

In the most straightforward application, the purpose of well logging is to provide measurements which can be related to the volume fraction and type of hydrocarbon present in porous formations. Measurement techniques are used from three broad disciplines: electrical, nuclear, and acoustic. Usually a measurement is sensitive primarily to the properties of the rock or to the pore-filling fluid.

The first technique developed was a measurement of electrical conductivity. A porous formation has an electrical conductivity which depends upon the nature of the electrolyte filling the pore space. Quite simply, the rock matrix is nonconducting, and the usual saturating fluid is a conductive brine. Therefore, contrasts of conductivity are produced when the brine is replaced with nonconductive hydrocarbon. Electrical conductivity measurements are usually made at low frequencies. A dc measurement of spontaneous potential is made to determine the conductivity of the brine.

Another factor which affects the conductivity of a porous formation is its porosity. Brine-saturated rocks of different porosity will have quite different conductivities; at low porosity the conductivity will be very low, and at high porosity it can be much larger. Thus in order to correctly interpret conductivity measurements as well as to establish the importance of a possible hydrocarbon show, the porosity of the formation must be known.

A number of nuclear measurements are sensitive to the porosity of the formation. The first attempt at measuring formation porosity was based on the fact that interactions between high energy neutrons and hydrogen reduce the neutron energy much more efficiently than other formation elements. However, it will be seen later that a neutron-based porosity tool is sensitive to all sources of hydrogen in a formation, not just that contained in the pore spaces. This leads to complications in the presence of clay-bearing formations, since the hydrogen associated with the clay minerals is seen by the tool in the same way as the hydrogen in the pore space. As an alternative, gamma ray attenuation is used to determine the bulk density of the formation. With a knowledge of the rock type, more specifically the grain density, it is simple to convert this measurement to a fluid-filled porosity value.

The capture of low energy neutrons by elements in the formation produces gamma rays of characteristic energies. By analyzing the energy of these gamma rays, a selective chemical analysis of the formation can be made. This is especially useful for identifying the minerals present in the rock. Interaction of higher energy neutrons with the formation permit a direct determination of the presence of hydrocarbons through the ratio of C to O atoms.

Nuclear magnetic induction, essentially an electrical measurement, is sensitive to the quantity of free protons in the formation. Thus nuclear magnetic logging furnishes another determination of porosity. Other aspects of nuclear magnetic logging have been applied, with some success, to the prediction of permeability.

Acoustic measurements can yield determinations of the rock elastic parameters. One important factor which determines the propagation of compressional waves is the porosity, and thus acoustic measurements provide yet another measurement of porosity. Acoustic measurements are also useful for detecting fractures. The amplitude of low frequency acoustic waves traveling in the borehole have been linked to permeability.

The one impression which should be gleaned from the above description is that logging tools measure parameters related to but not the same as those actually desired. It is for this reason that there exists a separate domain associated with well logging known as interpretation. Interpretation is the process which attempts to combine a knowledge of tool response with geology, to provide a comprehensive picture of the variation of the important petrophysical parameters with depth in a well.

HOW IS LOGGING VIEWED BY OTHERS?

As the first exhibit, refer to Table 1-1, taken from Serra.[5] It is an abbreviated genealogy of the geological parameters of interest concerning the depositional environment. Bed composition is the only item which is considered in any detail here. It is broken down into the framework and the fluid. The framework must be identified in terms of its mineralogical family. The clay, if present, needs to be quantified. Notice that the common term *matrix* refers, in logging, to the rock formation. Clay, or shale, is treated separately. The fluid content must be separated into water and hydrocarbon. A variety of logging measurements provide quantitative information regarding the final items of the table. In the original table, dozens of logging measurements are shown to be linked to the geological parameters.[5]

The second exhibit, from Pickett is shown in Table 1-2.[6] It indicates some of the applications for borehole measurements in petroleum engineering. The thirteen different applications fall into three fairly distinct categories: identification, estimation, and production. Identification concerns subsurface mapping or correlation. Estimation is the more quantitative aspect of well logging, in which physical parameters such as water saturation or pressure are needed with some precision. The final category consists of well logging measurements which are used to monitor changes in a reservoir during its production phase.

The third and final exhibit, Table 1-3, is a list of well log uses prepared

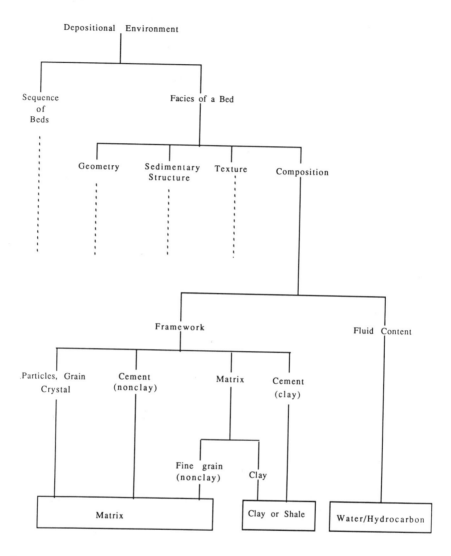

Table 1–1. Geological parameters of interest concerning depositional environment. Only the compositional family is shown in any detail. The final categories are accessible by a wide variety of logging measurements. Adapted from Serra.[5]

Logging Applications for Petroleum Engineering

Rock typing

Identification of geological environment

Reservoir fluid contact location

Fracture detection

Estimate of hydrocarbon in place

Estimate of recoverable hydrocarbon

Determination of water salinity

Reservoir pressure determination

Porosity/pore size distribution determination

Water flood feasibility

Reservoir quality mapping

Inter-zone fluid communication probability

Reservoir fluid movement monitoring

Table 1–2. Uses of well logging in petroleum engineering. Adapted from Pickett.[6]

by a commercial education firm. If taken literally, it demonstrates that everyone needs well logs. To test the validity of this hypothesis, we need to look at the measurements in more detail. To start this analysis, we turn to the historical origins of logging to discover why it was called electrical coring.

USES OF LOGS

A set of logs run on a well will usually mean different things to different people. Let us examine the questions asked — and/or answers sought by a variety of people.

The Geophysicist:

As a Geophysicist what do you look for?
- Are the tops where you predicted?
- Are the potential zones porous as you have assumed from seismic data?
- What does a synthetic seismic section show?

The Geologist:

The Geologist may ask:
- What depths are the formation tops?
- Is the environment suitable for accumulation of Hydrocarbons?
- Is there evidence of Hydrocarbon in this well?
- What type of Hydrocarbon?
- Are Hydrocarbons present in commercial quantities?
- How good a well is it?
- What are the reserves?
- Could the formation be commercial in an offset well?

The Drilling Engineer:

- What is the hole volume for cementing?
- Are there any Key-Seats or severe Dog-legs in the well?
- Where can you get a good packer seat for testing?
- Where is the best place to set a Whipstock?

The Reservoir Engineer:

The Reservoir Engineer needs to know:
- How thick is the pay zone?
- How Homogeneous is the section?
- What is the volume of Hydrocarbon per cubic metre?
- Will the well pay-out?
- How long will it take?

The Production Engineer:

The Production Engineer is more concerned with:
- Where should the well be completed (in what zone(s))?
- What kind of production rate can be expected?
- Will there be any water production?
- How should the well be completed?
- Is the potential pay zone hydraulically isolated?

Table 1–3.   Questions answered by well logs, according to someone trying to sell a well log interpretation course.

References

1. Allaud, L., and Martin, M., *Schlumberger, The History of a Technique*, John Wiley, New York, 1977.
2. Segesman, F. F., "Well Logging Method," *Geophysics*, Vol. 45, No. 11, 1980.
3. Jordan, J. R., and Campbell, F., *Well Logging I - Borehole Environment, Rock Properties, and Temperature Logging*, Monograph Series, SPE, Dallas, 1984.
4. Collins, R. E., *Flow of Fluids through Porous Materials*, Reinhold, New York, 1961.
5. Serra, O., *Fundamentals of Well-Log Interpretation*, Elsevier, Amsterdam, 1984.
6. Pickett, G. R., *Formation Evaluation*, unpublished lecture notes, Colorado School of Mines, 1974.

2
INTRODUCTION TO WELL LOG INTERPRETATION: FINDING THE HYDROCARBON

INTRODUCTION

This chapter presents a general overview of the problem of log interpretation and examines the basic questions concerning a formation's potential hydrocarbon production which are addressed by well logs. The borehole environment is described in terms of its impact on the electrical logging measurements, and all of the qualitative concepts necessary for simple log interpretation are presented.

Without going into the specifics of the logging measurements, the log format conventions are presented, and an example is given which indicates the process of locating possible hydrocarbon zones from log measurements. Although the interpretation example is an exercise in the qualitative art of well log analysis, it raises a number of issues. These relate to the extraction of quantitative petrophysical parameters from the logging measurements. This extraction process is the subject of subsequent chapters. Once these relationships are established, more quantitative procedures of interpretation are described.

RUDIMENTARY WELL SITE INTERPRETATION

Log interpretation, or formation evaluation, requires the synthesis of logging tool response physics, geological knowledge, and auxiliary measurements or information to extract the maximum petrophysical information concerning subsurface formations. In this section, a subset of this procedure is considered: well site interpretation. This subset refers to the rapid and somewhat cursory approach to scanning an available set of logging measurements, and the ability to identify and draw some conclusion about zones of possible interest. These zones, probably hydrocarbon-bearing, will warrant a closer and more quantitative analysis, which is possible only by the inclusion of additional knowledge and measurements.

The three most important questions to be answered by well site interpretation are:

1. Does the formation contain hydrocarbons?
2. If so, what is the quantity present?
3. Are the hydrocarbons recoverable?

In order to see how logging measurements can provide answers to these questions, a few definitions must first be set out. *Porosity* is that fraction of the volume of a rock which is not matrix material and may be filled with fluids. Fig. 2–1 illustrates a unit volume of rock. The pore space has a fractional volume denoted by ϕ, and the matrix material occupies the remaining fraction of the volume, $1 - \phi$.

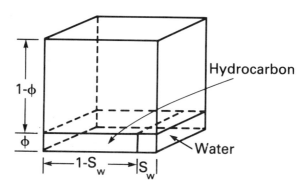

Figure 2–1. A unit volume of formation showing the porosity ϕ and the fractional pore volume of water S_w. The fractional volume of hydrocarbons is $\phi \times (1-S_w)$.

In addition to these fractional volumes, it is useful to use fractions in describing the contained pore fluids. *Water saturation*, S_w, is the fraction of the porosity ϕ which contains water. This fractional volume is also indicated in Fig. 2–1. In an oil/water mixture, the oil saturation, S_o, is given by $1 - S_w$. Note that the fractional volume occupied by the water is given by the product $\phi \times S_w$, and the total fraction of formation occupied by the oil by $\phi \times S_o$.

Since one of the principal logging measurements used for the quantification of hydrocarbon saturation is electrical in nature, it is necessary to mention some of the terminology used to describe these measurements. Electrical measurements are natural for this determination, since current can be induced to flow in a porous rock which contains a conductive electrolyte. The resistivity of a formation is a measure of the ease of electric conduction. Resistivity, a characteristic akin to resistance, is discussed in much more detail later. Replacing the conductive brine of a porous medium with essentially nonconducting hydrocarbons can be expected to impede the flow of current and thus increase its resistivity.

The resistivity of the undisturbed region of formation, somewhat removed from the borehole, is denoted by R_t, or true resistivity. As is implied, the formation resistivity R_t is derived from measurements that yield an apparent resistivity. These measurements can then be corrected, when necessary, to yield the true formation resistivity. In the region surrounding the well bore, where the formation has been disturbed by the invasion of drilling fluids, the resistivity can be quite different from R_t. This zone is called the invaded or flushed zone, and its resistivity is denoted by R_{xo}.

Two other resistivities will be of interest: the resistivity of the brine, R_w, which may be present in the pore space, and the resistivity of the filtrate of the drilling fluid, R_{mf}, which can invade the formation near the well bore and displace the original fluids.

Returning to the three questions which must be addressed by well site interpretation, refer to Fig. 2-2, which attempts to show the interrelationships

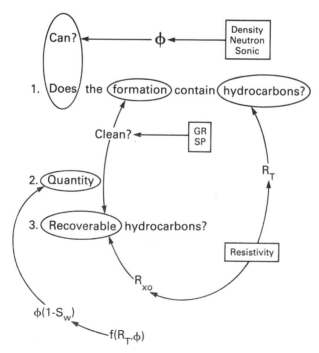

Figure 2-2. A schematic representation of the logging measurements used and the petrophysical parameters determined for answering the basic questions of well site interpretation.

implicit in the questions. Regarding Question 1, the selection of an appropriate zone must be addressed. It is known that shale-free, or clean, formations are much more likely to produce accumulated hydrocarbons. Thus the first task is to identify the clean zones. This task is routinely accomplished through two measurements: the gamma ray, GR, and the spontaneous potential, SP.

The second step is to answer the question: "Can the formation contain hydrocarbons?" This condition will be possible only if the formation is porous. Three curves from three different types of measurements will give porosity information. They are commonly referred to as the density, neutron, and sonic curves.

Once a porous, clean formation is identified, the analyst is faced with deciding whether it contains hydrocarbons or not. This analysis is done in quite an indirect way, using the resistivity R_t of the formation. Basically, if the porous formation contains conductive brine, its resistivity will be low. If, instead, it contains a sizable fraction of nonconducting hydrocarbon, then the formation resistivity will be rather large.

Another common resistivity measurement, R_{xo}, corresponds to the resistivity of the flushed zone, a region of formation close to the borehole, where drilling fluids may have invaded and displaced the original formation fluids. The measurement of R_{xo} is used to get some idea of the recoverability of hydrocarbons in the following way. If the value of R_{xo} is found to be the same as the value of R_t, then the original formation fluids are present in the so-called invaded zone, indicating that no formation fluid displacement has taken place. However, if the resistivity of the invaded or flushed zone corresponds to the resistivity expected for the formation invaded with the drilling fluid, then the drilling fluid has displaced the original fluid (some of which may be hydrocarbon). Therefore the formation fluids are movable and will probably be producible.

In order to determine the quantity of hydrocarbon present in the formation, the product of porosity and saturation ($\phi \times S_w$) must be obtained. For the moment, all that need be known is that the water saturation S_w is a function the both formation resistivity R_t and porosity ϕ.

THE BOREHOLE ENVIRONMENT

The borehole environment in which logging measurements are made is of some interest from the standpoint of logging tool designs and the operating limitations placed upon them. Furthermore, it is important in terms of the disturbance it causes in the surrounding formation in which properties are being measured.

Some characterization of the borehole environment can be made using the following set of generalizations. Well depths are ordinarily between 1,000' and 20,000', with diameters ranging from 5" to 15". Of course, larger ones can exist. A truly vertical hole is rarely encountered, and generally the

Introduction to Well Log Interpretation   19

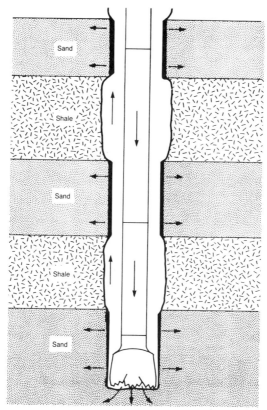

Figure 2-3. Degradation of the formation during and after drilling. Overpressured mud is indicated to be invading porous and permeable sand formations with the formation of a mudcake. The mud circulation also causes borehole washout in the shale zones. From Dewan.[1]

deviation of the borehole is between 0° and 5° onshore, and between 20° and 40° offshore. The temperature, at full depth, ranges between 100° and 300°F.

The drilling fluid, or mud, ranges in density between 9 and 16 lb/gal; weighting additives such as barite ($BaSO_4$) or hematite are added to ensure that the hydrostatic pressure in the well bore exceeds the fluid pressure in the formation pore space to prevent disasters such as blow-outs. The salinity of the drilling mud ranges between 3000 and 200,000 ppm of NaCl. This salinity, coupled with the fact that the well bore is generally overpressured, causes invasion of a porous and permeable formation by the drilling fluid. The general result of the invasion process is conveyed by Fig. 2-3. In the permeable zones, due to the imbalance in hydrostatic pressure, the mud begins to enter the formation but is normally rapidly stopped by the buildup of a mudcake of the clay particles in the drilling fluid.

To account for the distortion which is frequently present with electrical measurements, a simplified model of the borehole/formation has evolved. It considers the invaded formation of interest, of resistivity R_t, to be surrounded by "shoulder" beds of resistivity R_s. The invasion is represented by a step profile which is shown schematically in Fig. 2–4, along with the regions and

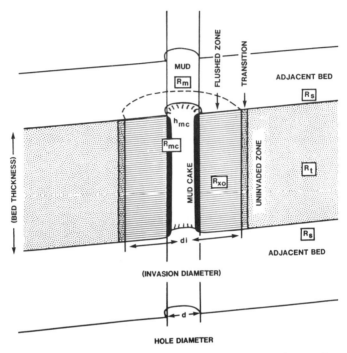

Figure 2–4. Schematic model of the borehole and formation used to describe electric logging measurements and corrections.

parameters of interest, starting with the mudcake of thickness h_{mc} and resistivity R_{mc}. The next annular region of diameter d_i is the flushed zone whose resistivity is denoted by R_{xo}, determined principally by the resistivity of the mud filtrate. Beyond the invaded zone lies the uninvaded or virgin zone with resistivity R_t. A transition zone separates the flushed zone from the virgin zone. Fig. 2–5 indicates schematically the distribution of pore fluids in the uninvaded, transition, and flushed zones.

QUALITATIVE INTERPRETATION

In this rapid overview, the basic rules for interpretation are given, without explanation. The reasons behind them are presented in later chapters, as the measurements made by each type of tool are discussed in detail. For the moment, consider this to be a recipe for answering the three basic questions of Fig. 2–2, from the phenomenological behavior of the measurement curves presented next.

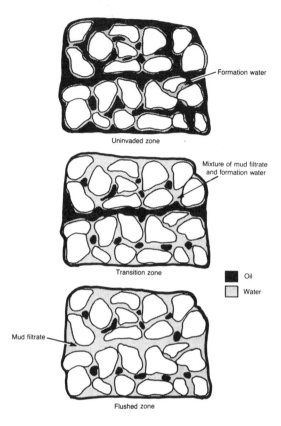

Figure 2–5. Distribution of pore fluids in zones around a well which initially contained hydrocarbons. From Dewan.[1]

In order to assess the formation for shaliness, two indicators can be used: the gamma ray (GR) and the spontaneous potential (SP). The qualitative behavior of the SP (a voltage measurement reported in mV) is to become less negative with increases in formation shale content. The GR signal will generally increase in magnitude according to the increase in shale content. The shale content, or V_{shale}, is quoted as the volume fraction of the formation consisting of shale.

Three logging devices yield a value of porosity. In the case of the *density tool*, the measured parameter is the formation bulk density ρ_b. As porosity increases, the bulk density ρ_b decreases. The *neutron tool* is sensitive to the presence of hydrogen. Its reported measurement is the neutron porosity ϕ_n, which reflects the value of the formation porosity. The *sonic tool* measures the compressional wave transit time Δt (reported in μsec/ft). It will increase for increases in porosity.

The presence of hydrocarbon is inferred from the value of the formation resistivity R_t. As the water saturation increases, R_t decreases because of the presence of relatively conductive water. Conversely, as the oil saturation increases, R_t increases. However, there is also an effect of porosity on the resistivity. As porosity increases, the value of R_t will decrease if the water saturation remains constant.

Movable hydrocarbons are inferred by comparing the resistivity of the flushed zone with that of the virgin zone. If the resistivity of the two zones are the same, there has been no invasion. However, if the ratio of the resistivity in the nearby zone to that in the virgin zone is not the same as the ratio of the mud filtrate resistivity to that of the formation water resistivity, then some type of formation fluid has been displaced.

A summary of these relations is found in Table 2–1.

QUALITATIVE INTERPRETATION

DESCRIPTOR	MEASUREMENT	FUNCTIONAL BEHAVIOR		
CLEAN/SHALEY	SP	$V_{Shale} \uparrow$	$\rightarrow$	SP $\uparrow$
	GR	$V_{Shale} \uparrow$	$\rightarrow$	GR $\uparrow$
POROSITY (ϕ)	DENSITY	$\phi \uparrow$	$\rightarrow$	$\rho_b \downarrow$
	NEUTRON	$\phi \uparrow$	$\rightarrow$	$\phi_N \uparrow$
	SONIC	$\phi \uparrow$	$\rightarrow$	$\Delta t \uparrow$
HYDROCARBON	R_t	$S_w \uparrow$	$\rightarrow$	$R_t \downarrow$
		($S_o \uparrow$	$\rightarrow$	$R_t \uparrow$)
		$\phi \downarrow$	$\rightarrow$	$R_t \uparrow$
RECOVERABLE/ MOVEABLE	R_{xo} vs. R_t (shallow vs. deep)	$R_{xo} = R_t$	$\rightarrow$	no invasion
		$\dfrac{R_{xo}}{R_t} \neq \dfrac{R_{mf}}{R_w}$	$\rightarrow$	Moved fluid

Table 2–1. A summary of phenomenological interpretation.

READING A LOG

Reading a log with ease requires familiarity with some of the standard log formats, which are shown in Fig. 2–6. These are seen generally to contain

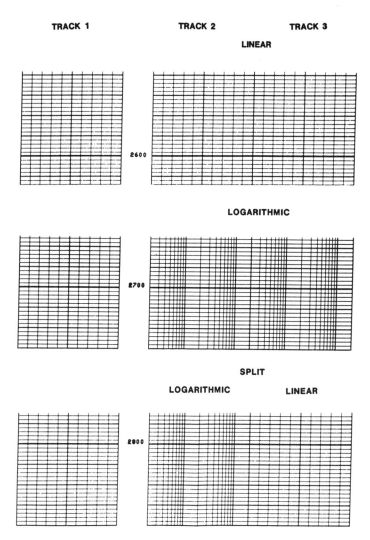

Figure 2–6. Standard log presentation formats.

three tracks. A narrow column containing the depth is found between track 1 and tracks 2 and 3. The latter two are generally contiguous.

The top illustration shows the normal linear presentation, with the grid lines in all three tracks having linear scales. Each track contains ten divisions, and a vertical line is indicated for each 2' of depth. The middle figure shows the logarithmic presentation for tracks 2 and 3. Four decades are drawn to accommodate the electrical measurements, which can have large dynamic ranges. Note that the scale begins and ends on a multiple of two rather than unity. The bottom illustration is a hybrid scale with a logarithmic grid on track 2 and a linear one in track 3. Electrical measurements that may spill over from track 2 into track 3 will still be logarithmic even though the indicated scale is linear.

Fig. 2–7 shows the typical log heading presentation for several of the

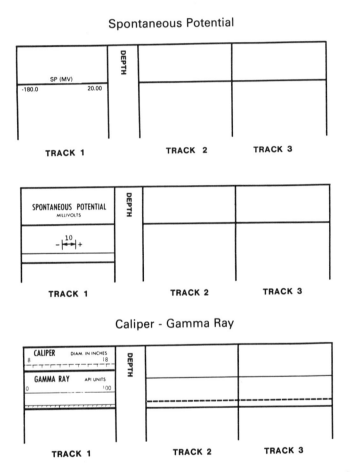

Figure 2–7. Presentation of SP and GR log headings used for clean formation determination.

basic logs that will be used shortly. The upper two presentations show two variations for the SP, which is always presented in track 1. The bottom presentation shows the caliper, a one-axis measurement of the borehole diameter, and the GR, which are also generally presented in track 1. Note that the SP decreases to the left. As the rule given for finding clean sections was that the SP becomes less negative for increasing shale, this will be seen to correspond to deflections of the SP trace toward the right for increasing shale content. The GR curve, as it is scaled in increasing activity (in API units) to the right, will also produce curve deflections to the right for increasing shale content. Thus the two shale indicators can be expected to follow one another as the shale content varies.

One of the resistivity log headings is shown in Fig. 2-8, along with a

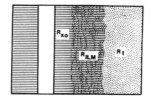

Figure 2-8. The induction log heading and schematic of the formation, with three zones corresponding approximately to the simultaneous electrical measurements of different depths of investigation.

schematic indication of the zones of investigation. The particular tool associated with this format is referred to as the dual induction and will normally show three resistivity traces (the units of which are $\Omega \cdot m$; see Chapter 3). The trace coded for ILD (Induction Log Deep; see Chapter 6) corresponds to the deepest resistivity measurement and will correspond to the value of R_t when invasion is not severe. The curve marked ILM (Induction Log Medium; see Chapter 6) is an auxiliary measurement of intermediate depth of penetration and is highly influenced by the depth of invasion. The third curve, in this case marked SFLU (Spherically Focused Log; see Chapter 7), is a measurement of very shallow depth of investigation and corresponds to the resistivity of the invaded zone R_{xo}. This curve is made with an electrode device and is not properly an integral part of the induction tool but an auxiliary piece of equipment. By combining the three resistivity measurements, it is possible to compensate for the effect of invasion on the ILD reading.

In Fig. 2-9, three typical headings for the three types of porosity devices are indicated. The top heading shows the format for porosities derived from

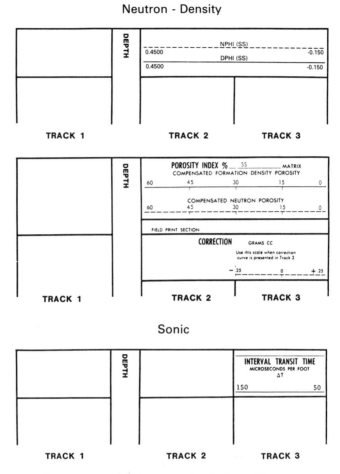

Figure 2-9. Log headings for three porosity devices. The top two correspond to two possible formats for simultaneous density and neutron logs. The bottom is the sonic log format.

neutron and density measurements simultaneously. Sometimes the porosity is expressed as a decimal or in porosity units (PU), each of which corresponds to 1% porosity. In this example, although the scale may vary depending upon local usage, porosity is shown from −15 PU to 45 PU. The middle example shows, in addition, the correction curve for the density log, which can be used to get some idea of the mudcake and rugosity of the borehole encountered during the density measurement. Another convention is to present the density measurement in g/cm^3 usually on a scale of 2.00 to 3.00. The bottom heading of Fig. 2–9 is for the sonic trace with the apparent transit time Δt increasing to the left. In all three presentations, the format is such that increasing porosity produces curve deflections to the left.

For the neutron and density logs, another point to be aware of is the

Introduction to Well Log Interpretation 27

matrix setting. This setting corresponds to a rock type assumed in a convenient preinterpretation that establishes the porosity from the neutron and density device measurements. In both examples shown in Fig. 2–9, the matrix setting is listed as SS, which means that the rock type is taken to be sandstone. If the formations being logged are indeed sandstone, then the porosity values recorded on the logs will correspond closely to the actual porosity of the formation. However, if the actual formation matrix is different, say limestone, then the porosity values will need to be shifted or corrected in order to obtain the true porosity in this particular matrix.

EXAMPLES OF CURVE BEHAVIOR AND LOG DISPLAY

In this section, each of the primary curves to be used in a later section is shown individually, to provide more familiarity with their presentation and behavior with expected changes in lithology and porosity. The first example is the SP, which is shown over a 150' interval in Fig. 2–10. The intervals of

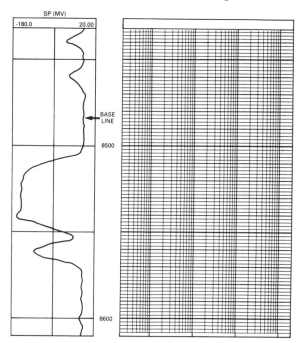

Figure 2–10. An SP log over a clean section bounded by shales.

high SP above 8500' and below 8580' are generally identified with shale sections. The value of the typical flat response is called the shale base line, as indicated on the figure. Sections of log with greater SP deflection (i.e., with a more negative value than the shale base line) are taken as clean, or at least cleaner, zones. One clean section is the zone between 8510' and 8550'.

28   Well Logging for Earth Scientists

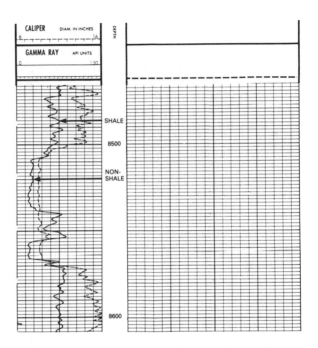

Figure 2–11. A GR and caliper log over the same section as Fig. 2–10.

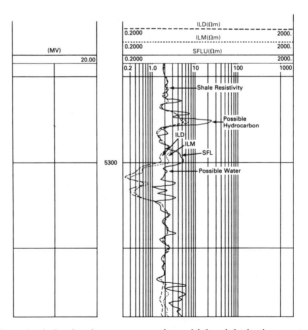

Figure 2–12. An induction log over a section which might be interpreted as a water zone with a hydrocarbon zone above it.

In Fig. 2–11, the caliper (solid) and GR (broken) traces are shown for the same section of the well. Note the similarity between the GR trace of Fig. 2–11 and the SP trace of Fig. 2–10. In the clean sections, the GR reading is on the order of 15 to 30 API units, while the shale sections may read as high as 150 API units. Note also that the caliper, in this example, follows much of the same trend. This trend results from the fact that the shale sections can "wash out," increasing the borehole size compared to the cleaner sand sections that retain their structural integrity.

Fig. 2–12 shows a 150′ section of an induction log. The shallow, deep, and medium depth resistivity curves are indicated. In the zone below 5300′, a possible water zone is indicated. This zone is a possibility because of a number of tacit assumptions. First, it has been assumed that the resistivity of the formation water is much less (i.e., the water is much more saline) than the resistivity of the mud. The effect of the resistivity of the mud can be seen by sighting along the shallow resistivity curve, which for the most part stays around 2 Ω·m. At a depth of 5275′, a possible hydrocarbon zone is noted. It is clear that the deep resistivity reading (ILD) is much greater than in the

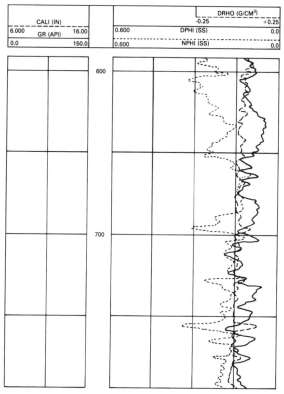

Figure 2–13. Sample neutron and density logs which have been converted to sandstone porosity. The auxiliary curve Δρ indicates little borehole irregularity.

supposed water zone. However, this increase in resistivity may not be the result of hydrocarbon presence. A decrease in porosity could produce the same effect for a formation saturated only with water. The real clue here is that even though the R_{xo} reading has also increased (this indicates that the porosity has decreased), there is less of a separation between the R_{xo} and R_t curves than in the water zone. This means that the value of R_t is higher than should be expected from the porosity change alone. By this plausible chain of reasoning, we are led to expect that this zone may contain hydrocarbons.

Fig. 2–13 shows a typical log of a neutron and density device in combination. In addition to the density porosity estimate (ϕ_d, or DPHI, on the log heading), in solid, and the dotted neutron porosity, the compensation curve $\Delta\rho$ (or DRHO) is also shown. This latter curve is the correction which was applied to the density measurement in order to correct for mudcake and borehole irregularities. It can generally be ignored if it hovers about zero, as is the case in Fig. 2–13 at certain depths. Note, once again, the built-in assumption that the matrix is sandstone. Where the density and neutron-derived porosity values are equal, the presence of liquid-filled sandstone is confirmed. This is the case for the 20′ section below 700′. Separation of the two curves can be caused by an error in the assumed matrix or by the presence of clay or gas.

The presence of gas may be extremely easy to spot from a comparison of the neutron and density logs. In the simplest of cases, gas is indicated in any

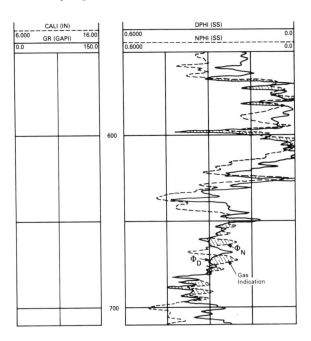

Figure 2–14. A neutron and density log exhibiting the characteristic crossover attributed to the presence of gas in the formation.

zone in which the neutron porosity is less than the density porosity. Fig. 2–14 shows sections which exhibit this behavior. Shale produces the opposite effect; the neutron porosity may far exceed the density porosity, as can be seen in the behavior in Fig. 2–15.

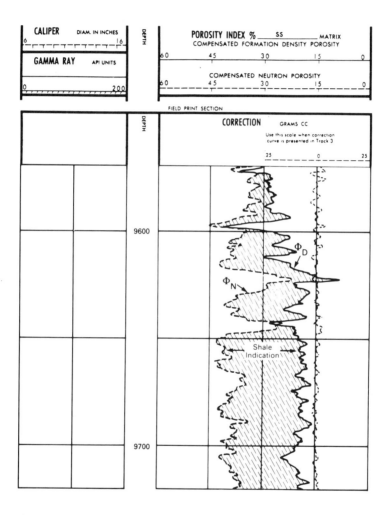

Figure 2–15. The signature of shale on a neutron and density combination log.

All of these generalities are true only if the principal matrix corresponds to the matrix setting on the log. The effect of having the wrong matrix setting on the log (or having the matrix change as a function of depth) is shown in Fig. 2–16. Several sections show negative density porosity. These are probably due to anhydrite streaks, which, because of their much higher density, are misinterpreted as a negative porosity.

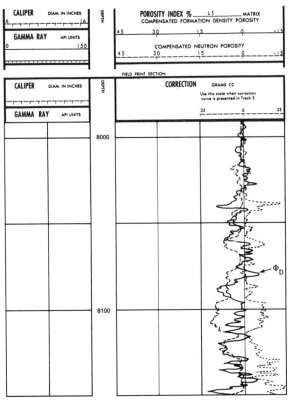

Figure 2–16. Neutron and density crossover caused by changes in lithology.

A SAMPLE RAPID INTERPRETATION

In this section the step-by-step process of identifying interesting zones for possible hydrocarbon production is traced. In analyzing the set of basic logs available, the first step is to identify the clean and possibly permeable zones. This is done by an inspection of the SP and GR curves. In Fig. 2–17, the SP curve has been used to delineate four clean, permeable zones which have been labeled A through D. For further confirmation that these zones are relatively clean, an inspection of the GR curve also shows a minimum of natural radioactivity associated with them.

In the next step, the resistivity readings in the four selected zones are examined. These curves are contained in the second track of Fig. 2–17. The first thing that is obvious is that the resistivity readings are roughly constant in the delineated zones, except in zone C, where a difference occurs between the lower and upper portions. For this reason, zone C is further subdivided into a zone of very low resistivity (≈ 0.2 $\Omega \cdot$m) at the bottom and a deep

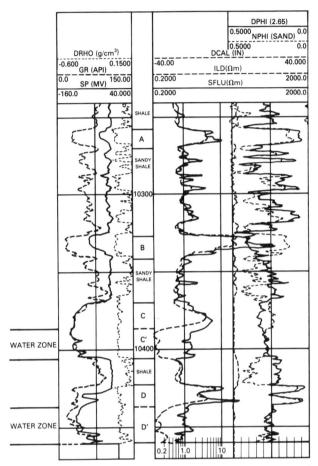

Figure 2–17. A basic set of logs for performing a well site interpretation.

resistivity of about 4.0 Ω·m in the upper portion. A similar delineation can be made for zone D.

A first estimate of fluid content can be established by looking at the lowest resistivity values and identifying them as water, as has been done in the figure. Then the zones such as A, B, C, and D may be suspected to be hydrocarbon-bearing. For the case of zone C, compared to C′, this seems clear. With reference to the porosity values in track 3, it is seen that the porosity over these two zones is approximately constant. In this case, the increase in resistivity in the upper zone, compared to the lower, suggests the presence of hydrocarbons. The case for zone D is not quite so clear. According to the neutron and density curves, the porosity has been considerably reduced in the transition between zone D′ and D. Perhaps the increased resistivity is due to a purely water-saturated low porosity formation and not hydrocarbon.

A careful look at the neutron and density curves in track 3 can yield some additional information. Notice the crossover between the neutron and density curves in zone C. This is indicative of the presence of gas. From this, it is now quite certain that the high resistivity in this zone is indeed the result of the presence of hydrocarbon, and that the hydrocarbon is gas. The same conclusion can be drawn for zone B, which shows an even greater neutron-density separation resulting from gas. The high resistivity streaks of zone D are still questionable. There is no evidence of gas from the neutron/density presentation in this zone, and thus the high resistivity value may simply be due to the reduced porosity. Any further speculation will depend on the ability to be more quantitative in the analysis.

A more quantitative approach to interpretation is developed through the next few chapters. One of the first questions which occurs to the observant analyst is how the porosity was actually determined in this example. For this quantity to be determined, some information is needed to identify the lithology. In the example just given, the matrix was specified as sandstone, from some prior knowledge perhaps. However, what would have been the conclusions if in fact the rock were mainly dolomite?

Another question which needs to be explored is the relationship between the resistivity of a water-saturated rock and its porosity. It has been noted that the resistivity of a porous rock sample can increase if the water is replaced by hydrocarbon or if the porosity is reduced. This relationship must be quantified in order to unscramble the effects of changing these two variables simultaneously.

References

1. Dewan, J. T., *Essentials of Modern Open-hole Log Interpretation*, PennWell Publishing Co., Tulsa, 1983.

2. Scholle, P. A., Bebout, D. G., and Moore, C. H., *Carbonate Depositional Environments*, AAPG Memoir 33, Tulsa, 1983.

Problems

1. Compute the porosity of a formation composed of uniform spherical grains of radius r arranged in the most "open" cubic packing. (The unit cube with side of length $2r$ spans eight grains; see Fig. 2–18.)

2. If the formation were composed of the nearly spherical plankton of Fig. 2–19, what would the porosity be for cubic packing? The spherical void at the center of each plankton seems to have a radius which is $\approx \frac{9}{10}$ of the total particle radius.

Introduction to Well Log Interpretation   35

Figure 2–18. Cubic or open packing of uniform-sized spherical particles.

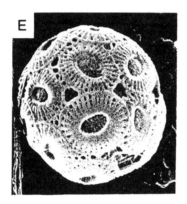

Figure 2–19. Photomicrograph of spherical plankton which contain a nearly spherical void. From Scholle et al.[2]

3. Most sandstone formations have porosities well below 30%. Can you suggest several reasons why this is the case?

4. What is the porosity (or liquid volume fraction) of an 11-lb/gal mud, assuming that it consists of water and clay particles of density 2.65 g/cm^3? The density of water is 8.3 lb/gal (1.00 g/cm^3).

5. In the formation of a mudcake, an annulus of lower porosity mud is formed by expelling some of the water in the mud into the formation. A typical mudcake density is 2.0 g/cm^3. What is its porosity, assuming that it was formed from the 11-lb/gal mud from the previous example?

6. The volume of water expelled from the mud, during the creation of the mudcake, will displace the formation fluid, creating the so-called invasion zone. The thickness of this zone, in which the formation fluid has been displaced by mud filtrate, will depend on the formation porosity. Show that the radius of invasion r_i is given by:

$$r_i^2 = \frac{1}{\pi}\left(\frac{dV}{\phi} + \pi r_{bh}^2\right),$$

where dV is the volume of mud filtrate/unit length displaced into the formation, ϕ is the porosity, and r_{bh} is the borehole radius.

7. Suppose that a mudcake of 40% porosity has been formed on the inside of a 6″ borehole, from a mud of 80% porosity. If the mudcake thickness is 1/2″, what is the diameter of invasion in a 20% porous formation? What is the diameter of invasion in a 2% porosity formation?

3
BASIC RESISTIVITY AND SPONTANEOUS POTENTIAL

INTRODUCTION

The preceding chapter showed through example that an important component of the well logging suite is the measurement of electrical properties of the formation. These measurements deal with the resistivity of the formation or the measurement of spontaneously generated voltages. These voltages are the result of an interaction between the borehole fluid and the formation with its contained fluids.

Historically, the first logging measurements were electrical in nature. The first log was a recording of the resistivity of formations as a function of depth and was drawn painstakingly by hand. Unexpectedly, in the course of attempting to make other formation resistivity measurements, "noise" was repeatedly noted and was finally attributed to a spontaneous potential. It seemed most notable in front of permeable formations. Both of these measurements are still performed on a routine basis today, and their physical basis will be explored in this chapter.

In this chapter the concept of a bulk property of materials, known as resistivity, is examined. It is a quantity related to the more familiar resistance. The contrast in resistivity between relatively insulating hydrocarbons and the conductive formation brines is the basis for hydrocarbon detection. The quantitative relationships between resistivity and hydrocarbon saturation are taken up in the next chapter. Here, the electrical characteristics of rocks and brines are reviewed, including the temperature and salinity dependence of electrolytic conduction, which is of great importance in hydrocarbon saturation determination. The final section of the chapter is an elementary presentation of the physical mechanisms responsible for the generation of the spontaneous potential observed in boreholes.

THE CONCEPT OF BULK RESISTIVITY

In order to understand the basic resistivity measurements used in standard logging procedures, the notion of resistivity is reviewed. It is a general property of materials, as opposed to resistance, which is associated with the geometric form of the material.

The familiar expression of Ohm's law:

$$V = IR$$

indicates that a current I flowing through a material with resistance R is associated with a voltage drop V. The more general form of this equation, used as an additional relationship in Maxwell's equations, is:

$$\overline{J} = \sigma \overline{E},$$

where $\overline{J}$ is the current density, a vector quantity; $\overline{E}$, is the electric field; and the constant of proportionality σ is the conductivity of the material. Resistivity, a commonly measured formation parameter, is defined as the inverse of conductivity:

$$\text{Resistivity}^* \equiv \rho = \frac{1}{\sigma}$$

and is an inherent property of the material.

To comprehend the concept of resistivity, consider the case of a very dilute ionized gas contained between two plates of area A, as illustrated in Fig. 3–1. The charge carriers are indicated to be moving under the influence of an applied electric field E, at an average drift velocity v_{drift}. The drift velocity can be estimated from the fact that the particles are accelerated in the applied electric field until they collide with another particle, at which time they are brought to rest and begin the process again. The mean time between collisions, τ, is the parameter of interest, since the drift velocity can be seen to be:

$$v_{drift} = \frac{F}{m} \tau, \qquad (1)$$

where the term $\frac{F}{m}$ represents the acceleration of the charge carriers of mass

---

[*] In this chapter, resistivity is denoted by ρ, and resistance by R. In later chapters, R will frequently be used to denote resistivity, as is done in most logging publications.

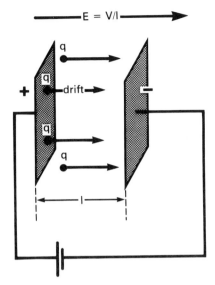

Figure 3-1. A dilute gas with particles of charge q, drifting under the influence of the electric field.

m, subject to a force F. In this case the force applied, F, is equal to the product of the charge and the electric field (qE).

A general expression for the drift velocity of a particle under the influence of an outside force F is:

$$v_{drift} = \mu F,$$

where the constant of proportionality μ is referred to as the mobility of the particle in question in a specified medium. By reference to Eq. (1) it can be seen that for the case of a dilute gas the mobility is given by:

$$\mu = \frac{\tau}{m}.$$

To illustrate the relationship between resistivity and resistance, an expression will be written for the current flowing in the system of Fig. 3-1, in a form which resembles Ohm's law. To compute the current, note that it is the charge collected per unit time. Fig. 3-2 illustrates the region of space containing charges that will reach the plate on the right during a time Δt; the thickness of this region is $v_{drift} \times \Delta t$. The number collected during the time interval Δt is $n_i v_{drift} \Delta t A$, where n_i is the particle density (number of charge carriers per unit volume) and A is the surface area of the electrode. The current is given by:

$$I = \frac{n_i v_{drift} \Delta t A}{\Delta t} q.$$

The relation for drift velocity is:

$$v_{drift} = \mu F = \mu q \frac{V}{l},$$

since the electric field strength is given by the voltage drop per unit length.

Combining these two relations results in the following expression for the current:

$$I = \frac{n_i \mu q \frac{V}{l} \Delta t A}{\Delta t} q,$$

which, when compared with Ohm's law, $I = \frac{1}{R} V$, indicates that the resistance of the geometry illustrated in Fig. 3–2 is given by:

$$R = \frac{1}{n_i \mu q^2} \frac{l}{A}.$$

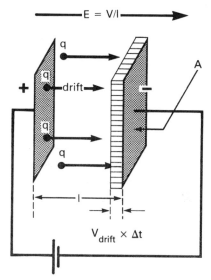

Figure 3–2. The region of space with indicated thickness $v_{drift} \times \Delta t$ is swept of charged particles in a time Δt, contributing to the current.

From this expression it is clear that the resistance R is composed of two parts, one which is material dependent ($\frac{1}{n_i \mu q^2}$) and a second which is purely geometric (the length of the sample divided by the surface area of the contact plates). Resistivity, ρ, is in fact, this first factor:

$$R = \frac{1}{n_i \mu q^2} \frac{l}{A} = \rho \frac{l}{A} \,. \tag{2}$$

It follows that the dimensions of resistivity are Ohms-m/m², or Ω·m. As the illustration of Fig. 3–3 indicates, a material of resistivity 1 Ω·m with

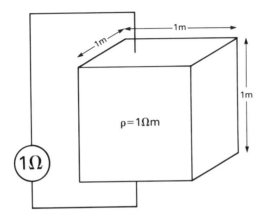

Figure 3–3. A 1–meter cube of characteristic resistivity 1 Ω·m has a resistance of 1 Ω face-to-face.

dimensions of 1 m on each side will have a total resistance, face-to-face, of 1 Ohm. Thus a system to measure resistivity would consist of a sample of the material to be measured contained in a simple fixed geometry. If the resistance of the sample is measured, the resistivity can be obtained from the relation:

$$\rho = R \times \frac{A}{l} \,,$$

which becomes, using Ohm's law:

$$\rho = \frac{V}{I} \frac{A}{l} = k \frac{V}{I} \,.$$

This constant k, referred to as the system constant, converts the measurement of a voltage drop V, for a given current I, into the resistivity of the material.

The practical exploitation of such a system is shown in Fig. 3–4, which shows the so-called mud cup into which a sample of drilling fluid can be placed for the determination of its resistivity. From the dimensions given in the figure, the system constant can be calculated to be 0.012 m. The resistivity ρ in Ω·m, is then obtained from the measured resistance R by:

$$\rho = R \frac{A}{l} = R \times 0.012 \,.$$

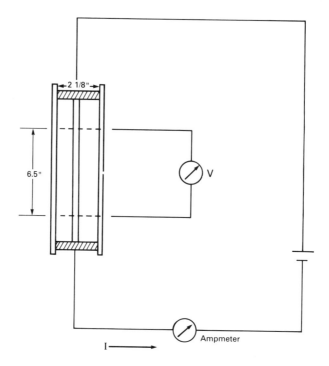

Figure 3-4. A schematic diagram of a mud cup, used for determining the resistivity of a mud sample. A current, I, is passed through the sample and the corresponding voltage, V, is measured.

For this particular measuring device, a sample of salt water with a resistivity of 2 Ω·m in the chamber would yield a total resistance of 166 Ω.

ELECTRICAL PROPERTIES OF ROCKS AND BRINES

There are two general types of conduction of interest to us: electrolytic and metallic. In electrolytic conduction, the mechanism is dependent upon the presence of dissolved salts in a liquid such as water. Examples of electronic conduction are provided by metals, which are not covered here.

The following table illustrates the resistivity of some typical materials. Notice the range of resistivity variation for salt water, which depends on the concentration of NaCl. Typical rock materials are in essence insulators. The fact that reservoir rocks have any detectable conductivity is usually the result of the presence of electrolytic conductors in the pore space. In some cases, the resistivity of a rock may result from the presence of metal, graphite, metal sulfides, or clays. The table shows that the resistivity of formations of interest may range from 0.5 to 10^3 Ω·m, nearly four orders of magnitude.

TYPICAL RESISTIVITY VALUES*	
MATERIAL	**RESISTIVITY ($\Omega \cdot m$)**
Marble	$5 \times 10^7 \rightarrow 10^9$
Quartz	$10^{12} \rightarrow 3 \times 10^{14}$
Petroleum	2×10^{14}
Distilled Water	5×10^3
Saltwater (15°C)	
2 Kppm	3.40
10	.72
20	.38
100	.09
200	.06
TYPICAL FORMATIONS	
Clay/Shale	$2 \rightarrow 10$
Salt-Water Sand	$0.5 \rightarrow 10$
Oil Sand	$5 \rightarrow 10^3$
"Tight" Limestone	10^3

The conductivity of sedimentary rocks is primarily of electrolytic origin. It is the result of the presence of water or a combination of water and hydrocarbons in the pore space in a continuous phase. The actual conductivity will depend on the resistivity of the water in the pores and the quantity of water present. To a lesser extent, it will depend on the lithology of the rock matrix, its clay content, and its texture (grain size and the distribution of pores, clay, and conductive minerals). Finally, the conductivity of a sedimentary formation will depend strongly on temperature.

Fig. 3-5 graphically presents the resistivity of saltwater (NaCl) solutions as a function of the electrolyte concentration and temperature. According to

---

* From Tittman.[6]

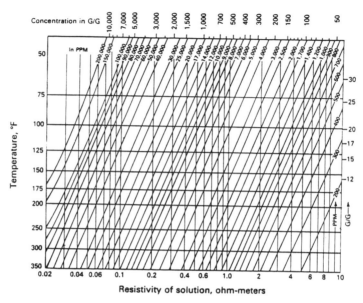

Figure 3–5. A nomogram for determining the resistivity of an NaCl solution as a function of the NaCl concentration and temperature. From Schlumberger.[1]

the preceding analysis, the resistivity is expected to depend inversely on the charge carrier concentration:

$$\rho \propto \frac{1}{nq^2\mu}.$$

To see that this is nearly the case, look at the figure to determine resistivity for concentrations of 4000 and 40,000 ppm. At a temperature of 100°F, the resistivities are 0.12 and 1.0 $\Omega \cdot$m , or nearly in the ratio expected. However, the temperature dependence, which is seen from the chart to be rather substantial, is not explicitly given by the simple expression for resistivity in Eq. (2). It was derived for the case of a dilute gas, a medium which is rather different from a saline solution. In the latter case, the interactions of the charge carriers with one another and with the medium in which they are found cannot be ignored.

An explanation for the temperature dependence comes from a consideration of viscosity. Fig. 3–6 shows a setup to measure the effects of viscosity. A film of liquid of thickness t is contained between two plates of surface area A. The bottom plate is fixed, and a force is applied to the top plate in order to move it parallel to the bottom plate. Experimentally it is found that for a given liquid film, the force F necessary to achieve the velocity v_o is directly proportional to the velocity, the surface area of the plate being dragged, and inversely to the thickness of the film. The constant

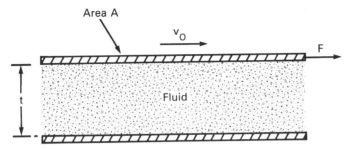

Figure 3–6. Relative motion between two parallel plates separated by a liquid film will require overcoming a drag force determined by the viscosity of the liquid. Adapted from Feynman.[7]

of proportionality η is the viscosity, and the experimental relationship is expressed as:

$$F = \eta \frac{v_o A}{t},$$

or

$$\frac{F}{A} = \eta \frac{v_o}{t}.$$

A practical application of this concept, known as Stoke's law, predicts that the viscous force on a spherical object of radius a is given by:

$$F = 6\pi \eta a v,$$

where v is the velocity of the object. In this case, it refers to the electrolytic particles in solution.

From the analysis of the ionized gas resistivity, it was seen how the mobility, μ, entered into the final expression:

$$v_{drift} = \mu F \rightarrow R = \frac{1}{n\mu q^2} \frac{l}{A}.$$

If the electrolytic particle is considered to be a sphere of radius a, then from Stoke's law the drift velocity would be given by:

$$v_{drift} = \frac{1}{6\pi \eta a} F,$$

and the resistivity should be given by:

$$R = \frac{6\pi \eta a}{nq^2} \frac{l}{A}. \tag{3}$$

The temperature dependence of resistivity for an electrolytic conductor comes from the viscosity factor in Eq. (3). The liquid's viscosity has a strong temperature dependence; unlike the case of ionized gas, it decreases with increasing temperature. In the case of a liquid, viscosity is the result of strong intermolecular forces which impede the relative motion of fluid layers. As the temperature increases, the kinetic energy of the molecules helps to overcome the molecular forces so that viscosity decreases. The experimental temperature dependence of viscosity for many liquids, such as water, can be described by an expression like:

$$\eta = \eta_o e^{\frac{C}{T}},$$

where C is characteristic of a given liquid.

SPONTANEOUS POTENTIAL

Spontaneous potential was shown in the last chapter to be of considerable practical use in the determination or identification of permeable zones. The origins of the spontaneous potential involve both electrochemical potentials and the cation selectivity of shales.

Electrochemical potentials of interest to the generation of the spontaneous potential are the *liquid junction potential* and the *membrane potential*. Fig. 3–7 schematically illustrates the situation for the generation of the liquid-junction potential. To the left is a saline solution of low NaCl concentration. To the right is one of a higher concentration, as indicated by the graph of electrolyte number density n(x) as a function of position. To add a note of realism, imagine the borehole, filled with a fluid of low salinity, to the far left of the figure. The first zone will then correspond to a permeable invaded zone, and the second region, to the undisturbed formation with water of greater salinity.

Because of the particle concentration gradient, there will be a diffusion of both Na and Cl ions from the region of higher concentration to that of lower concentration. An approximation of the diffusion process, known as Fick's law, is given by:

$$J_{diffusion} = -D\frac{dn}{dx},$$

where the current density of diffusing particles is $J_{diffusion}$. The diffusion constant D can be shown to be related to the mobility of the ions and the temperature, so that one can write:[8]

$$J_{diffusion} = -\mu kT \frac{dn}{dx}.$$

Because the Na and Cl ions have different mobilities, with $\mu_{Cl} > \mu_{Na}$, the diffusion current will tend to produce a charge separation. The higher mobility Cl ions will more readily migrate to the region of lower concentration and tend to create an excess negative charge to the left and a net excess positive charge to the right, as indicated in the lower half of the figure. The current which produces this charge separation can be written as:

Liquid Junction Potential

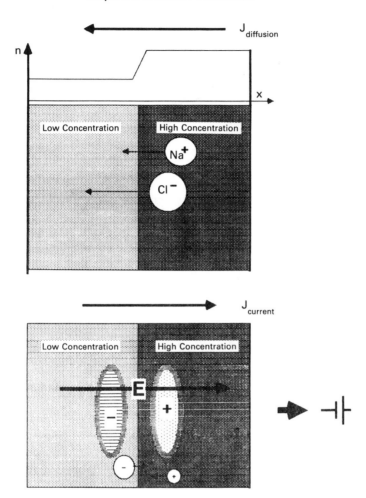

Figure 3–7. Schematic representation of the mechanism responsible for the generation of the liquid-junction potential. A concentration gradient results in diffusion. The mobility difference between Na⁺ and Cl⁻ causes a charge separation.

$$J_{sep} = -(\mu_{Cl} - \mu_{Na})kT\frac{dn}{dx} \ . \qquad (4)$$

This charge separation will induce an electric field of strength E, which in turn causes a current flow of the negative Cl ions to the right and positive Na ions to the left, as indicated in the figure. This electrical current can be expressed as:

$$J_{current} = \sigma_{Cl}\overline{E} + \sigma_{Na}\overline{E}. \tag{5}$$

The conductivities in Eq. (5) are proportional (κ is the constant of proportionality) to the number density of charge carriers and their mobilities. Thus the electrical current is:

$$J_{current} = \kappa n(\mu_{Cl} + \mu_{Na})E . \tag{6}$$

For this condition to remain stable, the two currents, the separation portion of the diffusion current, J_{sep}, and the electrical current, must balance. This leads to the following relation:

$$(\mu_{Cl} - \mu_{Na})kT\frac{dn}{dx} = \kappa n (\mu_{Cl} + \mu_{Na})E .$$

This expression can be rearranged and integrated to get a voltage drop from the electric field term:

$$\frac{(\mu_{Cl}-\mu_{Na})}{\kappa(\mu_{Cl} + \mu_{Na})} kT\int\frac{dn}{n} = \int E\, dx, \tag{7}$$

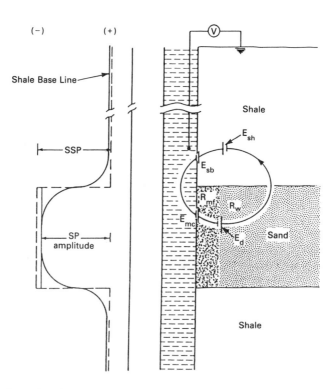

Figure 3–8. A schematic representation of the development of the spontaneous potential in a borehole. Adapted from Dewan.[2]

where the integration is performed over a dimension consistent with the particle density gradient. The liquid junction potential V_{l-j} is the expression on the right side of Eq. (7).

The integration results in the liquid junction potential being a logarithmic ratio of the particle concentrations in the two regions:

$$V_{l-j} = c\, T \ln \frac{n_{hi}}{n_{lo}}.$$

The equivalent electrical cell that such an effect generates is indicated to the right of Fig. 3–7. As is often the case, the resistivity of the drilling mud filtrate (R_{mf}) is greater than the resistivity of the formation water (R_W), so that the above equation can be written in the form:

$$V_{l-j} = -c' \log_{10} \frac{R_{mf}}{R_w},$$

which is widely used for estimating the water resistivity from the deflection of the SP caused by the liquid junction.

Fig. 3–8 is a schematic representation of the circuit producing the SP. The cell marked E_d corresponds to the liquid junction potential just discussed and is sketched with the polarity corresponding to a higher electrolyte concentration in the formation water than in the mud filtrate. As can be seen from the figure, an additional source of the spontaneous potential is associated with the shale. This second component of the SP is the result of the membrane potential generated in the presence of the shale, which has a large negative surface charge.

Fig. 3–9 shows a simplified setup for evaluating the membrane potential when a semipermeable shale barrier separates the solutions of two different salinities. The natural diffusion process is impeded because of the negative surface charge of the shale. The Cl ions which otherwise would diffuse more readily are prevented from traversing the shale membrane, whereas the less mobile Na ions can pass through it readily. The result is that the effective mobility of the chlorine in this case is reduced to nearly zero.

The diffusion current can thus be expressed as:

$$J_{diffusion} = -D\frac{dn}{dx} = -\mu_{Na}kT\frac{dn}{dx},$$

where only the Na mobility figures in the expression. As in the case of the liquid-junction potential, there will be a charge separation. However, this time there will be a positive charge accumulation to the left or low concentration side, which will tend to cause the Na ions to flow back to the region of higher concentration.

The potential generated from this effect can be computed in a manner similar to the previous case. It is obvious that this potential will be somewhat larger than the liquid-junction potential contribution, since the

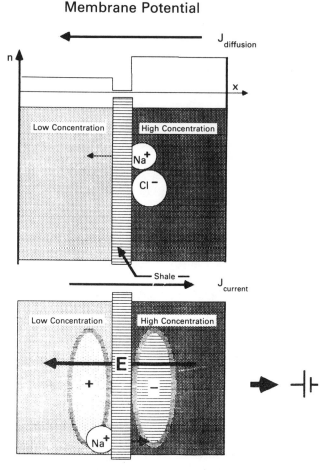

Figure 3-9. A schematic representation of the mechanism responsible for the generation of the membrane potential. The diffusion process is altered by the selective passage of Na⁺ through the shale membrane.

absolute value of the mobilities enters in, rather than a difference. Fig. 3-8 shows the two contributions to the SP and how in this case they are in addition. Numerous charts exist for the determination of R_w from a knowledge of R_{mf}, after correction of the SP reading for the effects of temperature, and for the presence of salts other than NaCl.[1] Reference 3 discusses other perturbations of the SP measurement.

In the best of cases, the measurement of the SP allows the identification of permeable zones and the determination of the formation water resistivity. A deflection indicates a permeable zone because of the establishment of the liquid junction potential, which implies that an invaded zone exists and that ionic conduction within the interconnected porosity is taking place.

LOG EXAMPLE OF THE SP

The measurement of the SP is probably the antithesis of the high-tech image of many of the logging techniques to be considered in subsequent chapters. The sensor is simply an electrode (mounted in the insulated "bridle" some tens of feet above any other measurement sondes) which is referenced to ground at the surface, as indicated in Fig. 3–8. The measurement is essentially a dc voltage measurement which is slowly time-varying as the electrode passes in front of various formations.

To illustrate some of the characteristic behavior to be anticipated by the SP measurement on logs, refer to Fig. 3–10. In the left panel of this figure, a sequence of shale and clean sand beds is represented, along with the idealized response. The shale base line is indicated, and deflections to the left correspond to increasingly negative values. In the first sand zone, there is no SP deflection since this case represents equal salinity in the formation water and in the mud filtrate. The next two zones show a development of the SP which is largest for the largest contrast in mud filtrate and formation water resistivity. In the last zone, the deflection is seen to be to the right of the shale base line and corresponds to the case of a mud filtrate which is saltier than the original formation fluid.

The second panel of Fig. 3–10 illustrates several cases, for a given

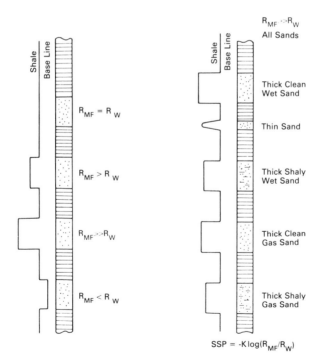

Figure 3–10. Schematic summary SP curve behavior under a variety of different logging circumstances commonly encountered. From Asquith.[4]

contrast in mud filtrate salinity and formation water salinity, where the SP deflection will not attain the full value seen in a thick, clean sand. (This latter value is referred to as the static SP, or SSP.) The first point is that the deflection will be reduced if the sand bed is not thick enough. Depending on invasion and the contrast between filtrate and water salinity, the bed thickness needs to be more than 20 times the borehole diameter to attain its full value. A second cause for reduced SP deflection is the presence of shale in the sand, an explanation of which is beyond the scope of this book. In addition to these effects on the SP behavior, elaborate interpretations have been developed based on connections established between SP curve shapes and geologicaly significant events. Some examples of using the SP curve to determine patterns of sedimentation are given in Reference 5.

References

1. *Schlumberger Log Interpretation Charts*, Schlumberger, New York, 1985, pp. 10-14.
2. Dewan, J. T., *Essentials of Modern Open-hole Log Interpretation*, PennWell Publishing Co., Tulsa, 1983.
3. Jorden, J. R., and Campbell, F. L., *Well Logging II - Resistivity and Acoustic Logging*, Monograph Series, SPE, Dallas (in press).
4. Asquith, G., with Gibson, C., *Basic Well Log Analysis for Geologists*, AAPG, Tulsa, 1982.
5. Pirson, S. J., *Geologic Well Log Analysis*, Gulf Publishing Co., Houston, 1977.
6. Tittman, J., *Geophysical Well Logging*, Academic Press, Orlando, 1986.
7. Feynman, R. P., Leighton, R. B., and Sands, M. L., *Feynman Lectures on Physics*, Vol. 2, Addison-Wesley, Reading, Mass., 1965.
8. Hearst, J. R., and Nelson, P., *Well Logging for Physical Properties*, McGraw-Hill, New York, 1985.

Problems

1. In the log example of Fig. 3–11, indicate the shale base line and zone it into three major units; label the shale and the two reservoir units.

Using the qualitative log interpretation guides of Chapter 2, assuming that the lower reservoir is water-filled, answer the following questions.

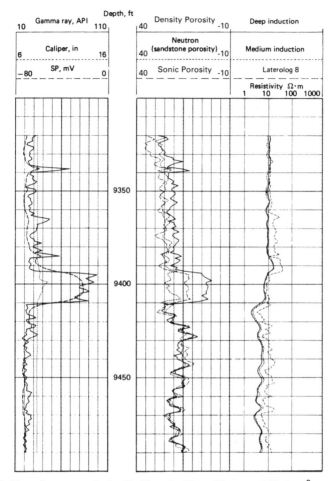

Figure 3-11. Log example for Problem 1. From Hearst and Nelson.[8]

a. Is the mud filtrate more or less saline than the formation water?

b. Is the average porosity of the upper reservoir greater or less than that of the lower reservoir?

c. In the upper reservoir, which curve(s) indicate(s) why the neutron porosity is greater than the density porosity?

d. On the basis of the resistivity curves alone, the upper reservoir may be split into two portions. Do they both contain hydrocarbons? Why?

e. Which of the two zones do you expect to be more permeable?

2. From the log of Fig. 3–12, determine the corrected value of the SP deflection in the one clean zone. Using the information at the bottom of the log, and chart SP-4 in the chartbook.[1] Note that the scale for the SP is 10 mV per division.

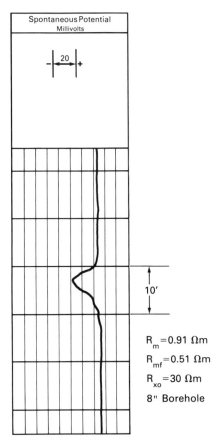

Figure 3–12.   SP log for Problems 2 and 3, showing effects of bed thickness.

3. Using the corrected deflection of the SP from Problem 2, estimate the water resistivity using the relation:

$$SSP = -70.7 \log_{10} \frac{R_{mf}}{R_w}.$$

What is the value of R_w if you use the uncorrected value of SP?

4. A 9" borehole is filled with mud at a constant temperature of 100°F. The resistivity of the mud is 0.9 Ω·m at 100°F.

a. What is the resistance of the mud column from surface to 7000'?

b. What is the resistance if the temperature is raised to 200°F?

c. In case b, what would be the resistance if the diameter of the borehole were increased to 1 m?

5. The log of Fig. 3–13 shows a measurement of the mud resistivity (in Ω·m) as a function of depth. Ignore the SP curve since it was measured 9 months prior to the mud measurements.

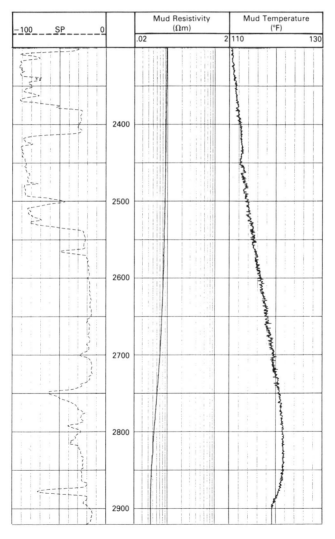

Figure 3–13. A log of SP, borehole mud resistivity, and temperature.

a. Assuming constant mud salinity for the bottom 300' of log, what temperature variation would be required to produce such a resistivity change? Sketch a log of it in track 3.

b. What salinity variation could produce a similar change in resistivity in accordance with the temperature over this zone? Plot a few points of the concentration of NaCl as a function of depth in track 3.

c. What is a good explanation for the resistivity behavior in this example?

4
EMPIRICISM: THE CORNERSTONE OF RESISTIVITY INTERPRETATION

INTRODUCTION

Before considering the details of measuring the resistivity of earth formations, let us look at the usefulness of such measurement. The desired petrophysical parameter from resistivity measurements is the water saturation S_w. In the previous chapter, the resistivity of various materials, including brines, is discussed. Here the focus is on the resistivity of a porous rock sample filled with a conductive brine in order to relate this measurable parameter to formation properties of interest for hydrocarbon evaluation.

In this chapter the empirical basis for the interpretation of resistivity measurements is reviewed. For many years, at the outset of well logging, it was not possible to address the water saturation question any more precisely than whether the resistivity of a formation was high or low. It was through the work of Leverett[1] and Archie[2] that it became possible to be more quantitative about the interpretation of a formation resistivity measurement.

After a review of the basis for the famous Archie relation, which links the measurement of the formation resistivity to that of the formation water resistivity, porosity, and water saturation, the principle of the simplest electrical logging measurements is presented. For this application, a review of some basic notions of electrostatics is made to indicate how, in a very idealized situation, the measurement of formation resistivity might be made.

EARLY ELECTRIC LOG INTERPRETATION

Fig. 4-1 shows a log of spontaneous potential and formation resistivity made

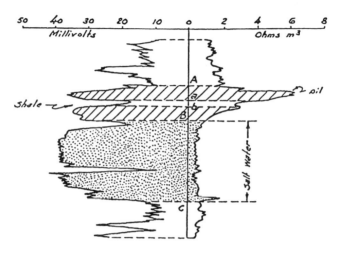

Figure 4-1. An early resistivity-SP log. The scale (Ohms m³) presumably refers to Ω·m. From Martin et al.[3]

prior to 1935. The notations on the figure make clear which zones are oil-bearing and which are water-bearing. It seems possible, noting the higher resistivity, that zone a-A contains more oil (has a lower S_w) than zone B-b. But how can this be verified?

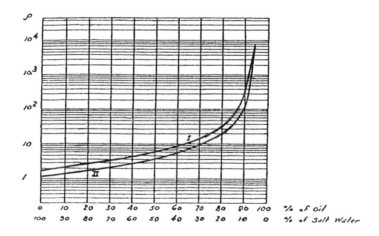

Figure 4-2. Resistivity measurements of two core samples as a function of water saturation for use in electric log interpretation. From Martin et al.[3]

The "standard" procedure at the time was to take a core sample, representative of the zones in question, and to make laboratory measurements of its resistivity under different conditions of water saturation. Fig. 4-2 is an example of two such core sample measurements. Presumably the core was saturated with water of the same resistivity as the undisturbed formation water for the resistivity determination. In the laboratory, the water was progressively displaced by hydrocarbon, and the measured resistivity of the sample was plotted as a function of the water saturation.

At about the same time, M. C. Leverett was conducting experiments with unconsolidated sands, to determine the relative permeability of oil and water as a function of the water saturation.[1] As a by-product of his research, he measured the conductivity of the material in a sample chamber (see Figs. 4-3 and 4-4, and note similarity to the mud cup of Fig. 3-4), after a calibration of the system constant, in order to conveniently determine the fraction of kerosene and water in his permeable samples. Fig. 4-5 is a summary of his calibration data. The fractional water saturation (S_w) is plotted versus the normalized conductivity. The normalizing point for this latter scale was taken to be the conductivity of the sample in the chamber when it was completely saturated with saltwater. Appropriately normalized points from the core measurements of Fig. 4-2 can be shown to clearly track Leverett's measurements and indicate the possibility of a general method for relating the resistivity of a porous sample to the water saturation. (See Problem 2.)

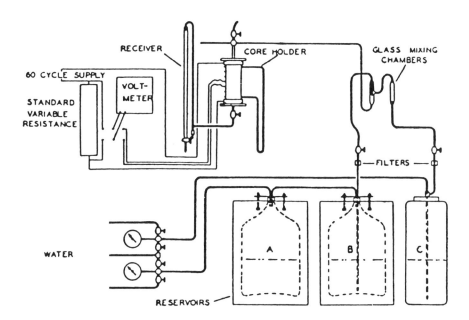

Figure 4-3. Schematic of Leverett's experimental setup for measuring the relative permeability of sand packs. From Leverett.[1]

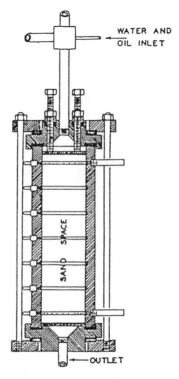

Figure 4–4. Detail of the core holder from Leverett's experiments. Note the similarity to the mud cup. From Leverett.[1]

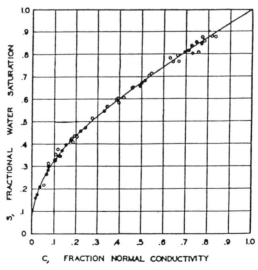

Figure 4–5. Calibration curve of Leverett's core holder with sand pack, showing variation of relative conductivity as a function of water saturation. From Leverett.[1]

EMPIRICAL APPROACHES TO QUANTIFYING R_t

Archie's Formation Factor

Shortly after the publication of Leverett's work, G. E. Archie of Shell was making electrical measurements on core samples, with the aim of relating them to permeability. His measurements consisted of completely saturating core samples with saltwater of known resistivity R_w and relating the measured resistivity R_o of the fully saturated core to the resistivity of the water. He found that, regardless of the resistivity of the saturating water, the resultant resistivity of a given core sample was always related to the water resistivity by a constant factor F. He called this the formation factor, and his experiments are summarized by the following relation:

$$R_{sample} \equiv R_o = F\, R_w .$$

Fig. 4-6 is an example of his work on cores from two different locations,

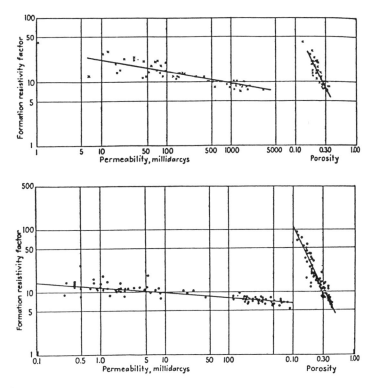

Figure 4-6.   Examples of the attempts to correlate the electrical formation factor with permeability and porosity for water-saturated rock samples from two regions. From Archie.[2]

where the formation factor F is plotted as a function of permeability and, almost as an afterthought, porosity (on a much compressed scale). Although he was searching for a correlation with permeability, he finally admitted that a generalized relationship between formation factor and permeability did not exist, although one seemed to exist for porosity. His summary graph (Fig. 4–7) shows the hopelessness of a formation factor/permeability

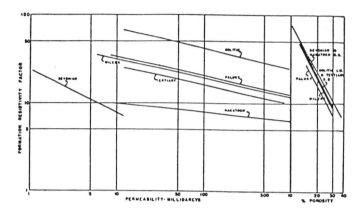

Figure 4–7. A summary of an exhaustive set of measurements of formation factor, concluding with a strong correlation with porosity and an unpredictable one with permeability. From Archie.[2]

correlation. However it indicates that the formation factor is a function of porosity and can be expressed as a power law of the form:

$$F \approx \frac{1}{\phi^m}$$

where the "cementation" exponent m is very nearly 2 for the data considered. This empirical observation can be used to describe the variation in formation resistivity for a fixed saturation and water resistivity when the porosity changes. Under conditions of constant saturation, the lower the porosity, the higher the apparent resistivity will be.

Archie's Synthesis

The practical application of resistivity measurements is for the determination of water saturation. This was made possible by another observation of Archie. He noticed that the data of Leverett and others could be conveniently parameterized after having plotted the data in the form shown in Fig. 4–8. On log-log paper, the data of water saturation versus relative resistivity

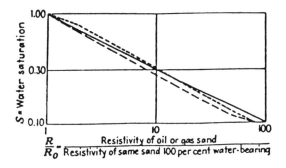

Figure 4-8. A synthesis of various resistivity/saturation experiments, indicating a general power law relationship. From Archie.[2]

plotted as a straight line, suggesting a relationship of the form:

$$S_w = \left[\frac{R_t}{R_o}\right]^{-\frac{1}{n}}.$$

The exponent n, called the saturation exponent, is very nearly 2 for the data considered.

From this, an approximate expression for the water saturation is:

$$S_w \approx \sqrt{R_o/R_t}.$$

However, the fully saturated resistivity R_o (which is not usually accessible in formation evaluation), can be related to the water resistivity using the previously discovered Archie relationship. So the expression becomes:

$$S_w \approx \sqrt{F\frac{R_w}{R_t}},$$

and with the porosity dependence, the final form is:

$$S_w \approx \sqrt{\frac{1}{\phi^2}\frac{R_w}{R_t}}, \qquad (1)$$

which can be used for purposes of estimation. The general form, however, is:

$$S_w^n = \frac{a}{\phi^m} \frac{R_w}{R_t}, \qquad (1a)$$

where the constants a, m, and n need to be determined for the particular field or formation being evaluated.*

From the above analysis it is clear that, in order to interpret a resistivity measurement in terms of water saturation, two basic parameters need to be known: the porosity ϕ and the resistivity of the water in the undisturbed formation R_w. To illustrate the basic procedures of resistivity interpretation, it is of some interest to turn back to the log example in Chapter 2 (Fig. 2–17) to make use of the empirical observations. As a starting point, the value of the water resistivity R_w can be estimated. This can be done in either zone D′ or zone C′, which have tentatively been identified as water zones. In either case, the porosity is about 28 PU, so the formation factor F is $\frac{1}{.28^2}$, or 12.8. Thus the apparent resistivity of about 0.2 $\Omega\cdot$m in these zones, which is assumed to be the fully water-saturated resistivity R_o, corresponds to a water resistivity of 0.2/12.6, or 0.016 $\Omega\cdot$m.

It is clear that the increase in deep resistivity in zone C to about 4 $\Omega\cdot$m must correspond to a decrease in water saturation compared to zone C′; the porosity seems to be constant at 28 PU over both zones. The saturation in zone C can be estimated from:

$$S_w = \sqrt{\frac{R_o}{R_t}} = \sqrt{\frac{.2}{4.0}} = 22\%,$$

so the hydrocarbon saturation is about 78%.

Another zone of hydrocarbon (A) indicates the same resistivity value as zone C. However, in the upper zone the porosity is much lower and can be estimated to be about 8 PU. Thus the formation factor in zone A is $\frac{1}{(.08)^2}$, or 156. If it were water-filled, the resistivity would be expected to be about 1.25 $\Omega\cdot$m compared to the 4 $\Omega\cdot$m observed. Thus the zone may contain hydrocarbons, but the water saturation can be expected to be higher than in zone C. The water saturation in this zone can be estimated from Eq. (1) to be:

---

* There has been much speculation and some study of the nature of these constants, in particular the Archie or cementation exponent m. Reference 7 reviews some of the models of tortuosity and constriction of regular pore systems developed to suggest Archie relation behavior for porous rock conductivity. Notable among the work on the cementation exponent is a paper by Roberts and Schwartz, which clearly relates the exponent to the size of restrictions in the porous medium and leads to a porosity dependence of m for a model of grain consolidation.[4] That m is not constant can be verified by noting the options for calculating formation factor in chartbooks published by logging service companies.[5]

$$S_w = \sqrt{156 \frac{.016}{4.0}} = 79\% ,$$

so it appears to be only about 21% hydrocarbon-saturated. The limit of confidence in the estimate of saturation can be determined from Eq. (1a) and is left as an exercise. In addition to measurement uncertainty, an important component in the uncertainty is the variability of the cementation exponent.

Now that the basic ideas of resistivity interpretation have been explored, it is appropriate to consider the question of how resistivity measurements of sedimentary formations are made in-situ. First we make a rapid review of some basic notions of electrostatics, which forms the basis for resistivity measurements.

A REVIEW OF ELECTROSTATICS

One concept of considerable use is that of the electrostatic potential, which follows directly from Coulomb's law. To arrive at an understanding of the electrostatic potential and to derive a simple expression for it, consider the case (Fig. 4-9) of two charges (q_1 and q_2) at a distance r from one another.

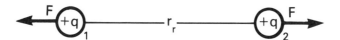

Figure 4-9.   Two charged particles separated by a distance r, exhibiting a repulsive force F.

Coulomb's law states that the force of repulsion between the two charges is inversely proportional to the square of the separation and varies directly with the product of the magnitudes of the charges. This can be expressed as:

$$F = \frac{1}{4\pi\varepsilon_o} \frac{q_1 q_2}{r^2} .$$

This leads directly to an expression for the electric field vector $\overline{E}$, which is defined as the force per unit charge, from which it follows that,

$$\overline{E} = \frac{1}{4\pi\varepsilon_o} \frac{q}{r^2} \hat{r} . \qquad (2)$$

Here $\hat{r}$ is the unit vector in the direction from the charge producing the field to the point of observation. Eq. (2) gives the electric field strength at any point r from a charge of magnitude q.

From the definition of work W, which is the integral of the opposing force over the distance traveled, one can write:

$$W = -\int_a^b \overline{F} \cdot d\hat{s} ,$$

which for a unit of charge in an electric field is:

$$W = -\int_a^b \overline{E} \cdot d\hat{s} = -\frac{q}{4\pi\varepsilon_o} \int_a^b \frac{dr}{r^2}$$

$$= \frac{q}{4\pi\varepsilon_o}\left[\frac{1}{r_a} - \frac{1}{r_b}\right].$$

It is to be noted that the amount of work done in moving from point a to point b is independent of the path taken. It depends only on the value of the two endpoints. Thus in analogy with the notion of potential energy, the electrostatic potential $\phi(P)$ is defined as:

$$\phi(P) = -\int_{P_o}^{P} \overline{E} \cdot d\hat{s} ,$$

or

$$\overline{E} = -\overline{\nabla}\phi .$$

The reference point P_o is usually taken to be at a distance infinitely removed from the charge producing the potential, and $\phi(P_o)$ is set to zero. In this case $\phi(P_0)$ is also called the voltage V. For a point charge, this results in:

$$\phi(r) = \frac{q}{4\pi\varepsilon_o}\frac{1}{r} = V(r) . \qquad (3)$$

A THOUGHT EXPERIMENT FOR A LOGGING APPLICATION

Fig. 4-10 shows the setup for measuring the resistivity of a homogeneous formation. It consists of a current source of intensity I and a voltage measurement electrode M at some distance r from the current emission at point A. The resistivity of the homogeneous medium is R_t, so its conductivity σ is given by $\sigma = \frac{1}{R_t}$.

One way to determine the relationship between the potential at M and the current I is to use some of the relationships from electrostatics. The current I, being a continuous source of charge, can be thought of as producing a potential V, just as would be expected from some equivalent point charge q:

$$V(r) = \frac{1}{4\pi\varepsilon_o}\frac{q}{r} .$$

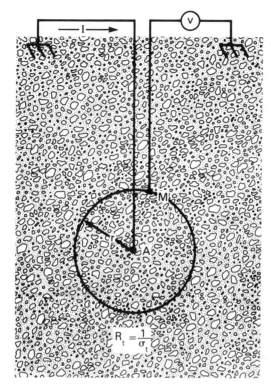

Figure 4–10. Idealized experiment for determination of the resistivity of an infinite uniform medium of conductivity $\sigma(= \frac{1}{R_t})$. It consists of the injection of a current at point A, and measurement of the potential at point M at a distance r from the current electrode.

The problem is to relate the equivalent charge q to the current I.
At any point in the system there will be a current density $\overline{J}$ given by:

$$\overline{J} = \sigma\overline{E} = -\sigma\frac{\partial}{\partial r}V(r)$$

$$= \frac{\sigma}{4\pi\varepsilon_o}\frac{q}{r^2}\hat{r} ,$$

where $\hat{r}$, the unit vector, is directed radially outward from the current source. In order to put the expression for potential in terms of the total current, I, the current density is integrated over the surface of a sphere enclosing the current source:

$$I = \int \overline{J} \cdot dS = \frac{\sigma q}{4\pi\varepsilon_o r^2} 4\pi r^2 = \frac{\sigma q}{\varepsilon_o} ,$$

and q is solved for in terms of I:

$$q = \frac{\varepsilon_o I}{\sigma} = \varepsilon_o I \, R_t \; .$$

This expression for q is now put back into the potential for a single point charge, Eq. (3), to obtain the voltage at a distance r from the current source:

$$V(r) = \frac{\varepsilon_o I R_t}{4\pi\varepsilon_o r} \; .$$

A less tortuous determination of the potential is obtained from Ohm's law in spherical geometry. For the source of current I, the current density on the surface of a sphere of radius r centered on the source is:

$$J = \frac{I}{4\pi r^2} \; .$$

The relation between current density and electric field E implies that:

$$E = \frac{R_t I}{4\pi r^2} \; .$$

From this expression, the voltage at a distance r from the current source is obtained from:

$$V(r) = \phi(r) = -\int_\infty^r \frac{R_t I}{4\pi r^2} dr = \frac{R_t I}{4\pi r} \; .$$

Thus the value of R_t is found to be:

$$R_t = 4\pi r \frac{V}{I} = k \frac{V}{I} \; . \qquad (4)$$

The setup of Fig. 4–10 can be considered as a rudimentary monoelectrode measurement device for determining formation resistivity. For this device the tool constant k is seen to be $4\pi r$, where r is the spacing between the current electrode and the measurement point. Knowing the injected current and the resultant voltage, the resistivity of the homogeneous medium R_t may then be found.

As an exercise, it is interesting to determine the sensitivity to resistivity variations of such a device following the treatment of Tittman.[6] This question can be examined by considering the current electrode to be at the center of a number of concentric spheres of differing resistivities, as indicated in Fig. 4–11. The object is to find the sensitivity of the measurement to the layers beyond the measurement electrode.

This can be found from the differential form of the basic tool response, that is:

Empiricism and Resistivity Interpretation

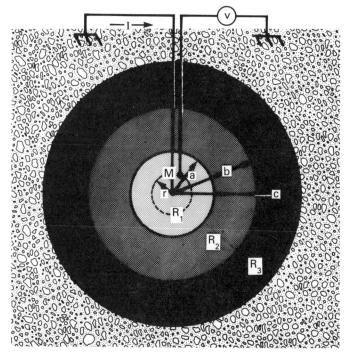

Figure 4-11. Geometry for determining the sensitivity of the two-electrode device to concentric layers of different resistivities. Adapted from Tittman.[6]

$$V(r) = \frac{1}{4\pi r} IR_t \rightarrow dV = -\frac{IR_t}{4\pi r^2} dr .$$

From this expression of the incremental potential, the voltage at point r can be found by integrating the effect of all the layers, starting at the outermost, up to the point r:

$$V(r) = \int_\infty^r dV = -\frac{I}{4\pi} \int_\infty^r \frac{R(r)}{r^2} dr = \frac{I}{4\pi} \int_r^\infty .$$

This integration can be broken up into sums over the various regions,

$$V(r) = \frac{I}{4\pi} \left[R_1 \int_r^a + R_2 \int_a^b + R_3 \int_b^c + \cdots \right] .$$

This can finally be simplified to

$$V(r) = \frac{IR_1}{4\pi r} \left[(1 - \frac{r}{a}) + \frac{R_2}{R_1}(\frac{r}{a} - \frac{r}{b}) + \cdots \right] .$$

For the case of r≪a this expression reduces to the result of Eq. (4) for a homogeneous medium of resistivity R_1.

References

1. Leverett, M. C., "Flow of Oil-Water Mixtures Through Unconsolidated Sands," *Trans.* AIME 132, 1938.

2. Archie, G. E., "Electrical Resistivity Log as an Aid in Determining Some Reservoir Characteristics," *Trans.* AIME 146, 1942.

3. Martin, M., Murray, G. H., and Gillingham, W. J., "Determination of the Potential Productivity of Oil-bearing Formations by Resistivity Measurements," *Geophysics*, Vol. 3, 1938.

4. Roberts, J. N., and Schwartz, L. M., "Grain Consolidation and Electrical Conductivity in Porous Media," *Phys. Rev. B* 31, 5990, 1985.

5. *Schlumberger Log Interpretation Charts*, Schlumberger, New York, 1985, p. 15.

6. Tittman, J., *Geophysical Well Logging*, Academic Press, Orlando, 1986.

7. Jordan, J. R., and Campbell, F., *Well Logging II - Resistivity and Acoustic Logging*, Monograph Series, SPE, Dallas (in press).

Problems

1. Table 4–1 is a copy of some of Archie's original data. For the samples listed, plot the formation factor versus porosity, and graphically determine an expression for a reasonable fit to the data.

   What is the maximum percentage error in F if the approximate form $F = \dfrac{1}{\phi^2}$ is used?

2. Fig. 4–2 shows resistivity measurements versus saturation for two core samples. Using information derived from curve I, plot the equivalent values of relative conductivity (at the following values of S_w: 0.1, 0.2, 0.4, 0.6, and 0.9) on Fig. 4–5, which presents equivalent data from Leverett's experiments on sand packs.

3. The original text accompanying Fig. 4–2 states that the salinity of the saturating water for the two samples was identical and that the porosity of sample I is 45% and that of sample II is 25%. What two inconsistencies are indicated by the data shown in the figure?

Core Analysis Data*			
Porosity	Formation Factor	Porosity	Formation Factor
30	9	27	10
32	7	30	9
25	13	28	10
27	11	20	8
34	14	28	8
27	11	27	11
30	9	28	9
31	9	27	10
25	9	25	13
30	9	21	20
20	14	23	15
25	14	24	14
27	11	25	13
26	12		

Table 4–1

4. Consider a two-electrode device, such as that shown in Fig. 4–10, where the spacing, A–M, between the current source and voltage monitor is 1 m.

   a. What is the resistance seen by this device in a completely water-saturated 20% (20 PU) porous limestone formation? The formation water is seawater (20 Kppm NaCl), and the temperature is 100°F.

   b. What resistance does it see in a zero porosity limestone (marble)?

5. The water resistivity R_w in the log example of Fig. 2–17, was estimated to be 0.016 $\Omega \cdot$m, based on the assumption that a water zone had been identified.

   a. What is the estimate of R_w, if 10% residual oil saturation is present in the "water" zone?

   b. Assuming that the water contains only dissolved NaCl, what is a reasonable value for the concentration at a temperature of 200F?

6. The water saturation in zone A of Fig. 2–17 was estimated to be 79% assuming a cementation exponent of 2.

   a. What value of cementation exponent produces a water saturation of 50%?

   b. What value of porosity would be required to yield a value of S_w of 50% using a cementation exponent of 2?

7. Suppose a series of core measurements on reservoir rocks in the range of 20–30% porosity has established that the cementation exponent is between 1.8 to 1.9, i.e., $F = \dfrac{1}{\phi^{1.8}}$ or $F = \dfrac{1}{\phi^{1.9}}$.

   a. What percentage error in the logging measurement of R_t can be tolerated so that its influence on the saturation estimate is smaller than that induced by the possible variation in cementation exponent? Assume a saturation exponent of 2.

   b. Show that a 20% error in R_t can be tolerated if the porosity is 10%.

5
RESISTIVITY: ELECTRODE DEVICES AND HOW THEY EVOLVED

INTRODUCTION

We have seen the utility of knowing formation resistivity and an idealized approach to making the measurement. This chapter focuses on the evolution of one type of electrical logging tool: electrode devices, so named because the measurement elements are simply metallic electrodes. In general, these electrode devices utilize very low frequency (<1000 Hz) current sources. The historical progression from the normal device to state-of-the-art focused laterolog arrays will be traced. An indication of the measurement limitations for each of these types of tools will be given and related to their design. Methods used for the prediction of their response will be discussed.

UNFOCUSED DEVICES

The Short Normal

The earliest commercial device, the *short normal*, is illustrated in Fig. 5–1. It bears a strong resemblance to the thought experiment of the preceding chapter. The differences include the presence of a borehole and a sonde (on which the current electrode A and measure electrode M are located). As indicated in the figure, the spacing between the current electrode and voltage

74 Well Logging for Earth Scientists

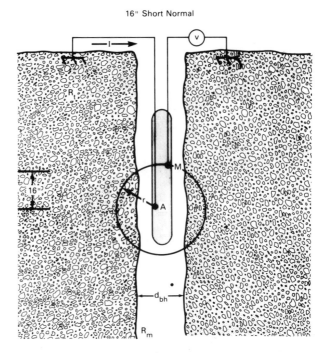

Figure 5–1. A schematic representation of the short normal. A 16" spacing is indicated between current electrode A and measure electrode M.

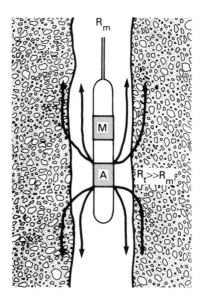

Figure 5–2. Idealized current paths for the short normal in a very conductive borehole mud.

electrode was 16", and thus the designation "short."

Two basic problems are associated with the short normal, both related to the presence of the borehole, which is normally filled with a conductive fluid. There is a sensitivity of the measurement to the mud resistivity and hole size, as indicated in Fig. 5–2. In a borehole filled with very conductive mud, the current tends to flow in the mud rather than the formation. In this case, the apparent resistivity as deduced from the injected current, and resultant voltage will not reflect the formation resistivity very accurately.

The second difficulty with this measuring technique is illustrated in Fig 5–3. Once again, the conductive borehole fluid provides an easy current path for the measure current into adjacent shoulder beds of much lower resistivity (R_s) than the formation (R_t) directly opposite the current electrode. In this case, again the apparent resistivity (from the measurement of the voltage of electrode M and the current I, in combination with the tool constant) will be representative not of the resistive bed, but, more likely, of the less resistive shoulder bed.

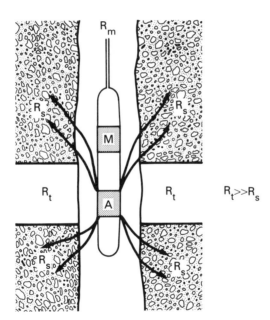

Figure 5–3.   Idealized current paths for the short normal in front of a thin resistive bed.

Estimating the Borehole Size Effect

To get an idea of the effect of the borehole size on the short normal and to gain an appreciation for the need for computational methods to attack such questions, a simple approach is investigated.

For estimating the borehole size effect, first assume that the potential distribution caused by the current source is spherical. This means ignoring the presence of the borehole, on the one hand, and on the other, considering that the borehole represents a small current loss from the injected measure current. In this way, a simple model for the tool and formation can be used as indicated in Fig. 5–4. The measure current is presented with two

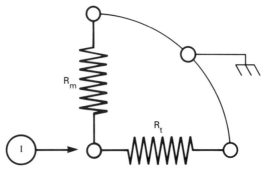

Figure 5–4. A simple equivalent circuit for estimating the short normal borehole effect. The effective resistance of the mud is R_m, and the effective resistance of the formation is given by R_t.

equivalent resistance paths: R_t, which represents the resistance presented by the formation of resistivity ρ_t, and R_m, the effective resistance of the borehole between the current electrode and the voltage measure electrode.

Taking advantage of the first assumption, that the equipotential surfaces are spherical, the response equation of the short normal Eq. (4), in Chapter 3, can be used to define the formation resistance, out to a distance r, in terms of the formation resistivity ρ_t:

$$V = \frac{I \rho_t}{4\pi r}.$$

This yields the effective resistance of the formation (if the borehole were not present):

$$R_t = \frac{\rho_t}{4\pi r}.$$

The borehole resistance can be estimated from its geometry, using the analysis developed earlier in conjunction with the mud cup. The radius of the borehole and measurement sonde are given by r_{bh} and r_s, respectively. As Fig. 5–5 indicates, the mud resistance R_m, is given, in terms of its resistivity, ρ_{mud}, by:

$$R_m = \rho_{mud} \frac{l}{A},$$

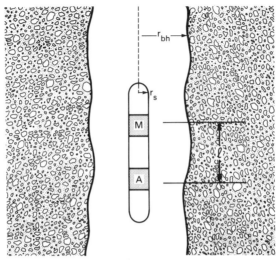

Figure 5–5. Geometry for estimating the borehole resistance between electrodes A and M.

where l, in this case, is the electrode spacing r. So:

$$R_m = \rho_{mud} \frac{r}{\pi(r_{bh}^2 - r_s^2)},$$

where no electrical interaction has been assumed between the borehole and the formation.

To evaluate the sensitivity of the model to the borehole mud, an expression will now be derived for the ratio of the apparent resistivity (ρ_{16}) to the mud resistivity ρ_m. First, the apparent resistivity is expressed in terms of the formation resistance and the mud resistance of the equivalent circuit of Fig. 5–4:

$$\frac{1}{\rho_{16}} = \frac{1}{4\pi r}\left[\frac{1}{R_t} + \frac{1}{R_m}\right].$$

This can be rewritten as:

$$\frac{1}{\rho_{16}} = \frac{1}{4\pi r}\left[\frac{4\pi r}{\rho_t} + \frac{\pi(r_{bh}^2 - r_s^2)}{r\,\rho_m}\right]$$

$$= \frac{1}{\rho_t} + \frac{1}{\rho_m}\left[\frac{1}{4}\frac{(r_{bh}^2 - r_s^2)}{r^2}\right] = \frac{1}{\rho_t} + \frac{1}{\rho_m^*}.$$

This expression is then inverted to get the desired form:

$$\frac{\rho_{16}}{\rho_m} = \frac{\rho_t}{\rho_m} \frac{\rho_m^*}{\rho_t + \rho_m^*} . \qquad (1)$$

Evaluating the second part of Eq. (1) for the case of an 8" borehole and a sonde of 4" diameter, letting $x = \dfrac{\rho_t}{\rho_m}$, yields

$$\frac{\rho_{16}}{\rho_m} \approx x \left[\frac{1}{1 + .01x} \right],$$

which is plotted in Fig. 5–6, along with the standard presentation of the

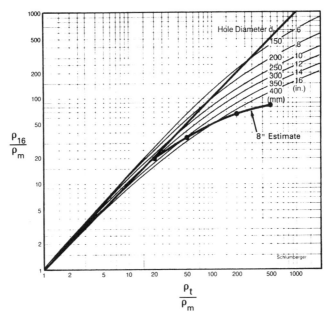

Figure 5–6. Borehole correction chart for the 16" short normal. Indicated is the approximate model correction for an 8" borehole, showing the need for a careful evaluation. Adapted from Schlumberger.[1]

borehole size effect for the short normal device. It is clear that this simple analysis has indicated a trend but is considerably in error in predicting the actual perturbation that results from hole size and mud resistivity contrast. The figure shows that in front of a formation with a resistivity 100 times that of the mud, the simple model predicts an error of a factor of 2, whereas in fact there will be none. At higher contrasts the difference between this simple model and the actual tool behavior becomes even greater.

A look at the correction chart data shows that for an 8" hole size, the short normal does a fairly good job of measuring the correct formation resistivity, except for very large mud/formation resistivity contrasts. However, for the 16″ borehole size, this is not the case. For a mud/formation contrast of 100, the measurement will be in error by a factor of 2. Thus the need for such correction charts. But how are they constructed?

Borehole correction charts for electrical logging tools are constructed by obtaining a solution of Laplace's equation:

$$\nabla^2 V = 0,$$

subject to the boundary conditions imposed by the borehole and tool configuration. There are three approaches to obtaining solutions to this equation: analog simulation, analytic solutions, and computer modeling. Fig. 5–7 is a sketch of the situation to be modeled, indicating the zones of

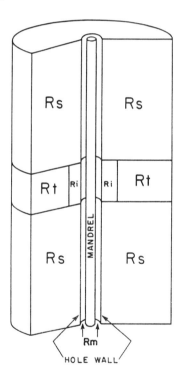

Figure 5–7. A geometric and electrical model of the borehole and formation used for generating electrical tool response to layered beds with step profile invasion. The centered tool is referred to as a *mandrel*.

interest, including an invaded zone of resistivity different from the formation resistivity. For the analog simulation, the axially symmetric rings of formation about the borehole axis are replaced by sets of resistors, as indicated in Fig. 5-8. The construction of such analog computers calls for

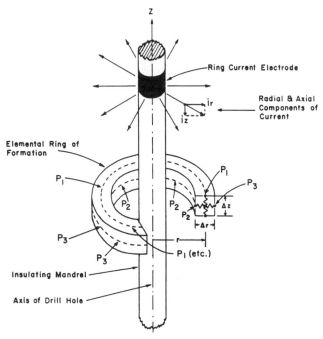

Figure 5-8. Analog simulation of the borehole/formation replaces axial rings of formation by a network of resistor pairs.

hundreds of thousands of individual resistors to be soldered into place. Such simulators were constructed in the 1950s and have been used until relatively recently. Advances in analytical solutions and high-speed digital computers have begun to supplant this technique.

The geometry for the analytical solution of Laplace's equation for the logging problem is shown in Fig. 5-9. The three components of current are indicated at a point $P(\rho,\phi,z)$ some distance from the current source. For this axially symmetric situation, the equation reduces to:

$$\frac{1}{\rho}\frac{\partial}{\partial \rho}\left[\rho \frac{\partial V}{\partial \rho}\right] + \frac{\partial^2 V}{\partial z^2} = 0,$$

and the two components of current are related to the potential by:

$$J_\rho = -\sigma \frac{\partial V}{\partial \rho}$$

and

$$J_z = -\sigma \frac{\partial V}{\partial z}.$$

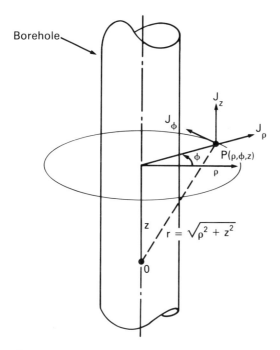

Figure 5–9. Geometry for the analytic solution of Laplace's equation in the cylindrical symmetry of the borehole.

The axial component of current is zero. The solution for the potential $V(\rho,z)$ is found by assuming a form which is separable:

$$V(\rho,z) = R(\rho)Z(z) .$$

The solution is found, after using boundary conditions of potential and normal current continuity across the borehole interface, to be expressed as infinite integrals of Bessel functions.[6] These can be evaluated numerically to give good predictive behavior for various borehole sizes, mud contrasts, and depth of invasion.

For the more complicated case of tool response to bed boundaries, powerful computer modeling techniques can be used. One is known as the *finite elements method*.[8,9] In this technique a grid is set up to represent the borehole and formation. A solution of Laplace's equation is then sought, subject to the boundary conditions, using a trial function for the potential, which is then evaluated at each one of the node points. The final solution is obtained by determining the potential at each point, which minimizes the energy of the system. It is obtained through the solution of a set of simultaneous difference equations.

A continuing examination of the shortcomings of the short normal, in

Fig. 5–10 reveals the kind of response problems encountered for large contrasts between the shoulder beds and the bed of interest. Note that in the upper part of the figure some idea of the actual tool implementation is given.

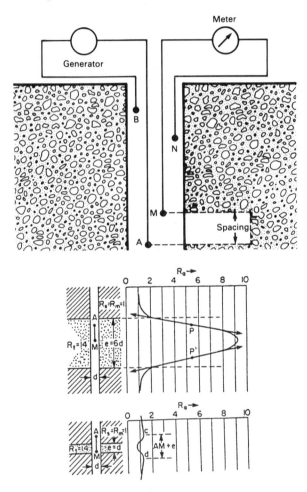

Figure 5–10. A schematic of the short normal and its response in two common logging situations. Adapted from Doll et al.[2]

The electrode B is at the surface, whereas the electrode N, to which the potential measurement is referenced, is actually located down-hole on the measurement sonde. In this particular case, the resistivity contrast between beds is 14, and the borehole diameter is half the spacing between current source and voltage electrode. Even for a bed 3′ thick, it is seen that the central value of resistivity does not attain the desired value. If the bed is only 6″ thick, then the behavior becomes bizarre, with the apparent resistivity

Electrode Devices and How They Developed   83

dipping below the value of the shoulder bed.

When attempts were made to improve this bed-boundary resolution, the normal device evolved to the *lateral* device, illustrated in Fig. 5–11. The

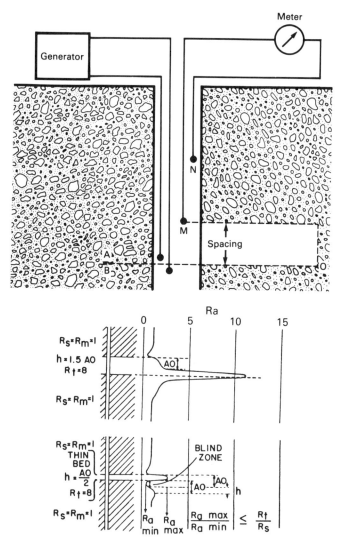

Figure 5–11. A schematic representation of a lateral device which uses a differential voltage measurement to define its response. Two cases of its response in common logging situations are shown. Adapted from Schlumberger.[10]

lateral sonde is much like the normal sonde except that there are two voltage electrodes, and the potential difference between them is used to indicate the resistivity of the formation layer between them. This will be nearly the case for beds whose thickness exceeds the spacing between the electrodes marked A and N. The bottom of the figure shows the response to two beds, whose thickness is given in terms of the electrode spacing. It is clear that there has been some improvement for bed resolution, but the response is still quite complicated because of current flow through the mud to zones other than the one directly in front of the measuring points.

FOCUSED DEVICES

Laterolog Principle

The next step in the evolution of electrical tools was the implementation of current focusing. Fig. 5–12 illustrates, on the left half of the diagram, the

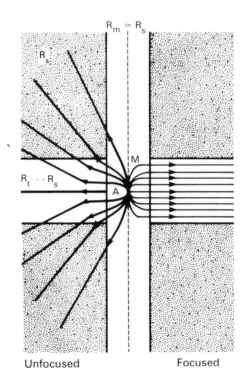

Unfocused   Focused

Figure 5–12. Idealized patterns of current flow in the borehole and formation from a central electrode. On the left the pattern is altered from the expected radial pattern because of the presence of a highly resistive bed. On the right is the desired flow, so that the resistivity of the bed of interest is sampled properly. Adapted from Schlumberger.[10]

current paths for the normal device in the case of a resistive central bed. The current tends to flow around it, through the mud, into the less resistive shoulders. The desired current path is shown on the right half of the figure, where the measure current is somehow forced through the zone of interest.

The principle of focusing is shown in Fig 5–13, where there are now three current-emitting electrodes. This type of array is known as a *guard focusing* device and is commonly referred to as a Laterolog–3, or LL3 device. The potential of the electrodes marked A_1 and A_1' is held constant and at the same potential as the central electrode A_0. Since current flows only if a potential difference exists, there should, in principle, be no current flow in the

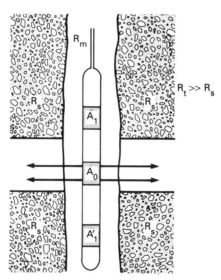

Figure 5–13. The current focusing principle. Two additional electrodes, A_1 and A_1', held at the same potential as the central current electrode, provide a bucking current to impede current flow along the mud column.

vertical direction. The first of these devices is illustrated in Fig. 5–14, with a schematic representation of the current lines. A sheath of current is shown to be emanating horizontally from the central measurement electrode. The current emitted from the focusing, or "guard" electrodes is often referred to as the "bucking" current, as its function is to impede the measure current from flowing in the borehole mud. It is the continuous adjustment of the bucking current which keeps A_1 and A_1' at the same potential as A_0.

Despite these good intentions, the LL3 device still showed some difficulty with bed boundaries. This is illustrated in Fig. 5–15, which shows cases of large contrast between the shoulder bed resistivity and the value of R_t. In the upper portion of the figure which shows a thick resistive bed, the principal measure current is seen to be escaping through the mud and into the shoulder.

86 Well Logging for Earth Scientists

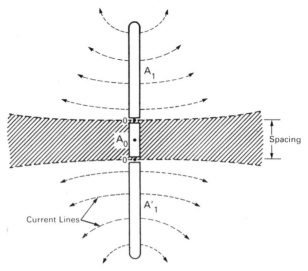

Figure 5-14. Idealized current distribution from the Laterolog-3 device, with current focused into the formation. From Serra.[3]

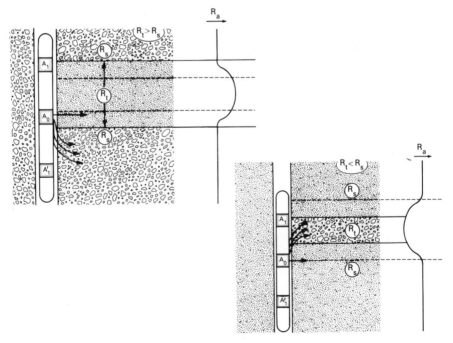

Figure 5-15. The effects of shoulder bed resistivity on the behavior of an LL3 device. The top sketch indicates current passing through the mud into a highly conductive shoulder. The bottom sketch indicates the effect of a thin conductive bed.

In the lower example, for a thin conductive streak, current is seen to seek it out sooner than expected, giving a broader apparent bed thickness than in the previous case.

Another approach to focusing the measure current is the seven electrode device, or LL7. The electrode configuration of one such device, the Laterolog-7, is sketched in Fig. 5–16. The additional monitoring electrodes

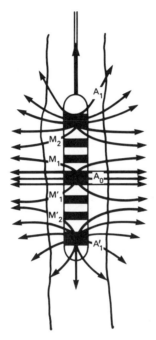

Figure 5–16. The electrode configuration of the Laterolog-7. Monitor electrodes drive the bucking current in the guard electrode to maintain a differential voltage of zero. Adapted from Serra.[3]

have been introduced in order to impede the flow of current parallel to the sonde though the borehole mud. This is achieved by varying the bucking current of electrodes A_1 and A_1' so that the potential drop between the pairs of monitor electrodes (i.e., M_1–M_1' and M_2–M_2') is zero. Since the potential drop is zero along this vertical direction, the current will be focused into the formation.

Spherical Focusing

Another approach to compensating for the effect of the borehole is the concept of spherical focusing. In this technique, which has been adopted for medium and shallow resistivity measurements, bucking currents are used to attempt to establish the spherical equipotential surfaces which would exist if no borehole were present. Fig. 5–17 is a rough sketch of the equipotential

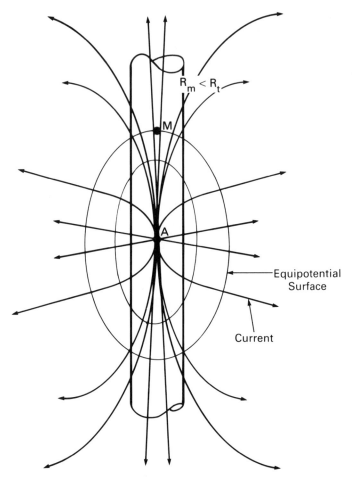

Figure 5–17. Current lines and equipotential surfaces for the short normal in a borehole.

surfaces which surround the current electrode in a normal device, as a result of the presence of the conductive mud in the borehole. Instead of spherical surfaces, they are of elongated shape. The objective of the spherical focusing is to provide a bucking current to force the equipotential lines to become spherical once again. Then the potential difference at two points along the sonde will be determined by the resistivity of a slice of formation in a spherical shell with radii equal to the two spacings. The depth of investigation can be controlled by the size of the shell. The idea is more clearly presented in Fig. 5–18.

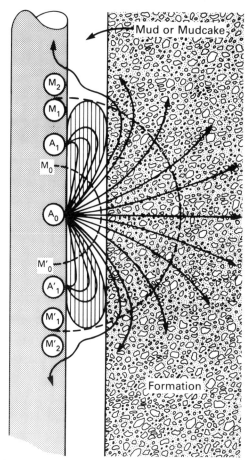

Figure 5–18. The electrode configuration of the spherically focused array.

Once again two sets of monitor electrodes (M_1–M_2 and M_1'–M_2') are used for controlling the bucking current, but the condition this time is that the potential difference between them remains constant. The electrode A_0 furnishes two sources of current; the measure current which is returned to a distant electrode, and the bucking current. The bucking current returned to the electrodes A_1 and A_1' is varied, so that the potential difference between the monitors is kept constant. The dotted lines then trace out approximately the surface of constant potential. The additional measure current injected by the central electrode then flows radially outward, at least until the outer potential surface is reached. The volume of formation investigated will be nearly the space between the two equipotential surfaces, with the exclusion of the region close to the borehole interface, which is "plugged" by the bucking current. The bucking current can be viewed either as setting up the equipotential surface or providing the current through the mud so that the actual measure current is forced into the formation.

Modern electrode devices use a dual focusing system. Those known as dual laterolog arrays combine the features of the LL3 and LL7 arrays, and the spherically focused device, in an alternating sequence of measurements.[6,7] By rapidly alternating the role of the various returns, a simultaneous measurement of deep and shallow resistivities is achieved. Fig. 5–19 shows the current paths computed for such a device. On the left side of the figure, the electrodes are in the deep configuration. The length of the guard electrodes, which uses parts of the sonde, is about 28′ to achieve deep penetration of a current beam of 2′ nominal thickness. On the right side, they are in the shallow (or medium) configuration. The current distributions, as calculated from the finite elements method, are shown in Fig. 5–20, for the case of a 5′ resistive bed surrounded by conductive shoulder beds in two borehole conditions. The current paths are much like those imagined for the focused tools.

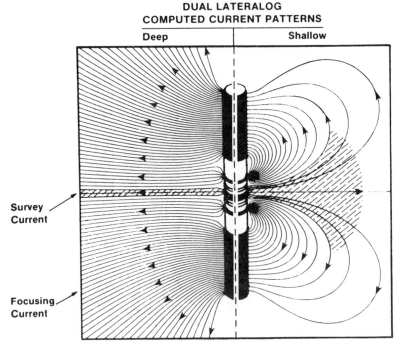

Figure 5–19. The current distributions computed for the dual laterolog in its two modes of operation. In the deep mode, the collection of central electrodes acts as the source of the measure current, and the two long electrodes as sources of bucking current. In the shallow mode, only the central electrode emits the measure current, and the spherical focusing technique is used to provide the bucking current which returns to the long guard electrodes. From Chemali.[6]

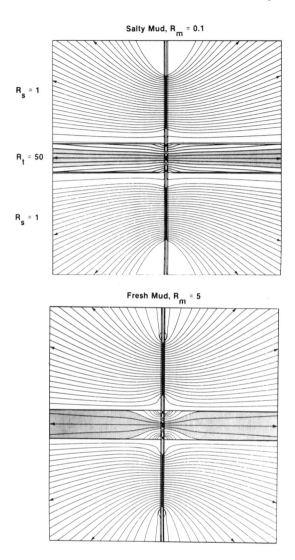

Figure 5–20. Computed behavior of the current distributions for the deep mode of the dual laterolog in the presence of a thin resistive bed for two conditions of borehole resistivity. From Chemali.[6]

PSEUDOGEOMETRIC FACTOR

For purposes of comparison of the different electrical measuring devices, it is convenient to think of the signal measured as being the result of the influence of three distinct regions of the measuring environment, as shown earlier in Fig. 2–4: the borehole, the invaded zone, and the undisturbed formation. Each of these zones is attributed its own characteristic resistivity: R_m, R_{xo}, and R_t. Generally the mud resistivity R_m is much less than either R_{xo} or R_t.

A model of these three zones, considered to be in series connection, is illustrated in Fig. 5–21. In this manner, the response of an electrode device

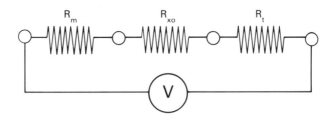

Figure 5–21. Model for the pseudogeometric factor development. It represents the effective resistances of the borehole, invaded zone, and formation as being in series.

can be thought of as being derived from a linear combination of the invaded zone and the true resistivity. This is expressed as:

$$R_a = J(d_i) R_{xo} + \left[1 - J(d_i)\right] R_t,$$

where R_a is the apparent resistivity. The pseudogeometric factor J is a normalized weighting factor which gives the relative contributions of the invaded zone (of diameter, d_i) and virgin zone to the final answer. It is referred to as the pseudogeometric factor (as opposed to a pure geometric factor, as will be seen later with the induction tool) since the weighting function will actually be influenced by the contrast between R_{xo} and R_t. Fig. 5–22 illustrates the pseudogeometric factor for several of the devices discussed, for the case of invaded zone resistivity that is greater than that of the virgin zone as well as the case of an invaded zone that is one tenth the resistivity of the virgin formation.

The pseudogeometric factors can be used to estimate the influence of the invaded zone on the measurement of resistivity when there is a contrast between R_t and R_{xo}. The response curves of Fig. 5–22 indicate how sensitive the deep and medium resistivity measurements are to the layer of R_{xo}. The shallow curve (marked LLs) rises steeply and indicates that in the case of a more conductive invasion zone ($R_{xo} = 0.1R_t$), half of the shallow signal comes from the first 8″ of invasion. Alternatively, 90% of the shallow signal comes from within a diameter of about 80″. The deep measurement (marked LLd) shows an insensitivity to the invaded zone since only about 15% of its signal comes from a diameter of 20″ (or the first 6″ of invasion in this calculation for an 8″ borehole).

It is important to note the laterolog's sensitivity to the borehole. Fig. 5–23 shows the correction chart for the deep and shallow measurement of a particular dual laterolog device, plotted in a manner similar to Fig. 5–6

Electrode Devices and How They Developed   93

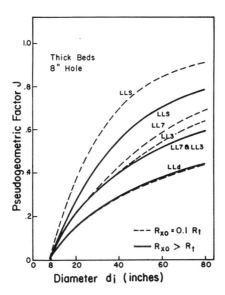

Figure 5–22. The comparison of calculated pseudogeometric factors for a number of common electrode devices. LLd and LLs refer respectively to the deep and shallow arrays of a dual laterolog device. From Schlumberger.[10]

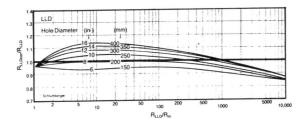

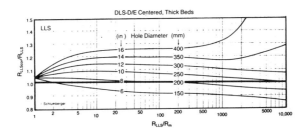

Figure 5–23. A borehole correction chart for the deep and shallow laterolog measurements. It is to be compared to Fig. 5–6, for the short normal, to appreciate the improved response due to focusing. From Schlumberger.[1]

for the short normal. It is seen that the deep reading is rarely in error by more than 10%, for a variety of borehole sizes and resistivity contrasts. The shallow measurement, however, may differ by as much as 30% from the value of R_t in the range of resistivity contrasts less than 1000, and far more when in excess of 1000, reflecting its sensitivity to R_{xo}. However, this measurement is used in comparison with the deep for the purposes of identifying invasion and providing a correction for the deep resistivity reading, rather than for saturation evaluation. Use of this technique for the correction of resistivity readings is covered in Chapter 7.

DUAL LATEROLOG EXAMPLE

Fig. 5-24 shows a typical dual laterolog presentation for a hypothetical reservoir (which is used in succeeding chapters to demonstrate the response of various logging tools). The reservoir consists of several water zones and a hydrocarbon zone of moderate porosity, and a section of tight (low porosity) limestone. Only two of the curves shown on the log are uniquely associated with the dual laterolog. They are indicated by the coding that designates LLs and LLd. These correspond to the shallow and deep resistivity measurements, respectively. The additional resistivity curve denoted by MSFL* is produced by a microresistivity device (indicating shallow depth of investigation, because of small electrode spacings), discussed in Chapter 7. The curve in track 1 is a gamma ray, which can be taken to indicate clean zones, as mentioned in Chapter 2.

Three water zones (A, B, D) indicated on the log are characterized, in this case, by the rather low resistivity readings and the lack of separation between the deep and shallow laterolog readings. The actual resistivity value will be determined by the formation water resistivity and the porosity, which at this stage is unknown. Assuming for the moment that the water resistivity is the same in all three zones, it would appear that zone D has the highest porosity since its resistivity is the smallest. The high resistivity readings in zone E are the result of the very low porosity in this limestone section.

The hydrocarbon section C indicates values of deep resistivity which are greater than those observed in the upper two indicated water zones. The shallow resistivity measurement however, is lower than in the two water zones. This may be the result of greater porosity in zone C than in zones A and B. Any further quantification of the contents of this formation will depend on further measurements or knowledge. As indicated, one of the most important pieces of information will be an estimate of the porosity.

Before any quantified interpretation can be made, it will be necessary to apply corrections to the resistivity readings. The pertinent correction charts can be found in References 1 and 4, among others. Reference 5 details the

---

* Mark of Schlumberger

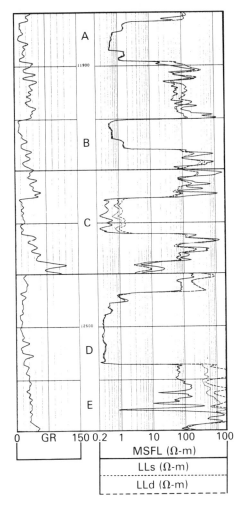

Figure 5-24. Section of a log showing the measurements from the dual laterolog and a gamma ray log.

steps involved in performing the corrections with which it may be helpful to become familiar. A summary of the major points of the procedure follows. Most of the corrections are made in terms of the mud resistivity (R_m) and mud filtrate (R_{mf}) resistivity, which can be obtained from the log heading (see Fig. 5-25, for example). The first step is to convert these two resistivity readings to the values they would have at formation temperature. This can be estimated from a recorded bottom hole temperature or from typical geothermal gradients for the region. The next step, for zones of interest, is to correct the resistivity readings for the influence of the borehole. This

```
                    DUAL LATERDLOG / GR
    Schlumberger
                         CSU   Field Log

    COMPANY:      USA-WEST
   *WELL:         STATE D-3                        OTHER SERVICES-
                                                     DIT
    FIELD:        OIL                                EPT
    COUNTY:       EASTERN                            TDT
    STATE:        COLORADO                           CNT
    NATION:       USA                                LDT
    LOCATION:                                        SGT

      SEC:      21        TWP: 28N       RGE: 24W

                                                   PROGRAM
    PERMANENT DATUM:       GL        ELEVATIONS-   TAPE NO:
    ELEV. OF PERM. DATUM:              KB:           26.2
    LOG MEASURED FROM:     KB          DF:         SERVICE
                   ABOVE PERM. DATUM   GL:         ORDER NO:
    DRLG. MEASURED FROM:   KB                        NONE

    DATE:                7 JUN 84
    RUN NO:

    DEPTH-DRILLER:
    DEPTH-LOGGER:
    BTM. LOG INTERVAL:
    TOP LOG INTERVAL:

    CASING-DRILLER:
    CASING-LOGGER:
    CASING:          0            0

    BIT SIZE:        0            0


    TYPE FLUID IN HOLE:    MUD
    DENSITY:               13.5 LB/G
    VISCOSITY:             67.0 S
    PH:                    12.1
    FLUID LOSS:            2.7 C3
    SOURCE OF SAMPLE:      CIRC
    RM:                    .078 OHMM AT 92.0 DEGF
    RMF:                   .035 OHMM AT 93.0 DEGF
    RMC:                        AT
    SOURCE RMF/RMC:        CUP/PRESS
    RM AT BHT:             .030 OHMM AT 243. DEGF
    RMF AT BHT:            .014 OHMM AT 243. DEGF
    RMC AT BHT:                      AT 243. DEGF

    TIME CIRC. STOPPED:
    TIME LOGGER ON BTM.:

    MAX. REC. TEMP:        243.0 DEGF

    LOGGING UNIT NO:
    LOGGING UNIT LOC:
    RECORDED BY:
    WITNESSED BY:
```

Figure 5–25. A sample log heading for the dual laterolog.

correction will depend on the size of the borehole, the stand-off of the tool, and the ratio of the resistivity reading to the mud resistivity. The final step is the correction for invasion, if present, which will be covered in Chapter 7.

References

1. *Schlumberger Log Interpretation Charts*, Schlumberger, New York, 1985, p. 59.
2. Doll, H. G., Tixier, M. P., Martin, M., and Segesman, F., *Electrical Logging*, Vol. 2 of *Petroleum Production Handbook*, SPE, 1962.
3. Serra, O., *Fundamentals of Well-Log Interpretation*, Elsevier, Amsterdam, 1984.
4. *Well Logging and Interpretation Techniques, The Course for Home Study*, Dresser Atlas, Dresser Industries, 1983.
5. Asquith, G. B., and Gibson, C. R., *Basic Well Log Analysis for Geologists*, AAPG, Tulsa, 1982.
6. Chemali, R., Gianzero, S., Strickland, R., and Tijani, S. M., "The Shoulder Bed Effect on the Dual Laterolog and Its Variation with the Resistivity of the Borehole Fluid," Paper UU, Trans. SPWLA Annual Symposium, 1983.
7. Suau, J., Grimaldi, P., Poupon, A., and Souhaite, P., "The Dual Laterolog–R_{xo} Tool," Paper SPE 4018, SPE Annual Meeting, 1972.
8. Anderson, B., and Chang, S-K.,: "Synthetic Deep Propagation Tool: Response by Finite Element Method," Paper T, Trans. SPWLA Annual Symposium, 1983. paper T.
9. Zienkiewicz, O. C., *The Finite Element Method in Engineering Sciences*, McGraw-Hill, New York, 1971.
10. *Log Interpretation: Principles*, Schlumberger, New York, 1972.

Problems

1. Fig. 5–22 shows, among other things, the pseudogeometric factor for the deep and shallow laterolog (LLd & LLs). Using this information, what is the apparent resistivity that you would expect for the LLd and LLs in a 30-PU water-bearing formation which has a diameter of invasion of 30″? The borehole is filled with relatively fresh water of 2.0 $\Omega\cdot$m resistivity, and the formation water resistivity is 0.1 $\Omega\cdot$m at the same temperature.

2. Using the log values of Fig. 5–24 and assuming the porosity of water zone D to be 20%:
   a. What do you estimate the porosity of zone E to be assuming that it is also water-filled?

b. What value of R_w would produce the resistivity observed in zone C if it were also a 20% porosity water zone rather than a hydrocarbon zone?

3. Using the data of Fig. 5–25, what value of R_{mf} would be appropriate for the evaluation of zone C, assuming that the bottom hole temperature (BHT) corresponds to the final depth of the log of Fig. 5–24?

4. Which of the zones of Fig. 5–24 indicate the presence of invasion?

6
INDUCTION DEVICES

INTRODUCTION

The presence of a conductive mud in the borehole is somewhat of a nuisance for electrode devices, as was illustrated in the last chapter. Many improvements have been made in electrode tool design to compensate for the problems. However, conductive borehole mud does provide one advantage: It effectively places the current and voltage measurement electrodes into electrical contact with the formation whose resistivity is to be measured.

What about those cases in which the mud is nonconductive (oil-base mud) or nonexistent (air-filled hole), or in which a plastic liner has been inserted into the borehole? It is for these cases that the induction tool was designed originally, although it has since found widespread use even in conductive muds. Induction devices use medium frequency alternating current to energize transmitter coils in the sonde; they, in turn, induce eddy currents in the formation. The strength of the induced eddy current is proportional to the formation conductivity. The magnitude of the induced currents is detected by receiver coils in the tool.

Before discussing the principles involved in the use of the induction tool, this chapter reviews some of the basics of electromagnetic theory. This review will serve as the basis for analyzing the characteristics of a two-coil device in detail. The analysis will develop the notion of the geometric factor, which is used to predict the radial and vertical tool response. The development of multi-coil focused devices follows directly from geometric factor theory. A modification to this simple theory is shown to be necessary to account for attenuation of the magnetic induction field, known as the skin effect.

REVIEW OF MAGNETOSTATICS AND INDUCTION

Induction devices employ alternating currents in transmitter coils to set up an alternating magnetic field in the surrounding conductive formation. This changing magnetic field induces current loops in the formation that are detectable by a receiver coil in the sonde. The details of the relationships between electric currents and magnetic fields, both steady-state and time-varying, are reviewed in this section to provide the basis for geometric factor theory that is used to demonstrate the response of induction logging tools.

Ampere's law states that a magnetic field will be associated with the flow of an electric current and directed at right angles to it. The strength of the magnetic field B is related to the current I. In particular the integral of the tangential component of B around any closed path Γ is proportional to the current piercing the area enclosed by Γ. This is expressed as:

$$\int_\Gamma \overline{B} \cdot \overline{dl} = \frac{I}{\varepsilon_o c^2} ,$$

where $\overline{dl}$ is a unit vector directed along the path Γ. Through the use of a vector identity, this is often written as:

$$\nabla \times \overline{B} = \frac{\overline{j}}{\varepsilon_o c^2} ,$$

where $\overline{j}$ is the current density, or the normal component of the current I divided by the surface area enclosed by Γ.

One simple application of this relation is the calculation of the magnetic field associated with the current flowing in a long wire, shown in Fig. 6–1. At a radial distance r from the wire, the path integral is just $B \cdot 2\pi r$, since the magnetic field B is in the form of closed circles around the current-carrying wire, and since B and I are known to be at right angles to one another. Thus the magnetic field strength relation may be determined from:

$$B \cdot 2\pi r = \frac{I}{\varepsilon_o c^2} ,$$

or

$$B = \frac{1}{4\pi \varepsilon_o c^2} \frac{2I}{r} .$$

The generalized expression for calculating the magnetic field from a current element is called the law of Biot-Savart and resembles the preceding expression:

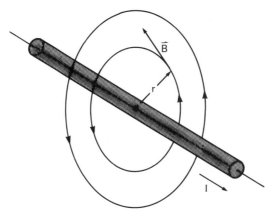

Figure 6–1. Circular lines of magnetic flux B, surrounding a very long straight wire carrying a current. From Feynman et al.[3]

$$\bar{B} = -\frac{1}{4\pi\varepsilon_o c^2} \int \frac{I \, d\hat{r} \times d\hat{l}}{r^2},$$

where $d\hat{l}$ is an elemental length along the current path Γ and $d\hat{r}$ is the unit vector in the direction of the observation point from the current element.

A simple application of the law of Biot-Savart which will be useful in the discussion of the induction device is the calculation of the component of the magnetic field perpendicular to the plane of a circular loop of current. As

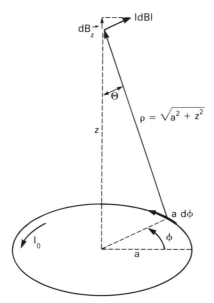

Figure 6–2. Geometry for the calculation of the vertical component of the magnetic field on the axis of a current-carrying circular loop of radius a.

shown in Fig. 6–2, the application is simple for reasons of symmetry. The vertical component B_z is to be calculated on the axis of the current loop. The sketch shows that the component $d\overline{B}$ is the result of one element of the current loop; it is oriented at right angles to the current element. The magnitude of this contribution to the magnetic field at a distance z above the loop of radius a is given from the law of Biot-Savart by:

$$dB \propto \frac{I_o a \, d\phi}{a^2 + z^2},$$

where the element of current is of length $a \times d\phi$.

It is clear that all but the z-component of the B field will be canceled when the whole current loop is considered. The component dB_z is seen to be:

$$dB_z = dB \, \sin\Theta = dB \, \frac{a}{\sqrt{a^2 + z^2}}.$$

The total contribution to the z-component is given by the integral of all the elements of current around the loop:

$$B_z = \int_0^{2\pi} \frac{I_o a^2}{(a^2 + z^2)^{3/2}} d\phi \propto \frac{m}{\rho^3}, \qquad (1)$$

where m, the magnetic moment (current in loop times area of loop), is given by:

$$m = I_o \pi a^2.$$

Another relation for which there will be a need is that of the vertical component of the magnetic field of a small current loop off-axis. For this problem, recourse is made to the vector potential $\overline{A}$, which is defined by:

$$\overline{B} = \nabla \times \overline{A}$$

and can be related to a current distribution in a fashion analogous to the relation between the electrostatic potential and a charge distribution:

$$\overline{A} = \frac{1}{4\pi\varepsilon_o c^2} \int \frac{\overline{j} \, dV}{r}.$$

Here, the current density distribution must be integrated over the volume (dV) which contains it. Once the vector potential is obtained, then the z-component of the magnetic field is simply obtained from:

$$B_z = (\nabla \times \overline{A})_z$$

$$= \frac{\partial A_y}{\partial x} - \frac{\partial A_x}{\partial y}.$$

The vector potential of a small current loop can be written in analogy with the electrostatic potential at a distance r from a dipole which is given by:

$$\phi(r) = \frac{1}{4\pi\varepsilon_o} \frac{p \cos \theta}{r^2},$$

where p is the dipole moment (the charge times separation distance) and θ is the angle between the orientation of dipole and the observation point. For the current loop in the x-y plane shown in Fig. 6-3, we will write an expression for the vector potential at the point indicated. It will consist of only two components, A_x and A_y, since there is no current distribution in the z-direction. To find the x-component of $\overline{A}$, only the current in the x-direction is considered, as shown in Fig. 6-3. The two parallel current paths are

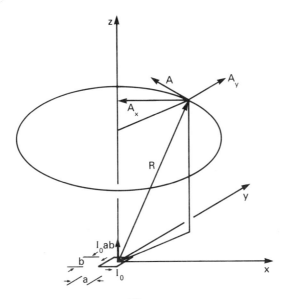

Figure 6-3. The vector potential $\overline{A}$, from a small current-carrying loop of rectangular cross section. Adapted from Feynman et al.[3]

equivalent to the concept of an electric dipole. By analogy with two charged rods, each with charge per unit length λ, the dipole moment would be the total charge times the separation or:

$$p = \lambda a\, b,$$

and the cosine of the angle between the point P and the dipole moment is $-\frac{y}{R}$. Continuing the analogy,

$$A_x = -\frac{I\, ab}{4\pi\varepsilon_o c^2} \frac{y}{R^3}.$$

The y-component can be found in the same manner to be:

$$A_y = I \frac{ab}{4\pi r \varepsilon_o c^2} \frac{x}{R^3} .$$

From the two components of the vector potential the spatial dependence of the vertical component of the magnetic field can be determined:

$$B_z \propto \frac{\partial}{\partial x}\left[\frac{x}{R^3}\right] - \frac{\partial}{\partial y}\left[\frac{-y}{R^3}\right]$$

$$B_z \propto \frac{1}{R^3} - \frac{3z^2}{R^5} . \qquad (2)$$

The final review item for the following discussion is that of Faraday's law of induction. From experimental observations, Faraday deduced that a changing magnetic field would set up a current in a loop of conductor present in the field. He also demonstrated that a changing current in one loop of wire could induce a current in another loop of wire, as illustrated in Fig. 6–4. The induced electromotive force associated with the induced

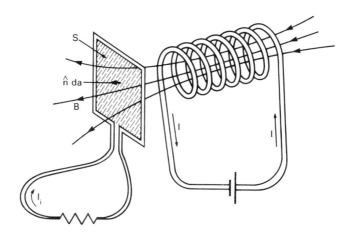

Figure 6–4. One aspect of Faraday's law of induction. An alternating current in the primary loop produces an induced current in the receiver loop.

current was found to be proportional to the rate of change of magnetic flux linking the circuit. This is most compactly expressed as:

$$\nabla \times \overline{E} = -\frac{\partial \overline{B}}{\partial t} .$$

Using Stoke's theorem, we can write this as:

$$\oint_\Gamma \overline{E} \cdot \overline{dl} = \int_S (\nabla \times \overline{E}) \cdot \hat{n} \, da = -\int_S \frac{\partial \overline{B}}{\partial t} \cdot \hat{n} \, da .$$

This last expression is seen to be the time rate of change of the normal component of magnetic flux through a surface S. The integral on the left is just the voltage seen at the terminals of a coil with cross-sectional area S.

THE TWO-COIL INDUCTION DEVICE

Fig. 6–5 shows the essential features of an induction logging device. It

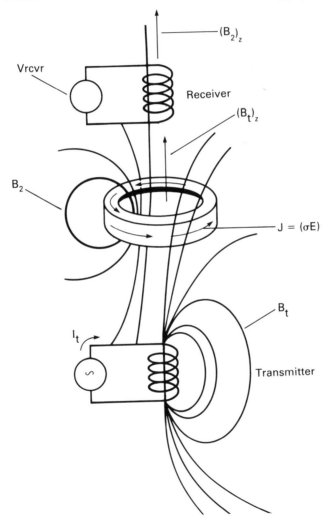

Figure 6–5. The principle of the induction tool. The vertical component of the magnetic field from the transmitting coil induces ground loop currents. The current loops in the conductive formation produce an alternating magnetic field detected by the receiver coil.

consists of a transmitter coil, excited by an alternating current of medium frequency (≈ 20 KHz) and a receiver coil. The two coils, contained in a nonconductive housing, are presumed to be surrounded by a formation of conductivity σ. One axially symmetric ring of current-bearing formation is indicated in the figure. Before analyzing the geometric sensitivity of such a device, it is worthwhile to step through the sequence of physical interactions which produce, finally, a signal at the receiver. In this way we will be able to see the dependence of the detected signal on excitation frequency and formation conductivity, as well as the phase relation between received and transmitted signal.

The first step to consider is the excitation of the transmitter coil by the transmitter current I_t:

$$I_t = I_0 e^{-i\omega t}.$$

The transmitter coil, which can be considered as an oscillating magnetic dipole, sets up throughout the formation a magnetic field B_t, whose vertical component is of interest. The vertical component will have a time dependence given by:

$$(B_t)_z \propto I_0 e^{-i\omega t}.$$

If a ring of formation material which is axially symmetric with the tool axis is considered, it forms the perimeter of a surface through which passes a time-varying magnetic field. From Faraday's law, an electric field E will be set up which is proportional to the time derivative of the vertical component:

$$E \propto -\frac{\partial (B_t)_z}{\partial t} \propto i\omega I_0 e^{-i\omega t}.$$

This electric field, which curls around the vertical axis, will induce a current density in the loop of formation sketched. It will be proportional to the formation conductivity:

$$J \propto \sigma E \propto i\omega\sigma I_0 e^{-i\omega t}.$$

The current in the ground loop considered will behave like the transmitter coil; that is, it will set up its own magnetic field B_2. The vertical component of the secondary magnetic field $(B_2)_z$ has the same time dependence as the current density in the loop:

$$(B_2)_z = i\omega\sigma I_0 e^{-i\omega t},$$

and its time dependence will induce a voltage V_{rcvr} at the receiver coil:

$$V_{rcvr} \propto -\frac{\partial (B_2)_z}{\partial t} \propto -\omega^2 \sigma I_0 e^{-i\omega t}.$$

This final result indicates that the voltage detected at the receiver coil will vary directly with the conductivity of the formation and with the square of the excitation frequency. It is also seen to be 180° out of phase with the transmitter current driving signal, whereas the voltage induced by the direct flux linkage from the transmitter, which has been ignored in this discussion, will be 90° out of phase. Phase-sensitive detection is used to separate these two signals.

Geometric Factor for the Two-coil Sonde

In order to determine the geometric sensitivity of the two-coil induction sonde, following the treatment by Moran and Kunz,[1] we now make use of the relations derived in the review of magnetostatics. The first expression to be derived is the component of the driving magnetic field, which sets up the ground current indicated in Fig. 6–6. The driving coil is considered to be a magnetic dipole source which produces a vertical component of magnetic field at a distance z above the transmitter. In this case, the dipole moment of the transmitter is given by the product of the current I_o, the winding area A, and the number of transmitter winding turns n_T ($I_o n_T A$). At any position identified by the coordinates (ρ_t, z), the vertical component, from Eq. (2), is given by:

$$B_z \propto I_o e^{-i\omega t} A n_t \left[\frac{1}{\rho_t^3} - \frac{3z^2}{\rho_t^5} \right].$$

The left side of Fig. 6–6 shows the geometry to be considered for determining the magnitude of the current density set up in the indicated ground loop. Dropping, for convenience, the time-dependent terms and other constants, the relation for the induced electric field is:

$$\int E \cdot d\bar{s} = -\frac{\partial}{\partial t} \int_s B_n \, dA ,$$

where the surface integral is over the element of area:

$$dA = 2\pi r dr$$

and the normal component of the magnetic field, B_n, is B_z. This results in:

$$\int E \cdot d\bar{s} = E \cdot 2\pi r \propto \int_0^r \left[\frac{1}{\rho_t^3} - \frac{3z^2}{\rho_t^5} \right] r dr \propto \frac{r^2}{\rho_t^3} ,$$

or

$$E \propto \frac{r}{\rho_t^3} .$$

This electric field then causes a current density J which is given by:

$$J = E \sigma ,$$

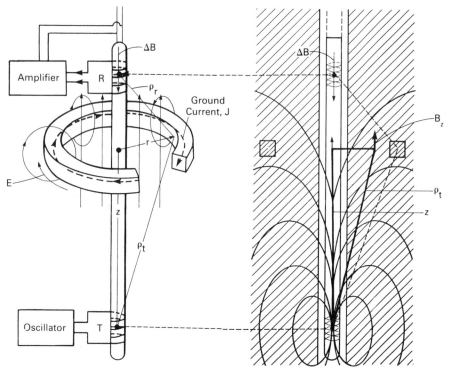

Figure 6–6. Geometry for the development of the geometric factor for a three-coil induction sonde. Adapted from Doll.[5]

where σ is the formation conductivity. Thus the geometric dependence of the induced current is given by:

$$J \propto \frac{\sigma r}{\rho_t^3} .$$

The induced voltage in the receiver coil will be proportional to the vertical component ΔB, indicated in Fig. 6–6, which passes through the receiver coil. From Eq. (1), it is seen to be:

$$\Delta B \propto J \frac{r^2}{\rho_r^3} ,$$

where J is the current density in the ground loop in question of radius r, and ρ_r is the distance from any point along the current loop to the receiver coil.

The final result, then, is that the geometric dependence of the detected signal is proportional to:

$$g(r,z) \propto \frac{r}{\rho_t^3} \frac{r^2}{\rho_r^3} .$$

The above expression is known as the differential geometric factor, since it gives the contribution of a single ground loop of unit cross-section at position z and radius r to the final receiver output.

It is convenient to define two other geometric factors which give information on the tool response in a cumulative sense. The first is the differential radial geometric factor, which is defined to be:

$$g(r) = \int_{-\infty}^{\infty} g(r,z)dz .$$

It predicts the relative importance of each of the cylindrical shells of radius r to the overall response. This factor and a sketch of its radial dependence are shown in Fig. 6-7. A peaking in the relative importance is shown for

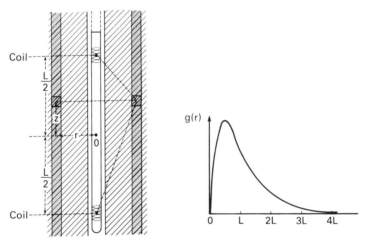

Figure 6-7. Integration of the geometric factor with respect to z at a constant radial value r produces the differential radial geometric factor.

cylindrical formation layers with radius somewhat less than the dimension of the coil separation.

In a similar fashion, the differential vertical geometric factor is defined as:

$$g(z) = \int_{0}^{\infty} g(r,z)dr$$

and gives the response of a unit-thickness slice of formation, located at position z, to the overall tool response. The geometry corresponding to the integration and the response curve are shown in Fig. 6-8. A fairly flat response is obtained from a slice of formation contained between the two coils, but the tapering-off of the response above and below the coils will produce distortion of bed boundaries.

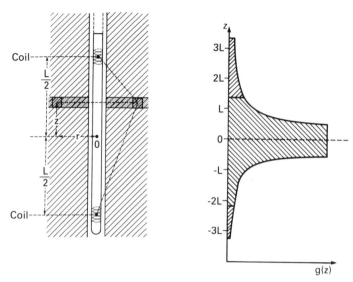

Figure 6–8. The differential vertical geometric factor produced by integration with respect to r, at fixed z.

In order to get an idea of the bed boundary response, we can make an integration of the differential vertical geometric factor. Fig. 6–9 shows an example of the integrated vertical factor G_r for a two-coil device with a 40″ coil separation. It is seen that a sharp transition of formation resistivity will be reported by the tool reading to vary over a distance which is roughly two times the coil spacing. It is obvious that the conductivity readings in thin beds will be considerably affected by this type of coil arrangement.

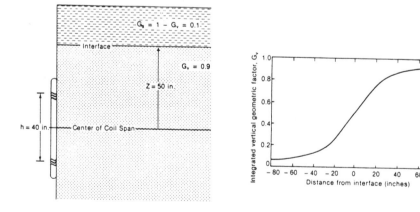

Figure 6–9. The integrated vertical geometric factor for estimating the influence of shoulder beds. Adapted from Dresser.[4]

The integrated radial geometric factor G_r for the two-coil device is shown in Fig. 6–10. There appears to be some sensitivity to the region nearest the borehole. It would be desirable to eliminate this sensitivity to a presumed invaded zone and to put more weight on the region farther from the borehole, where the true resistivity could be measured.

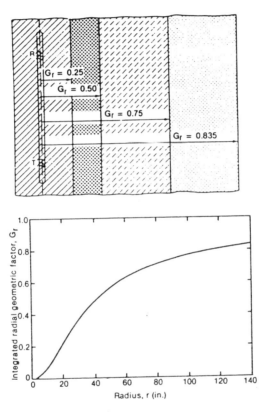

Figure 6–10.   The integrated radial geometric factor for estimating the importance of invasion. Adapted from Dresser.[4]

Focusing the Two-coil Sonde

The response of the two-coil device examined above can be altered to minimize the "tail" of sensitivity to beds above and below the measurement coils or to decrease the sensitivity to layers closest to the borehole. For an illustration of how the response is altered or focused, we examine the technique for changing the depth of investigation of the two-coil sonde. The idea is simply to add a second receiver coil which is a bit closer to the transmitter and to use its response, which will be somewhat shallower than the original receiver, to subtract from the response of the original. This subtraction, if properly normalized, should eliminate much of the signal from regions close to the borehole. This principle is shown schematically in Fig. 6–11. A similar procedure is used to sharpen the vertical resolution of the tool. This will change the sensitivity of the tool measurement to layers of different conductivity above and below the measurement coils.

Commercial induction devices employ focused arrays of coils, usually

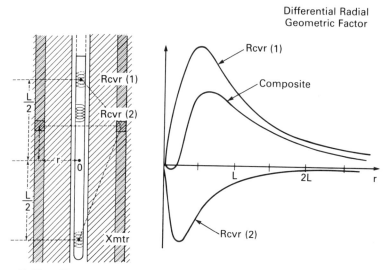

Figure 6–11. The principle of three-coil focusing. A second coil, wound with reverse polarity, produces a signal which cancels some of the signal from close to the borehole. Adapted from Doll.[5]

providing two measurements of conductivity (resistivity) at different depths of investigation. The improvement of the depth of investigation of one such device can be seen from a comparison of the integrated radial response functions of a two- and six-coil device shown in Fig. 6–12. Most of the response closer than about 30″ has been eliminated. For this reason the deep measurement* of the induction device is frequently referred to as R_t, since it is generally a measurement of that value without the perturbation of intervening layers.

Tailoring the response by the addition of coils may sound too good to be true. Of course there are limitations; the addition of focusing coils leaves some residual imbalance in the geometric factor, which will make itself known in logging situations. In the preceding discussion, we have considered

---

* The deepest measurement, i.e., the largest coil spacing, is indicated by the designation ILD on some log headings. The shallower-reading configuration may be designated by ILM.

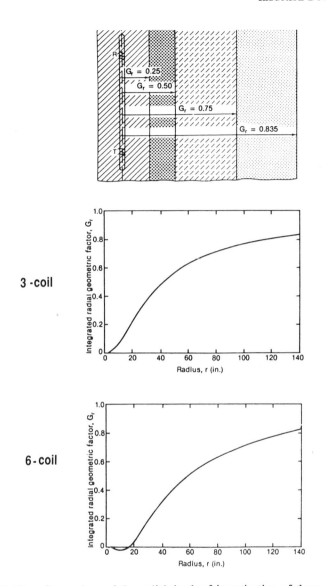

Figure 6–12. Comparison of the radial depth of investigation of three- and six-coil induction devices. Adapted from Dresser.[4]

only homogeneous formations. In reality, layered formations of differing conductivity will be the rule, not to mention radial conductivity profiles which are far from uniform because of invasion or the presence of dipping beds. How will these affect the response of the induction tool?

An idea can be obtained from a closer examination of the composite radial geometric factor of Fig. 6–11. Note the small undershoot. The impact of this imbalance is that the conductivity of the initial portion of the formation near the borehole will make a negative contribution to the total signal. This is no problem in a homogeneous formation. However, suppose there is a conductive anomaly near the borehole: Taken to the extreme, this could cause a negative reading!

Figure 6–12a shows the detailed geometric factors for the deep and medium arrays of one induction device. This three-dimensional display shows the contribution of all the significant rings of formation material; the portions in black correspond to negative contributions to the total signal. Note that in addition to being nonsymmetric, the medium array shows a negative contribution at the center of the array much like the six-coil device seen earlier. The deep array is symmetric but exhibits negative lobes above and below the center of the array. It is these lobes which can cause the appearance of "horns" at bed boundaries of sufficient conductivity contrast. Examples of this type of behavior can be found in Reference 6. Detailed studies of induction tool geometric factors and their influence on log response to specific situations such as caves, bed boundaries, dipping beds, and thin beds can be found in References 8, 9, and 10. Some modern induction devices even take advantage of the out of phase quadrature signal to improve bed boundary response and depth of investigation.[11,12]

SKIN EFFECT

One characteristic of electromagnetic waves has been overlooked in the discussion of induction devices. It is referred to as the skin effect and is simply the result of the fact that electromagnetic waves suffer an attenuation and phase shift when passing through conductive media. The parameters which govern this effect can be made apparent for the case of a time-varying electric field at the surface of a conductive formation, as shown in Fig. 6–13.

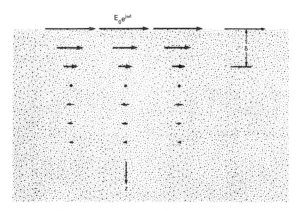

Figure 6–13. A one-dimensional model of a time-varying electric field at the surface of a conductor. Due to its conductivity, the intensity of the electric field diminishes with depth of penetration into the conductor.

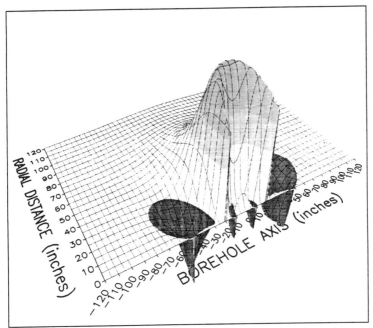
Deep Array

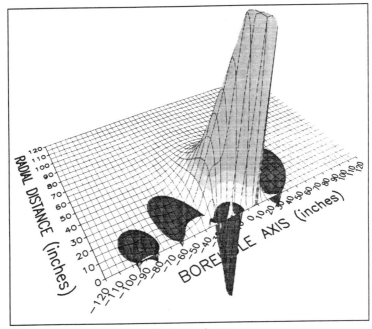
Medium Array

Figure 6–12a. Geometrical factor maps for medium and deep induction arrays. From Anderson.[10]

To simplify the situation, the electric field at the surface of this infinite half-space is taken to vary only in the z-axis.

From three of Maxwell's equations:

$$\nabla \cdot \overline{E} = 0, \qquad (3)$$

$$\nabla \times \overline{B} = \mu \overline{j}, \qquad (4)$$

$$\nabla \times \overline{E} = -\frac{\partial \overline{B}}{\partial t}, \qquad (5)$$

and the current relationship:

$$\overline{j} = \sigma \overline{E},$$

a wave equation can be derived by using a vector identity.

First the curl of Eq. (5) is taken:

$$\nabla \times \nabla \times \overline{E} + \nabla(\nabla \cdot \overline{E}) - \nabla^2 \overline{E} = \qquad (6)$$

$$-\frac{\partial}{\partial t} \nabla \times \overline{B}.$$

From Eq. (3), which implies that no free charges are present in this conductive medium, the left side of Eq. (6) reduces to:

$$-\nabla^2 \overline{E} = -\frac{\partial}{\partial t} \mu \sigma \overline{E}, \qquad (7)$$

after having used the relation (Eq. (4)) between $\nabla \times \overline{B}$ and the current density $\overline{j}$.

If a sinusoidal time dependence:

$$E = E_o e^{i\omega t}$$

is assumed for the electric field, then Eq. (7) reduces to:

$$\frac{\partial^2 \overline{E}}{\partial^2 z} = i\omega\mu\sigma \overline{E} = k^2 \overline{E}.$$

The solution to this equation is of the form:

$$E = E_o e^{-kz},$$

where k can be written as:

$$k = \sqrt{i} \sqrt{\omega\mu\sigma}.$$

Using the relation that:

$$\sqrt{i} = \frac{1+i}{\sqrt{2}},$$

k can be written as:

$$k = \frac{1+i}{\delta},$$

where δ, is given by:

$$\delta = \sqrt{\frac{2}{\omega\mu\sigma}}.$$

Thus the form of the electric field in the z-axis (into the conductive half-space) is given by:

$$E(x) = E_o e^{-\frac{x}{\delta}} e^{-\frac{i}{\delta}z},$$

which indicates both an attenuation and phase shift increasing with penetration into the conductive formation. The parameter δ is the skin depth, which is the distance over which the electric field will be reduced by a factor of 1/e. The magnitude of this distance is given in Fig. 6–14 for formation

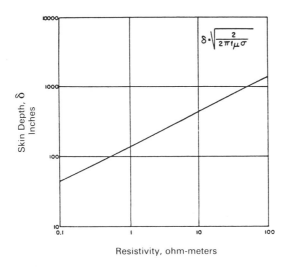

Figure 6–14. The numerical value of the skin depth in inches as a function of the formation resistivity in Ω·m. for operating frequencies around 20 kHz.

resistivity in Ω·m, evaluated for an induction tool operating at 20 kHz. It can be seen that there will be some noticeable effect on induction measurements if the resistivity is less than about 10 Ω·m. At this point the skin depth is on the order of the depth of investigation and can no longer be ignored. The skin depth will have some impact on the geometric factor. In Fig. 6–15, the integrated radial geometric factor is illustrated for several types of induction tools. Of particular note is the dotted curve, which is the geometric factor with skin effect taken into account, showing, in general, a reduced depth of investigation.

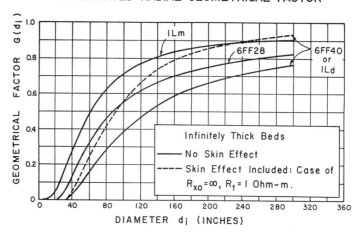

Figure 6–15. The integrated radial geometric factor of several commercial tools, indicating the effect of the skin effect. From Schlumberger.[2]

INDUCTION OR ELECTRODE?

When is an induction device used in preference to an electrode tool? It will depend on the conditions of mud and formation conductivity. Generally, in very conductive muds an electrode device will be preferred.

Fig. 6–16 gives some guidance as to when to use an electrode device over an induction device. This chart was devised for the case of obtaining the best value of R_t in hydrocarbon-bearing zones. It is presented in terms of the porosity and the ratio of the mud filtrate resistivity and the formation water resistivity. For the left side of the curve, it is clear that the laterolog is preferred, since the borehole is very conductive. This allows a good contact between the electrode and the formation but provides a very undesirable high conductivity value in the invaded zone, which will perturb the measured induction value. Using the geometric factor theory, the apparent conductivity seen by the tool is given by:

$$\sigma_a = G_r(d_i)\sigma_{xo} + (1-G_r(d_i))\sigma_t .$$

Thus the effect of the invaded zone conductivity σ_{xo}, can be evaluated. If this zone has a large enough conductivity, it can begin to obscure the effect of the deeper zone on the measurement.

On the right side of Fig. 6–16, the resistivity of the mud is several times greater than the water resistivity, and in most of the cases the induction tool is preferred. However, in cases where R_t is expected to be very high (several hundred $\Omega \cdot m$), the laterolog may be used.

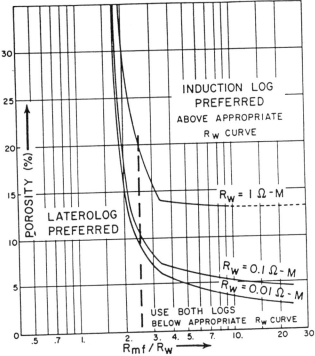

Figure 6–16. Ranges of application of induction and laterolog. From Schlumberger.[2]

INDUCTION LOG EXAMPLE

For the sample induction log, a simulated reservoir is used again. Two thick clean zones are indicated as A and B. Two much thinner clean streaks are shown as C and D. These four zones can be easily identified on the sample log presentation of Fig. 6–17. For the identification of the clean zones, the SP is shown in track 1 along with a gamma ray. Of the three resistivity curves shown in tracks 2 and 3, only two are properly associated with the induction. They are marked ILD (deep induction) and ILM (medium induction). The third curve (SFLU) is another variety of microresistivity (electrode) device to be considered in the next chapter.

As in the case of the laterolog curves, the induction curves must be checked for any necessary corrections before attempting quantitative interpretation. In addition to the same general types of borehole corrections, the induction may also require correction for bed thickness. The magnitude of this correction will depend on an estimate of the bed thickness and the resistivity of adjacent or shoulder beds.[7] This type of correction will certainly be necessary for the two thin streaks of zones C and D. Reference 6 gives a step-by-step example of correcting induction log reading for borehole effects, invasion, and bed thickness.

120 Well Logging for Earth Scientists

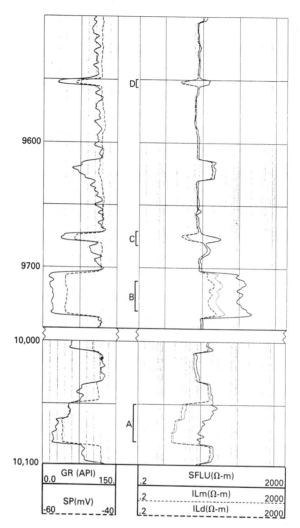

Figure 6–17. A sample induction log.

References

1. Moran, J. H., and Kunz, K. S., "Basic Theory of Induction Logging and Application to Study of Two-Coil Sondes," *Geophysics* Vol. 27, 1962.

2. *Log Interpretation:* Vol. 1: *Principles*, Schlumberger, New York, 1972.

3. Feynman, R. P., Leighton, R. B., and Sands, M. L.: *Feynman Lectures on Physics*, Vol. 2, Addison-Wesley, Reading, Mass., 1965.

4. *Well Logging and Interpretation Techniques: The Course for Home Study*, Dresser Atlas, Dresser Industries, 1983.

5. Doll, H. G., Pet. Trans. AIME 186, 1949.

6. Jorden, J. R., and Campbell, *Well logging II - Electrical and Acoustic Logging*, Monograph Series, SPE, Dallas (in press).

7. *Schlumberger Interpretation Charts*, Schlumberger, New York, 1985, p. 65-67.

8. Gianzero, S., and Anderson, B., "A New Look at Skin Effect," *The Log Analyst*, Vol. 23, No. 1, 1982, pp 20-34.

9. Anderson, B., and Chang, S. K., "Synthetic Induction Logs by the Finite Element Method," *The Log Analyst*, Vol. 23, No. 6, 1982.

10. Anderson, B., "The Analysis of Some Unsolved Induction Interpretation Problems Using Computer Modeling," Paper II, SPWLA Twenty-seventh Annual Logging Symposium, 1986.

11. Barber, T. D., "Real-Time Environmental Corrections for the DIT-E Digital Dual Induction Tool," *Trans.* SPWLA, 1986.

12. Barber, T. D., "Invasion Profiling with the Phasor Induction Tool," *Trans.* SPWLA, 1986.

Problems

1. You are logging with a rudimentary two-coil induction device through a 40"-thick water-bearing sandstone with very thick shale beds above and below. You know that $R_o = 0.5$ $\Omega\cdot$m and that the shale resistivity is 1.0 $\Omega\cdot$m. Using the integrated vertical geometric factor for the induction device given in the text (remembering that the geometric factors for induction devices apply to conductivity):

   a. Sketch the log response as the tool approaches and passes through the zone of sandstone.

   b. Calculate the minimum resistivity you would measure in the sandstone.

   c. Assuming that you can read the resistivity to 10% accuracy, what is the minimum bed thickness you could detect for the resistivity contrasts of part a?

2. In the log of Fig. 2–12, estimate the hydrocarbon saturation in the zone of interest, before and after correcting the deep induction reading for bed thickness. The correction can be made by use of appropriate charts from Reference 7 or others. The bed thickness appears to be 4'.

3. A formation is known to have a water saturation of 60% and water resistivity of 2 $\Omega\cdot$m. A well drilled through this formation was logged with the deep induction and shallow laterolog. The values observed were $R_{ILd} = 145$ $\Omega\cdot$m and $R_{LLs} = 180$ $\Omega\cdot$m. The mud filtrate had a resistivity of 3.8 $\Omega\cdot$m. The estimated diameter of invasion, d_i, was 60". What would you expect the residual oil saturation of this reservoir to be after water flooding?

4. Show that for large values of z the differential vertical geometric factor $g(z)$ varies as $\frac{1}{z^2}$. This is in contrast with an electrode tool which varies as $\frac{1}{z}$ and is the reason why induction tools are less influenced by shoulder beds.

7
ELECTRICAL DEVICES FOR MEASURMENTS OTHER THAN R_t

INTRODUCTION

The measurement of the resistivity of the invaded or flushed zone (R_{xo}) is of interest for several reasons; it is related to the identification of movable hydrocarbons and its value can be used to correct the observed resistivity of deeper reading devices. Historically, the first use of the invaded zone resistivity was, in the absence of any other measurement, to make an estimate of the formation porosity. By assuming that the saturation of the invaded zone was unity, and by making some assumption about the form of the formation factor-porosity relationship, a lower limit of porosity could be found.

Some of the log examples in earlier chapters make clear that R_{xo}, when compared with R_t, can be used used to obtain a visual indication of permeable zones and evidence of moved hydrocarbons. Using the measured values of R_{xo}, in conjunction with other information, it is possible also to determine the water saturation of the invaded zone S_{xo} and thereby estimate the efficiency of water-flood recovery. Another use of R_{xo} measurements, in combination with the medium and deep resistivity measurements, is to obtain a better estimate of the deep resistivity R_t. This correction is possible due to the dissimilar depths of investigation of the three measured resistivities involved.

Before presenting examples of the application of the R_{xo} measurement to the above problems, we will examine a few of the electrode devices which have been designed to make this measurement. Two additional topics conclude the review of resistivity measurements. The first is a brief presentation of the effect of clays on formation resistivity. Although this topic is taken up in more detail in Chapter 21, it serves here to make the student aware of the problem of practical resistivity interpretation. The second topic concerns the basis for the measurement of the formation dielectric constant and an introduction to some of the methods for its interpretation.

MICROELECTRODE DEVICES

Microelectrode devices, as their name implies, are electrical logging tools with electrode spacings on a much-reduced scale compared to the tools previously considered. A further distinction, a result of the smaller spacings, is that their depth of investigation is also much reduced. The electrodes are mounted on special devices, called pads, which are kept in contact with the borehole wall during the ascent phase of the tool to which they are attached.

The development of microelectrode devices followed the same evolution as seen previously for the electrode tools. The first was the *microlog* device

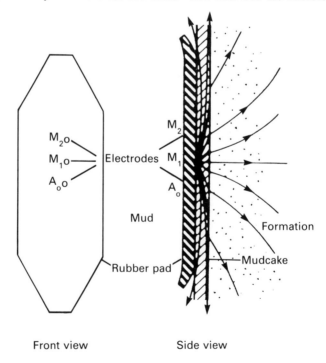

Figure 7–1. A microlog device: a pad version of the short normal. The spacing between the two voltage electrodes is 1″. From Serra.[1]

(Fig. 7–1), which was based on the same principle as the normal device. No focusing is used on this device. Current is emitted from the button marked A_o, and the potentials of the two electrodes M_1 and M_2 are measured. To ensure a shallow depth of investigation, the spacing between electrodes is 1″. Electrode M_1 has a shallower depth of investigation and is mostly influenced by the presence of mudcake. Electrode M_2, being farther from the current source, is influenced more by the flushed zone. Separation between the two curves presented on the log is indicative of permeable zones. Examples of many microelectrode device logs and their interpretation can be found in Reference 2.

As indicated by the current lines in Fig. 7–1, the microlog could be strongly influenced by the presence of mudcake, especially in the case of a resistive formation and a very conductive and thick mudcake. In order to deal with this problem, a focused or microlaterolog device was the next innovation. Fig. 7–2 is a schematic of this device, which shares many features of the laterolog, except for dimensions. As indicated in Fig. 7–2, the

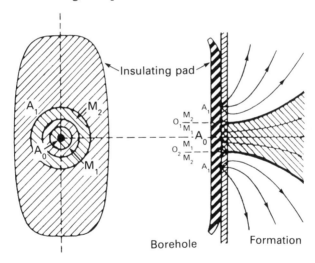

Figure 7–2. A microlaterolog device: a reduced scale and pad version of the laterolog. From Serra.[1]

bucking current from electrode A_1 focuses the measure current to penetrate the mudcake. The depth of investigation of this tool can be appreciated from a glance at the pseudogeometric factor in Fig. 7–3 which shows that 90% of the measured signal comes from the first 2 to 4 inches of formation (depending on R_{xo} and R_t contrast).

Another microresistivity device, based on the spherical focusing technique, is shown in Fig. 7–4. The two mudcake correction charts in Fig. 7–5 allow comparison between these two types of devices. The spherically focused device appears to be much less sensitive to the presence

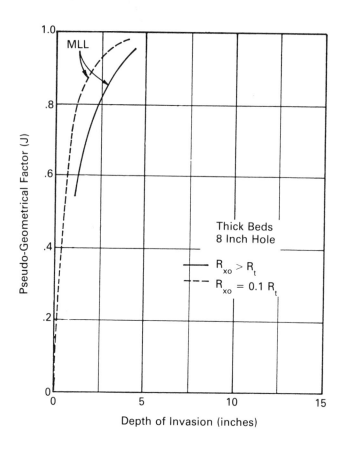

Figure 7–3. The pseudogeometric factors for a microlaterolog device. Adapted from Schlumberger.[3]

of mudcake. It is usually this type of device that is appended to a laterolog or induction tool to provide the third resistivity curve seen on all of the preceding resistivity log examples.

Another microresistivity device has been designed to determine the orientation of bedding planes. It is known as the dipmeter and consists of a sonde with three or four arms which press measurement "pads" against the borehole wall. Each pad contains one or several microelectrodes which sample (usually with a fine vertical resolution on the order of 0.1") the resistivity of the borehole wall. Although the measurements are not necessarily calibrated in terms of resistivity, the vertical sequence of resistivity anomalies is of interest for determining the three-dimensional orientation of strata intersecting the borehole. For a vertical well traversing horizontal layers of formation, the resistivity variations encountered by the measurement pads should correlate at the same depth. Depending on the

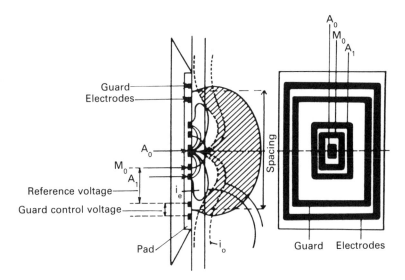

Figure 7-4. A micro spherically focused device. The upper portion shows the pad and electrode design with a schematic representation of the current paths. From Serra.[1]

orientation of the sonde (which is determined by an inertial platform or a magnetometer and pendulum), dipping beds will produce resistivity anomalies at different depths for each arm. The shift required to bring them into alignment will depend on the formation dip angle and borehole size.

The raw resistivity curves of the dipmeter are rarely used for interpretation but are subjected to various correlation or pattern recognition processing programs. These produce a summary log of the correlated events, which indicates the bedding orientation (dip angle and azimuth). The interpretation of the summary log, or "tadpole plot," in terms of structural geology and depositional environment, is beyond the scope of this book but is thoroughly treated in References 1, 4, 5, and 6.

A recent evolution from the dipmeter is a device which incorporates a large number of microelectrodes on several of the measurement pads.[7] The arrays of staggered electrodes are sampled at a high rate and processed to provide an electrical image of a significant portion of the borehole wall. Details on the scale of a few millimeters are resolved, and the electrical image is nearly indistinguishable from a core photograph.

Figure 7-5a compares a log from such a device with a core photograph over a 4' interval. The two images on the right were obtained from two microelectrode arrays on measurement pads at right angles to one another. The orientation of the core photograph corresponds to the electrical image on the right. The major features of the core photograph, corresponding to sedimentary features, are found in the electrical image. Additional features in the electrical image are due to conductivity variations (due to changes in

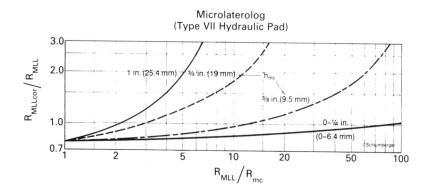

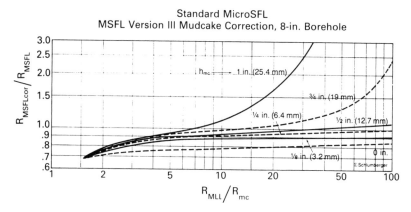

Figure 7-5. Mudcake corrections for two types of microresistivity devices. From Schlumberger.[1]

porosity and clay content) which are not visible on the core. In the image on the left, there are a number of vertical fractures which are easily identified. These images provide considerable enhancement to the conventional dipmeter measurements which are made simultaneously.

USES FOR R_{xo}

In the early years of resistivity logging, no porosity information was available from other logging devices. For this reason, the first use of R_{xo} to be considered, the estimation of porosity, is of historical interest only. This estimation procedure is based on knowledge of the mud filtrate resistivity R_{mf}

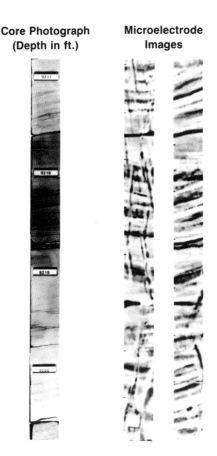

Figure 7–5a. An electrical image of the borehole produced by arrays of microelectrodes of a modified dipmeter. A scaled photograph of the core of the logged section is flanked by images from opposite sides of the hole. From Ekstrom et al.[7]

(obtained from a mud sample) and a very shallow resistivity measurement. The resistivity of a formation saturated with mud filtrate is related to R_{mf} by the formation factor, which is a function of porosity.

Following the definition of the formation factor F, which relates the fully water-saturated formation resistivity to the water resistivity,

$$R_o = F\, R_w ,$$

one can write an analogous expression for the invaded zone:

$$R_{xo} = F\, R_{mf} .$$

Here, it is supposed that the mud filtrate of known resistivity R_{mf} has displaced the connate water. Also, by analogy, an expression for the mud filtrate saturation of the invaded zone can be written:

$$S_{xo} = \sqrt{F \frac{R_{mf}}{R_{xo}}},$$

where the mud filtrate resistivity has replaced R_w in the usual formula, and R_{xo} has replaced R_t.

In order to get an estimate of the porosity, one can further make the assumption that the invaded zone is completely water-saturated and that the porosity dependence of F is $\frac{1}{\phi^2}$. From this, one obtains:

$$\frac{1}{\phi^2} = \frac{R_{xo}}{R_{mf}}.$$

Since the water saturation may not be complete, this can be used to obtain a lower limit to porosity, which is given by:

$$\phi \geq \sqrt{\frac{R_{mf}}{R_{xo}}}.$$

Although we have already seen the use of the R_{xo} curve for the identification of movable oil, it is worth a moment to quantify the separation often observed between the shallow resistivity curves, which correspond to R_{xo}, and the deep resistivity curves, which normally correspond to R_t.

From the generalized saturation equation:

$$S_w^n = \frac{a}{\phi^m} \frac{R_w}{R_t},$$

it is possible to write an expression to compare the initial value of the water saturation (that in the uninvaded zone) to the residual water saturation in the invaded zone. This is given by:

$$\left[\frac{S_w(\text{initial})}{S_w(\text{residual})}\right]^n = \frac{\frac{R_w}{R_t}}{\frac{R_{mf}}{R_{xo}}} = \frac{R_w}{R_{mf}} \frac{R_{xo}}{R_t}.$$

The utility of this expression can be appreciated by noting that the ratio R_{xo}/R_t is the separation between the deep and shallow resistivity readings on the logarithmic format usually used. Since the expression may also be rewritten as:

$$\frac{R_{xo}}{R_t} = \frac{R_{mf}}{R_w} \left[\frac{S_w(\text{initial})}{S_w(\text{residual})}\right]^n,$$

it is clear that in a water zone, where both saturations may be assumed to be identical (if not unity), the separation should just be equal to the ratio of the mud filtrate resistivity to the water resistivity. Any decrease in this ratio implies the presence of movable oil. This is based on the assumption that the mud filtrate saturation of the invaded zone will remain close to 100%. The decrease of the ratio can only mean that the initial saturation of the deeper zone has decreased.

An example of this type of behavior can be seen in the laterolog example of Fig. 7-6. Shown is a log of the bottom 800 feet of a hydrocarbon reservoir. At the bottom of zone 1 it can be assumed, in the absence of other information, that only water is present; the formation is fully water-saturated.

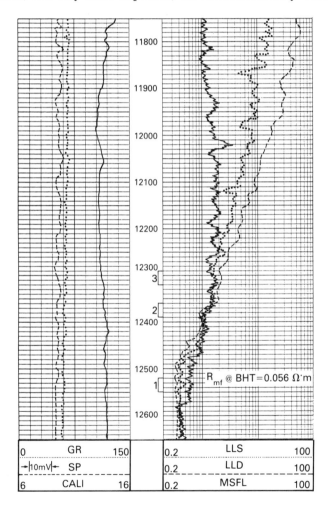

Figure 7-6. Idealized log to be expected from a dual laterolog with a microresistivity device in a thick reservoir. The bottom zone is a water zone and the uppermost portion is hydrocarbon. A long transition zone is apparent.

In the bottom interval, the shallow (MSFL) and deep resistivity separation is about a factor of 2. This ratio steadily decreases to about 1/80 in the upper part of the reservoir, indicating that the water saturation in the uninvaded zone gets progressively smaller. This indicates progressively greater hydrocarbon saturation. Thus the ratio R_{xo}/R_t decreases as hydrocarbon is displaced from the invaded zone by the mud filtrate.

In the preceding example, we looked at relative saturations between the invaded and deep zones. However, the saturation of the invaded zone is of some interest in its own right. For its determination, additional information is necessary. If the value of porosity is known from an additional measurement, then the residual oil saturation can be calculated from:

$$S_{xo}^n = \frac{a}{\phi^m} \frac{R_{mf}}{R_{xo}} .$$

This saturation can be used to determine the efficiency of water-flood

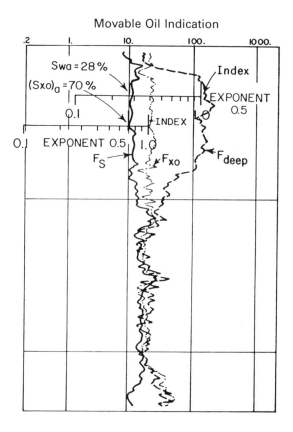

Figure 7–7. A graphical visualization of movable oil. This type of display was popular in the precomputer era and depends upon the simultaneous measurement of porosity (the F_s–curve). From Schlumberger.[8]

production, because it quantifies, to some extent, the residual hydrocarbon saturation after flushing with mud filtrate.

Fig. 7-7 shows a graphical application of R_{xo} as a movable oil indicator. In conjunction with a porosity log, three versions of the formation factor can be plotted: the porosity device formation factor F_s ($=1/\phi^2$), the apparent invaded formation factor F_{xo} ($= R_{shallow}/R_{mf}$), and the deep formation factor F_{deep} ($= R_{deep}/R_w$). As they are presented on a logarithmic scale, the deep and invaded formation factors simply track the respective resistivity curves and agree with the porosity-derived formation factor only in water-bearing zones. It is in these zones that the three curves are normalized, in the absence of another method. After this normalization, the porosity-derived

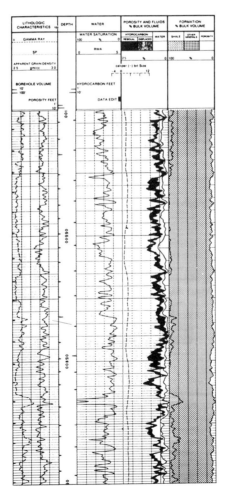

Figure 7-8. An example of a computerized interpretation which presents a display of the estimated movable oil. From Dresser.[9]

curve can be taken as the 100% water-saturation line. It traces the resistivity value to be expected from the variations in porosity for a fully water-saturated case. The other two curves can be read in water saturation directly, as indicated by the saturation scales on the figure (which change only 1 decade for each 2 decades of resistivity). Movable oil is readily identified in regions where $S_{xo} > S_w$. In the example shown, the apparent flushed zone saturation is 70%, and the apparent saturation from the deep device is 28%.

Fig. 7–8 is an example of one type of computerized interpretation presentation which combines porosity and resistivity measurements. The third track contains the information on movable oil. In this track, the curve farthest to the left is the value of the total porosity. The curve farthest to the right is the product of porosity and S_w from the deep measurement. It represents the total volume fraction of water. Between these two lines are two bands. The black strip to the left represents oil which has not moved, and the white band to the right is oil which has moved. The line separating the two is the product of porosity and S_{xo}.

The final use of the R_{xo} measurement, combined with the medium resistivity measurement, is the correction of the deep resistivity readings for the effects of invasion. Fig. 7–9 is an induction log taken from the simulated reservoir data set. At 9856', it is seen that the deep and medium resistivities are reading nearly the same value, which is quite separated from the invaded zone value. This indicates moderate invasion. Deep invasion would be signaled by a larger separation between deep and medium.

To quantify the invasion and to correct the deep resistivity to the true value (R_t), use is made of the invasion correction chart. These charts are parameterized in terms of resistivity ratios: the shallow resistivity (R_{sflu}) compared to the deep resistivity, and the medium resistivity to the deep. At the depth selected, the medium to deep resistivity ratio is found to be about 1.2, and the ratio R_{sflu}/R_{deep} is about 3.5. These two values are used to enter the correction chart, often referred to as a "tornado" chart, in Fig. 7–10.

Despite the clutter of curves in Fig. 7–10, there is a wealth of information. First, the intersection of the two ratios indicates that R_t/R_{ID} is about 0.98. This means that the deep measurement is only about 2% in error from the value of R_t because of invasion effects; R_t is 0.98 times the value of R_{deep}. Another of the parameterized sets of curves indicates a diameter of invasion of about 38" and that the value of R_{xo} is about 5.5 times the value of R_t. Thus, in this case, the deep induction measurement was little affected by the invasion. It was not deep enough for R_{sflu} to register the flushed zone value.

In Fig. 7–11, the laterolog response is shown in a portion of the simulated reservoir. At the indicated depth, 12435', the separation between medium and deep resistivity is a factor of 2, while the deep and shallow are separated by a factor of 30. The correction chart for the laterolog (Fig. 7–12) indicates for this case that the diameter of invasion is only about 25". However, the influence on R_t is much greater than in the last case; R_t is about

1.25 times the deep resistivity reading. In the zone below, 12455'–12466', the shallow and the deep laterolog readings overlay. This indicates, as can be confirmed from the correction chart, that there is little invasion and that the deep resistivity needs no correction.

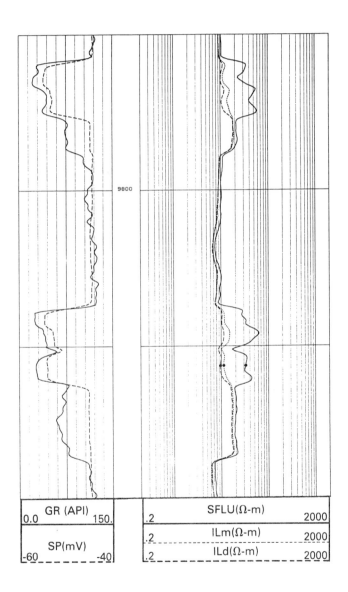

Figure 7–9.   An induction log of a section of a simulated reservoir model. Indicated is a zone which shows the effects of invasion on the resistivity measurement.

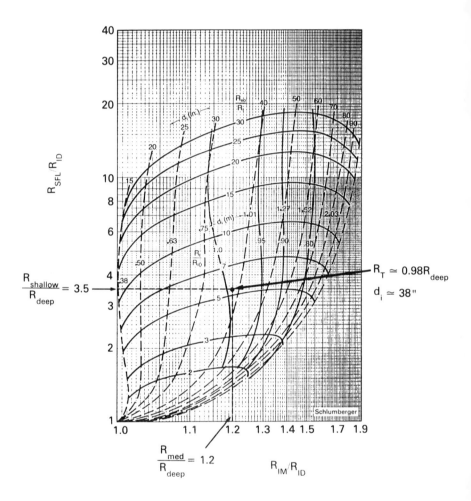

Figure 7–10. An invasion correction (or tornado) chart for a dual induction tool. Adapted from Schlumberger.[3]

Devices for Measurements other than R_t **137**

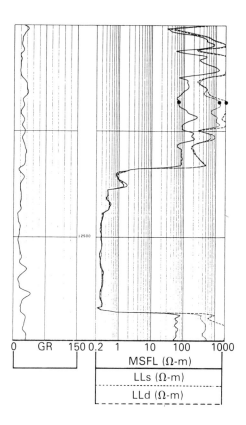

Figure 7–11. Idealized response of a laterolog in a portion of the simulated reservoir. The marked section shows need of invasion correction.

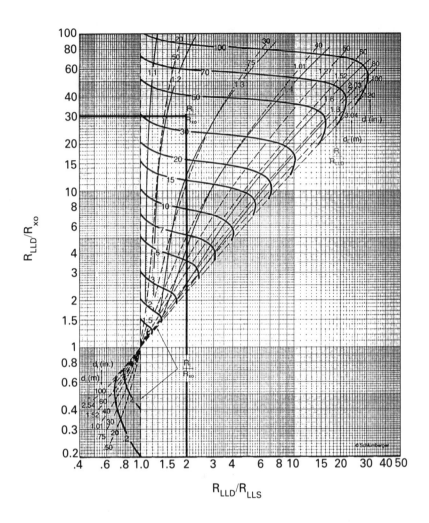

Figure 7–12. A dual laterolog invasion correction chart. Adapted from Schlumberger.[3]

A NOTE ON REALITY

Before leaving conventional resistivity measurements, we should point out that, for convenience, the subject of electrical measurements in anything other than clean (non-shale-bearing) formations has been ignored. However, the conventional resistivity tools studied up to this point are affected by the presence of clay in the formation. The effect of clay minerals and shales on resistivity and other log readings is considered in more detail in Chapters 19 and 20, but a short introduction is necessary here to temper the enthusiasm of the novice log interpreter.

The electrical effect of clay in core samples was studied by Hill and Milburn among others.[10] A typical resistivity response curve obtained by them is reproduced in Fig. 7–13. In this curve, where the resistivity of a fully saturated rock is being compared to the resistivity of the saturating

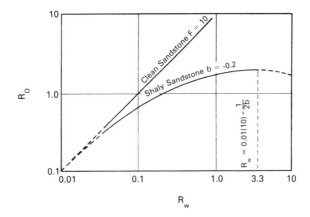

Figure 7–13. Schematic representation of the observation results of Hill and Milburn.[10] The Archie formation factor has been determined of a clay-free rock and one containing shale. From Lynch.[11]

water (Archie's experiment), two types of behavior are shown: the linear response, documented earlier by Archie, in the clean sandstone, and the curved response for a shaly sandstone. The presence of clay is seen to decrease the overall resistivity of the sample. This behavior is more clearly seen in Fig. 7–14 where the conductivity of a clay-bearing core sample is shown as a function of the saturating water conductivity (both of which are in units of Siemans/m, the standard unit for inverse $\Omega \cdot$m). A sample containing no clay minerals would be expected to have no conductivity when the value of the saturating water conductivity, C_w, falls to zero. This is seen not to be the case. The presence of the clay appears to provide additional conductivity.

To understand this phenomenon, it is necessary to recall that the structure of clay minerals produces a negative surface charge, because of substitution at the surface of the clay crystals of atoms of lower positive valence. The excess negative charge is neutralized by adsorption of hydrated cations which are physically too large to fit inside the crystal lattice. The neutralization occurs at locations referred to as exchange sites. In an ionic solution, these cations can exchange with other ions in solution. A measurement of this property is called the cation exchange capacity, or CEC.

Many models have been developed to explain the additional conductivity of clay-bearing rocks. They all have in common the idea of the presence of two parallel conduction paths. The first is the usual charge transport of ions of the electrolyte in the pore space. The second is conduction which occurs because of exchange of cations at negatively charged sites on the clay mineral

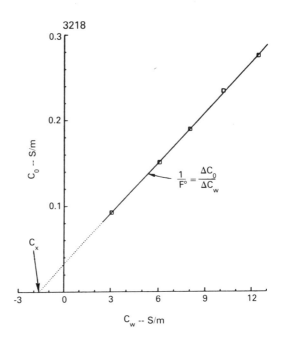

Figure 7–14. A representation of the measurements of Waxman and Smits on one core containing shale. The conductivity of the fully saturated rock has been measured for several saturating waters of different conductivities. The units (S/m or Siemans /m) are the inverse of $\Omega \cdot m$. Adapted from Clavier.[12]

particles. This secondary path of conduction is viewed differently by several current theories. In one model, it is held to be due to charge transport from the electrolyte to fixed exchange sites on the clay, by transport through the adjacent electrolyte from site to site, and between sites on different clay particles.[13] Regardless of the details, the magnitude of this secondary path of conduction will depend on the density of exchange sites available. This will be related to the specific surface area of the clay, the quantity of shale present in the core sample, and its distribution in the rock.

To show the type of variability of apparent formation factor that results from the presence of clay in core samples, see Fig. 7–15, which shows the data set used by Waxman et al.[13,14] These are summary plots of data similar to that of Fig. 7–14 for many rock samples with various clay contents. This is far from the expected Archie type of behavior. It is clear that at least another parameter, related to clay content, is necessary to explain the variance observed on the plot. One such parameter, Q_v, is the cation exchange capacity normalized to the pore volume. This parameter is related to the ability of the clay to form an agglomeration of positive ions near the clay surface, which provides an additional path of conduction. The measured values of Q_v for the core samples of the preceding figure are plotted on a third axis in Fig. 7–16, and the entire plot has been rotated. There is a clear

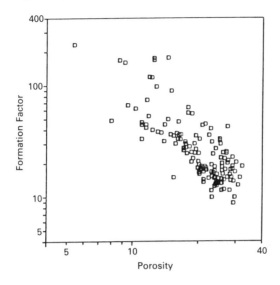

Figure 7–15. The results of the measurement of the formation factor for the ensemble of Waxman–Smits data for clay-bearing rocks. The simple Archie behavior is seen not to hold. Adapted from Clavier.[12]

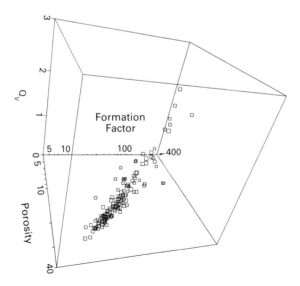

Figure 7–16. A rotated three-dimension view of the data of Fig. 7–15, with a third parameter, related to the electrical activity of the clay Q_v, now in view. Adapted from Clavier.[12]

reduction in scatter of the formation factor versus porosity plot when this new piece of information is present. The development of a direct measure of CEC to add to the suite of logging devices available would seem desirable.

The preceding discussion touches only on behavior of resistivity of fully water-saturated core samples which contain clay. The determination of saturation is yet another problem in the case of shaly samples. It is clear that resistivity measurements will have to be adjusted to offset the effect of the clay. Some of the techniques for dealing with this problem are presented in Chapter 20.

CHARACTERIZING DIELECTRICS

Another electrical property of material that is of interest for logging applications is one usually associated with insulators: the dielectric constant, or dielectric permittivity. The implication of the dielectric constant of a material can be best understood from a familiar application. It consists of the use of a dielectric material to increase the capacitance of a condenser such as that shown in Fig. 7–17. Without the dielectric material between the parallel plates, the capacitance C is given by:

$$C = \frac{\varepsilon_o A}{d},$$

where A is the area of the plates, ε_o is the dielectric permittivity of free space, and d the separation. The charge Q, stored on the plates, and the voltage V are related by:

$$Q = CV.$$

The observational fact is that when a dielectric material is placed between the plates of a parallel capacitor, and the charge is held constant, the voltage is

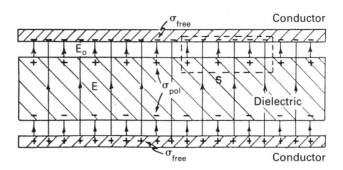

Figure 7–17. A parallel plate capacitor containing a dielectric material. From Feynman et al.[15]

seen to drop. Since the voltage is the integral of the electric field across the plates, it is apparent that the electric field somehow decreases.

The explanation for this lies in the polarizability of the atoms which make up the insulating dielectric. Under the influence of an applied electric field, positive charges are displaced with respect to the negative. In the presence of the applied electric field, the charge q can be imagined to physically separate a distance δ, thereby creating a number of dipoles, each of dipole moment $q\delta$. If N is the total number of atoms/cm^3, then the dipole moment per unit volume P is :

$$P = Nq\delta .$$

Despite the charge separation at each atom, there will be charge neutrality throughout the volume of the material, except for the outer layer of thickness δ, where a surface charge is present. The surface charge density can be found from the total number of excess charges in the layer of thickness δ, which is given by NδA, where A is the surface area. Since the charge associated with each dipole is q, the surface charge density (charge per unit area) is just Nδq. This is numerically equivalent to the dipole moment/unit volume. The reduction in electric field inside the dielectric can be seen with reference to the surface sketched by a dotted line in Fig. 7–17. From Gauss' law, the electric field outside of this enclosed volume is given by the contained net charge divided by ε_o. In this case, the volume contains two surface charges: a negative one of free charge, which has been stored on the capacitor plates, and a positive one on the surface of the dielectric, which has been induced. As the two are opposite in sign, the field is found to be:

$$E = \frac{\sigma_{free} - \sigma_{pol}}{\varepsilon_o} ,$$

from which it is apparent why the electric field decreases when a dielectric is placed between the plates.

The polarizability of materials varies considerably depending on their electronic configuration. For example, the O_2 molecule has a rather low polarizability, which might be surmised from its structure as sketched in Fig. 7–18. The centroids of the positive charges, from the nuclei, and the negative charges, from the electrons, perfectly overlap. An externally applied electric field would, however, produce some distortion of the charge distributions, but it would not produce nearly as great an effect as would be the case for a sample of water. As also shown, in Fig. 7–18, the water molecule is dipolar; there is a naturally occurring geometric separation between the centroids of the positive and negative charges due to the nonsymmetric shape of the molecule. Each molecule, then, is a tiny dipole. Due to thermal agitation, the orientation of these dipoles is at random in the absence of an applied field. However, upon application of an external electric field, the dipoles will tend to align and produce a rather large dipole moment per unit volume.

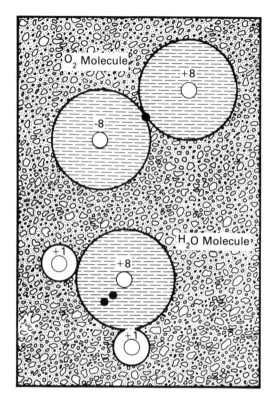

Figure 7–18. A schematic representation of the charge distribution of the O_2 and H_2O molecules. Because of the orientation of the hydrogen, water shows a large polarizability.

In most substances the polarization (number of induced dipoles/unit volume) is proportional to the applied electric field:

$$\overline{P} = \chi \varepsilon_o \overline{E},$$

where the constant of proportionality χ depends upon the substance.

In order to get to the defining equations for the dielectric constant, or rather the relative dielectric permittivity, we use the historical approach which was developed before the existence of the polarization charge was appreciated. In order to write Maxwell's equations in a simple manner, it was necessary to define a new vector $\overline{D}$, which is a linear combination of the electric field vector and the polarization vector:

$$\overline{D} = \varepsilon_o \overline{E} + \overline{P}.$$

In this case the divergence equation can be written as:

$$\nabla \cdot \overline{D} = \rho_{free},$$

and the relationship between D and E is given by:

$$\overline{D} = \varepsilon_o(1 + \chi)\overline{E} = \kappa\varepsilon_o\overline{E}.$$

This equation is usually written as:

$$\overline{D} = \varepsilon'\overline{E},$$

where ε' is the relative dielectric permittivity.

The relationship between polarizability and relative dielectric permittivity is given by:

$$\varepsilon' = \kappa\varepsilon_o = (1 + \chi)\varepsilon_o.$$

The relative dielectric permittivity (or constant) is given in Table 7–1 for a number of substances. Of all the substances listed on the chart, it is of note that water (a polar molecule) has the highest relative dielectric constant. There is a large difference between the relative permittivities of oil and water, which is one of the motivations for making a dielectric logging measurement.

MATERIAL	$\varepsilon = \varepsilon'/\varepsilon_0$	t_{pl}(nanosec/m)
Sandstone	4.65	7.2
Dolomite	6.8	8.7
Limestone	7.5 - 9.2	9.1 - 10.2
Anhydrite	6.35	8.4
Dry Colloids	5.76	8.0
Halite	5.6 - 6.35	7.9 - 8.4
Gypsum	4.16	6.8
Petroleum	2.0 - 2.4	4.7 - 5.2
Shale	5 - 25	7.45 - 16.6
Fresh Water at 25°C	78.3	29.5

Table 7–1. Relative dielectric permittivity and propagation time for various materials evaluated at 1100 MHz.

The dielectric constant is not really a constant but varies with the frequency of the polarizing field, as shown in Fig. 7–19. It is seen that up to about 10^9 Hz it is indeed constant, but beyond that the dipoles are not able to keep up with the rapidly changing electric field. The relative dielectric permeabilities shown in Table 7–1 were evaluated at a frequency of 1100 MHz.

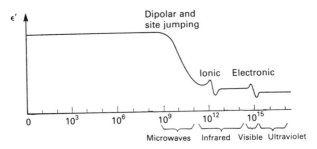

Figure 7–19. A schematic representation of the frequency dependence of the relative dielectric "constant." From Ramo et al.[16]

TOOLS TO MEASURE FORMATION DIELECTRIC PROPERTIES

Table 7–1 suggests why there is interest in the measurement of the dielectric constant; nearly an order of magnitude separates the values for water from other formation constituents. An equally important feature of a dielectric measurement in the borehole is that at microwave frequencies, the dielectric constant of water-saturated rocks is relatively independent of salinity. These two features combine to suggest an alternative means of distinguishing hydrocarbons from water in porous rocks. This is of particular interest in cases where the formation water is relatively fresh (resistivity greater than several $\Omega \cdot m$), making saturation estimation difficult because of very high formation resistivities. It is also of interest in evaluating zones for hydrocarbon reserves where the water salinity is completely unknown, as might be the case in secondary recovery projects where water flushing has altered the formation water.

In the discussion of the dielectric constant up to this point, the topic has been presented without any consideration as to how such a quantity might actually be measured for porous rocks saturated with conductive fluids. At low frequencies, conduction masks any presence of the dielectric effect. However, at very high frequencies, the dielectric properties will dominate. For the logging applications considered here, the measurement employs microwave radiation. The attenuation and velocity (phase shift) of the radiation are related to the dielectric constant and conductivity of the material through which it has passed.

This results from the well-known fact that the dielectric constant of a material plays an important role in the propagation of an electromagnetic (E–M) wave though it. The reciprocal of the velocity of propagation t_{pl}

depends upon the relative dielectric permittivity and conductivity of the material; as an example, the second column of Table 7–1 indicates the inverse velocity of E–M waves through each material. This dependence is reviewed in this chapter's appendix, through a consideration of E–M wave propagation in a conducting dielectric material. This is shown to result in a dielectric constant which is a complex quantity. The velocity of propagation is determined primarily by the real part of this quantity. The imaginary part determines, in large part, the attenuation of the E–M wave.

Measurement Technique

A device to measure dielectric constant properties of the formation is shown in Fig. 7–20 and is known as an electromagnetic propagation tool. One

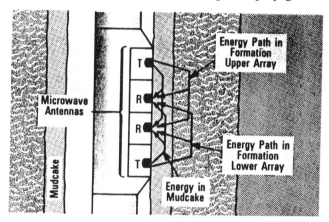

Figure 7–20. A schematic illustration of the configuration of a device to measure the dielectric properties of formations by the transmission of high frequency electromagnetic radiation.

version operates at a frequency of 1100 MHz and, because of the close spacing of the two receiver antenna (4 cm), has very good vertical resolution.[17,18] Its depth of investigation can be estimated to be on the order of the skin depth of these microwaves. This latter quantity will depend not only on the relative dielectric constant, but also on the conductive losses in the formation. For the ranges of signal attenuation observed with the tool at these frequencies, the depth of investigation is generally on the order of only several centimeters.* It makes a measurement in the shallow invaded zone.

---

* For increased depth of investigation, other tools operate in the range of 10–200 MHz.[19,20] In this frequency range, dielectric effects are still large, but the attenuation is much reduced, allowing transmitter-receiver spacings to be much larger, on the order of several feet, with a consequent increase in depth of investigation. The price to be paid, however, is a much-increased dependence on the salinity of the water.

The measurement consists of determining an attenuation and phase shift of the transmitted microwave signal from a comparison of signals at the two receiver antennae. The standard log format is to present the attenuation in dB/m and, rather than the phase shift, a quantity t_{pl}. This latter quantity, obtained from the phase shift, corresponds to the inverse of the microwave velocity measured in a lossy medium. It is presented in units of nanoseconds/m (ns/m).

Now, we have come almost full circle. We showed the interest in measuring ε', and a technique for doing so at microwave frequencies. The result, however, is a log of travel time t_{pl} and attenuation. What happened to ε'?

As shown in the appendix, both ε' and σ (at 1100 MHz) can be obtained from the phase shift and attenuation in the case of a plane wave. Certainly the tool configuration shown in Fig. 7–20 does not correspond exactly to this idealized case. The short source-detector spacing and the nonideal geometry results in additional attenuation or spreading loss. Thus the first step in recovering ε' and σ is to obtain the value of the formation attenuation A_f from the measured attenuation (EATT on the log heading). This can be estimated by using an empirical correction that has been developed. This correction chart is shown in Fig. 7–21 and takes into account that the spreading losses are not just geometric but also a function of conductivity and relative dielectric constant.

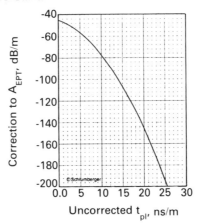

Figure 7–21. Correction charts for the attenuation and geometric spreading loss for a particular electromagnetic propagation tool. From Schlumberger.[3]

Then, for the particular tool design parameters (t_{pl} from the phase shift detected over 4 cm at a frequency of 1100 MHz), the following useful relations can be derived[21]:

$$\sqrt{t_{pl}^2 - \frac{A_f^2}{3604}} = t_{p-vac}\sqrt{\varepsilon'} = 3.33 \text{ ns/m } \sqrt{\varepsilon'}$$

and

$$\sigma_{EPT} = \frac{A_f t_{pl}}{5458},$$

where t_{p-vac} corresponds to the inverse velocity of light in a vacuum, and conductivity is in S/m. These two equations can be used to construct the graph of Fig. 7–22 which allows the determination of ε' and σ from the measured values of t_{pl} and the spreading-loss corrected attenuation A_f. But what can we do with these two values, once they are determined? The conductivity obtained from such a cross plot is not, by any means, the low frequency conductivity. Although the contrast of ε' between water and oil or rock matrix materials was seen to be large, how can this be used quantitatively? These questions can be addressed by looking at a log example.

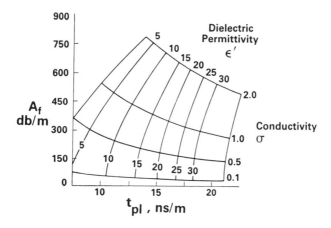

Figure 7–22. A graphical representation of the relation between the attenuation and the electromagnetic wave travel time, t_{pl}, for various values of the relative dielectric constant ε' and formation conductivity σ.

Log Example

Fig. 7–23, shows a typical electromagnetic propagation tool log presentation for a small section of the upper portion of the simulated reservoir. The zone of interest corresponds to the interval A noted earlier in the induction log example in Chapter 6. In track 1, the measured attenuation is shown along with the gamma ray. In tracks 2 and 3, the measured travel time is given in ns/m.

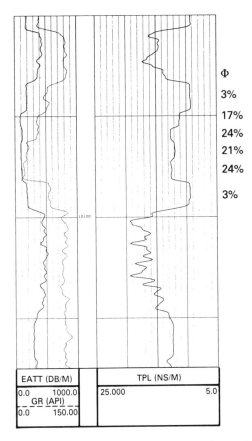

Figure 7-23. A high frequency electromagnetic propagation tool log of a section of the simulated reservoir model, showing the curves of t_{pl} and the attenuation.

The porosity over the sand section indicated contains several layers of moderate porosity (17–24%), bounded above and below by two tight sections of about 3% porosity. It is clear that t_{pl} is effectively reflecting the value of the porosity variations, which are noted also on the log. However, if we use the combination of attenuation and t_{pl} in conjunction with the chart of Fig. 7-22, the values of ε' can be deduced; ε' is around 8 or 9 in the two tight streaks and about 13 or 14 in the porous zones. But what have we learned? It seems likely that the invaded zone is water-filled because of the large value of ε', but what can be said further? To answer this we need to know how ε' changes for a mixture of matrix and water? This is the domain of log interpretation.

Rather than using the fundamental quantities ε' and σ, the interpretation has been developed in terms of the parameters actually measured, t_{pl} and attenuation. Since it is a relatively new measurement, the interpretation of the electromagnetic propagation tool is continually evolving. The initial method of interpretation was based on measurements made on rock samples. These indicated that the complex propagation constant k is linear in volume fractions. This can be expressed as:

$$k = k_w \phi + k_{ma}(1-\phi), \qquad (1)$$

where the subscripts w and ma refer to the fluid and matrix respectively. From Eq. (A–4), presented in the appendix that follows this chapter, it can be seen that this implies that the complex dielectric constant ε^* will have the following mixing law:

$$\sqrt{\varepsilon^*} = \sqrt{\varepsilon_{fl}^*}\,\phi + \sqrt{\varepsilon_{ma}^*}\,(1-\phi).$$

The relations developed in the appendix for the tool-measured parameters:

$$t_{pl} = t_{p-vac}\,\mathrm{Re}\,\sqrt{\varepsilon^*},$$

and

$$\mathrm{Att} = 200\,\mathrm{Im}\,\sqrt{\varepsilon^*}$$

can be equated with the real and imaginary parts of Eq. (1). This results in two rather simple relations:

$$t_{pl} = \phi t_{p-fluid} + (1-\phi)t_{p-ma}$$

and

$$\mathrm{Att} = \phi \mathrm{Att}_{fl}.$$

The second equation has only one term since it is assumed that the matrix term has no losses. These equations can be used together or independently to find porosity or the combination of porosity and water resistivity. This latter case is possible since both Att_{fl} and $t_{p-fluid}$ are somewhat salinity-dependent. Thus at the fundamental level, the electromagnetic propagation tool is a porosity device in the sense that it responds primarily to water.

As a quick interpretation, it is convenient to note that the value of propagation time t_p for oil is not much different from that of formation materials (see Table 7–1). By assuming that these values are identical, the porosity derived from the above expression will be only the water-filled porosity. This is because water has the greatest effect on the decrease of the E-M wave velocity. By comparing the porosity obtained from the electromagnetic propagation tool with other knowledge of the total formation porosity (i.e., water + hydrocarbon), we can estimate the water saturation in the invaded zone.

More recent experience with log measurements and laboratory investigations on core samples have revealed, in addition, that high frequency dielectric measurements have a dependence on rock texture.[22,23] Based on this observation,[22] depending on available accompanying logging measurements, three types of interpretations can be made. If the water resistivity and rock texture are assumed constant over a zone, then the output becomes similar to a shallow zone resistivity measurement R_{xo}. If the texture is assumed constant and another measurement of R_{xo} is available, then the model can be used to solve for porosity and water resistivity. The third option, in a zone where water resistivity is assumed constant, can be used to obtain the textural parameters from the model and thus predict the cementation exponent m, for use in the interpretation of the deep resistivity measurements.

APPENDIX

Propagation of E–M Waves in Conductive Dielectric Materials

To demonstrate the propagation characteristics for a real material which has properties of both conductivity and polarizability, we will proceed much as in the case for the derivation of the expression for the skin effect seen in Chapter 6. In that earlier example, we were only concerned with a material which was a pure conductor. For our more general case, Maxwell's equations are somewhat modified, to account for the displacement current associated with the polarization of the material. The first equation is now written as:

$$\nabla \times \overline{H} = \overline{J} - i\omega\overline{D} ,$$

where the time dependence of all vector quantities is represented by $e^{i\omega t}$. The constitutive equations which link the material properties σ, μ, and ε to the basic vector quantities are simply:

$$\overline{J} = \sigma\overline{E} ,$$

$$\overline{B} = \mu\overline{H} ,$$

and

$$\overline{D} = \varepsilon\overline{E} .$$

Using these relationships, the first Maxwell equation can be written as:

$$\nabla \times \overline{H} = \frac{1}{\mu} \nabla \times \overline{B} = \sigma\overline{E} - i\omega\varepsilon\overline{E} ,$$

or

$$\nabla \times \overline{B} = (\mu\sigma - i\omega\mu\varepsilon)\overline{E} . \qquad \text{(A-1)}$$

From the second Maxwell equation we have:

$$\nabla \times \overline{E} = -\frac{\partial \overline{B}}{\partial t} = i\omega\overline{B} .$$

As before, for the skin depth example, we take the curl of both sides of the equation:

$$\nabla \times \nabla \times \overline{E} = i\omega \nabla \times \overline{B} = i\omega(\mu\sigma - i\omega\mu\varepsilon)\overline{E}$$

$$= (\omega^2\mu\varepsilon + i\omega\mu\sigma)\overline{E}, \qquad (A-2)$$

from the result obtained in Eq. (A–1). The left side of Eq. (A–2) can be simplified by using the vector identity:

$$\nabla \times \nabla \times \overline{E} = \nabla(\nabla \cdot \overline{E}) - \nabla^2 \overline{E} = -\nabla^2 \overline{E},$$

since we also have the relation that:

$$\nabla \cdot \overline{E} = 0.$$

Thus the final result is:

$$\nabla^2 \overline{E} + (\omega^2\mu\varepsilon + i\omega\sigma\mu)\overline{E} = 0,$$

or for the much simpler one-dimensional case:

$$\frac{\partial^2 E}{\partial x^2} + (\omega^2\mu\varepsilon + i\omega\sigma\mu)E = 0.$$

This is the wave equation for an E–M wave traveling in the x-direction in a medium characterized by a magnetic permeability μ, conductivity σ, and dielectric constant ε.

A solution to this equation is the expression for a traveling wave:

$$E(x,t) = E_o e^{i(kx - \omega t)},$$

as can be verified by substitution. This results in the requirement that the wave number or propagation constant k satisfy:

$$k^2 = \omega^2\mu\varepsilon + i\omega\sigma\mu$$

$$= \omega^2\mu(\varepsilon + i\frac{\sigma}{\omega}). \qquad (A-3)$$

This relationship implies that the wave number k will be a complex number which can be represented as:

$$k = \alpha + i\beta.$$

This results in an attenuation and phase shift in the transmitted plane wave. This can be seen by substituting the expression for k into the traveling wave solution, which yields:

$$E(x,t) = E_o\, e^{-\beta x} e^{i(\alpha x - \omega t)}.$$

Thus an E–M wave traveling over a distance x will be attenuated by a factor $e^{-\beta x}$ and suffer a phase shift of αx radians.

For the case of a plane wave, it is possible to extract σ and the dielectric constant ε from the measured values of β and α. This is done by referring to the definition of α and β:

$$k^2 = (\alpha + i\beta)^2 = \omega^2 \varepsilon \mu + i\omega\mu\sigma ,$$

and, by expansion,

$$\alpha^2 + 2i\alpha\beta - \beta^2 = \omega^2 \varepsilon \mu + i\omega\mu\sigma .$$

By equating the real parts of the above equation, we have:

$$\alpha^2 - \beta^2 = \omega^2 \varepsilon \mu ,$$

and the imaginary parts yield:

$$2\alpha\beta = \omega\mu\sigma .$$

These can be rearranged to yield:

$$\varepsilon = \frac{\alpha^2 - \beta^2}{\omega^2 \mu}$$

and

$$\sigma = \frac{2\alpha\beta}{\omega\mu} .$$

Thus from a measurement of the attenuation and phase shift of an E–M plane wave, we can in principle obtain the original desired quantities, ε and σ.

Note that Eq. (A–3) can be rewritten as:

$$k^2 = \omega^2 \mu \left(\varepsilon + i\frac{\sigma}{\omega}\right) = \omega^2 \mu \varepsilon_o \left(\varepsilon' + i\frac{\sigma}{\varepsilon_o \omega}\right)$$

$$= \omega^2 \mu \varepsilon_o (\varepsilon' + i\varepsilon'') ,$$

where ε' is now the relative dielectric constant which corresponds to the values in Table 7–1. The imaginary part, ε'', is the loss term and corresponds to the second curve in Fig. 7–19. To denote the complex nature of the dielectric constant, the symbol ε^* is sometimes used and is defined as:

$$\varepsilon^* = \varepsilon' + i\varepsilon'' .$$

Thus one can represent the complex nature of k by treating ε as a complex quantity, the real part accounting for polarization and the imaginary part representing losses.

For the plane wave, a couple of facts can be deduced from the preceding equations. First, we can write:

$$k = \omega\sqrt{\mu\varepsilon_o} \sqrt{\varepsilon' + i\varepsilon''} ,$$

$$= \omega\sqrt{\mu\varepsilon_o} \sqrt{\varepsilon^*} \qquad\qquad (A-4)$$

For propagation in free space, where no conduction is possible, ε' becomes unity and ε'' is zero. We are left with:

$$k = \omega\sqrt{\mu\varepsilon_o}.$$

Thus from the traveling wave representation:

$$E = E_o e^{-i(kx - \omega t)} = E_o e^{-ik(x - vt)}, \qquad (A\text{-}5)$$

where v is the phase velocity of the traveling wave, we see that the velocity is simply given by $\dfrac{\omega}{k} = \dfrac{1}{\sqrt{\mu_o\varepsilon}}$, which is c, the velocity of light in a vacuum. The reciprocal of this expression, $\dfrac{k}{\omega}$, gives the wave travel time in units such as nanoseconds/meter. This quantity, when measured in a lossy formation, is referred to as t_{pl}. It is related to the real part of the wave number:

$$t_{pl} = \frac{k}{\omega} = \operatorname{Re}\sqrt{\mu\varepsilon_o}\sqrt{\varepsilon' + i\frac{\sigma}{\omega\varepsilon_o}},$$

so the travel time is just $\dfrac{1}{c}$ modulated by $\sqrt{\varepsilon'}$.

The attenuation is related to the imaginary part of the propagation constant k, defined in Eq. (A–4). Since the attenuation is exponential, it is necessary to use the factor 8.68 (20/log$_e$ 10) to convert to the usual units of dB. Thus we can write:

$$\text{Att} = 8.68\,\operatorname{Im}\,\omega\sqrt{\mu\varepsilon_o}\sqrt{\varepsilon^*}.$$

However, we note that the factor $\omega\sqrt{\mu\varepsilon_o}$ is equal to $\dfrac{2\pi}{\lambda}$. For the tool operating at 1100 MHz, the wavelength λ is 27.3 cm. Thus the attenuation in units of dB/meter is given by:

$$\text{Att} = 200\,\operatorname{Im}\,\sqrt{\varepsilon^*}.$$

References

1. Serra, O., *Fundamentals of Well-Log Interpretation*, Elsevier, Amsterdam, 1984.

2. Jordan, J. R., and Campbell, F., *Well Logging II - Resistivity and Acoustic Logging*, Monograph Series, SPE, Dallas (in press).

3. *Schlumberger Interpretation Charts*, Schlumberger, New York, 1985.

4. *Fundamentals of Dipmeter Interpretation*, Schlumberger, New York, 1970.

5. Serra, O., ed., *Sedimentary Environments from Wireline Logs*, Schlumberger, New York, 1985.

6. Doveton, J. H., *Log Analysis of Subsurface Geology, Concepts and Computer Methods*, John Wiley, New York, 1986.

7. Ekstrom, M. P., Dahan, C. A., Chen, M. Y., Lloyd, P. M., and Rossi, D. J., "Formation Imaging with Microelectrical Scanning Arrays," Paper BB, SPWLA Twenty-seventh Annual Logging Symposium, 1986.

8. *Log Interpretation: Vol 1, Principles*, Schlumberger, New York 1972.

9. *Well Logging and Interpretation Techniques: The Course for Home Study*, Dresser Atlas, Dresser Industries, 1983.

10. Hill, H. J., and Milburn, J. D., "Effect of Clay and Water Salinity on Electro-Chemical Behavior of Reservoir Rocks," Trans. AIME 207, 1956.

11. Lynch, E. J., *Formation Evaluation*, Harper & Row, New York, 1962, p. 213.

12. Clavier, C., Coates, G., and Dumanior, J., "Theoretical and Experimental Basis for the Dual Water Model for Interpretation of Shaly Sands," *Society of Petroleum Engineering Journal*, April 1984, 153–68.

13. Waxman, M. H., and Smits, L. J. M., "Electrical Conductivities in Oil-Bearing Shaly Sands," Trans. AIME 243, 1968.

14. Waxman, M. H., and Thomas, E. C., "Electrical Conductivities in Shaly Sands. I: The Relation Between Hydrocarbon Saturation and Resistivity Index II: The Temperature Coefficient of Electrical Conductivity," Trans. AIME 257, 1974.

15. Feynman, R. P., Leighton, R. B., and Sands, M. L., *Feynman Lectures on Physics*, Vol 2, Addison-Wesley, Reading, Mass, 1965.

16. Ramo, S., Whinnery, J. R., and Van Duzer, T., *Fields and Waves in Communication Electronics*, John Wiley, New York, 1965.

17. Wharton, R., Hazen, G., Rau, R., and Best, D., "Electromagnetic Propagation Logging: Advances in Technique and Interpretation," Paper SPE-9267, SPE Annual Meeting, Dallas, 1980.

18. Freedman, R., and Vogiatzis, J. P., "Theory of Microwave Dielectric Constant Logging Using the Electromagnetic Wave Propagation Method," *Geophysics*, Vol. 44, No. 5, 1979.

19. Huchital, G. S., Hutin, R., Thoroval, Y., and Clark, B., "The Deep Propagation Tool (A New Electromagnetic Logging Tool)," Paper SPE-10988, SPE Annual Meeting, San Antonio, 1981.
20. Janes, T. A., Hilliker, D. J., and Carville, C. L., "200 MHz Dielectric Logging System," Paper 28, Ninth International Formation Evaluation Transactions, SAID, 1984.
21. Delano, J. M., Jr., and Wharton, R. P., "An EPT Interpretation Procedure and Application in Freshwater, Shaly Oil Sands," *JPT*, Oct. 1984, p. 1763–1772.
22. Kenyon, W. E., and Baker, P. L., "EPT Interpretation in Carbonates Drilled with Salt Muds," Paper SPE-13192, SPE Annual Meeting, Houston, 1984.
23. Baker, P. L., Kenyon, W. E., and Kester, J. M., "EPT Interpretation Using a Textural Model," Paper DD, SPWLA Twenty-sixth Annual Logging Symposium, Dallas, 1985.

Problems

1. Using the SP and resistivity fundamentals, show that the following relation holds for clean formations:

$$SP = -K \left[\log_{10} \frac{R_{xo}}{R_t} + 2 \log_{10} \frac{S_{xo}}{S_w} \right].$$

2. In the log of Fig. 7–24, a scaled curve of $\frac{R_{xo}}{R_t}$ is presented in track 1 as an overlay to the SP curve. How does this quick-look technique relate to the expression derived above? How is it used to distinguish water from hydrocarbons? For the six zones of interest (A–D'), mark on the figure which are water zones and which are hydrocarbon zones. When might this technique fail or be misleading?

3. A section of sandstone reservoir was logged and found to have a porosity of 18%. The water resistivity is estimated to be 0.2 Ω·m, and R_t was measured to be 10 Ω·m.

   a. What is the water saturation?

   b. What error in S_w (in saturation units) is induced by a 10% relative uncertainty for each of the three parameters?

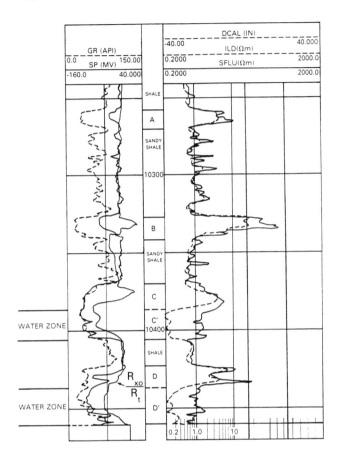

Figure 7–24. A resistivity log for Problem 2, showing the R_{xo}/R_t overlay.

4. Given the log of Fig. 7–6 with R_{mf} indicated at formation temperature, answer the following:

   a. Over the zone 11,800'–12,200', what is the average value of the lower limit to porosity which can be established?

   b. Evaluate S_w every 50' over the above interval and make a linear plot of S_w versus depth.

   c. The actual average porosity over the zone in question is 15 PU. How does this compare with your estimate? Is this discrepancy reasonable? How does this additional information impact the actual value of S_w along the zone (replot curve)?

5. In the bottom section of the well previously studied in question 4, assume that the porosity is constant at 15% over the entire interval and answer the following:

a. In the zones marked 1, 2, and 3, determine the corrected values of R_{LLd} and the diameter of invasion.

b. What can you conclude about the permeability of the three zones? Do any other measurements support this view?

c. What do you think is the cause of the resistivity jump which occurs just above zone 2?

d. Estimate the value of R_w in this reservoir.

6. You have logged a reservoir in which, from previous experience, the simple Archie relation ($1/\phi^2$, where the cementation exponent is 2) doesn't describe the measurements of F made on core samples. In the zone of interest, which has a porosity of 20%, a simple determination of the invaded zone saturation from the microresistivity device indicates a value of 50%. Fortunately an electromagnetic propagation tool has been run in this well. It indicates a porosity of 5%. What cementation exponent should be used to interpret R_t?

7. Suppose you are logging in a shaly sandstone whose clay content is constant and whose electrical effects have been evaluated from measurements on fully water-saturated samples, as shown in Fig. 7–14.

a. What value of R_t will you expect to find if the reservoir is 75% oil-saturated when $R_w = 0.1\Omega\cdot m$?

b. What value of R_t do you expect for the same saturation but when $R_w = 1\Omega\cdot m$?

c. How much higher would R_t be for case b if the formation were entirely clay free?

8
BASIC NUCLEAR PHYSICS FOR LOGGING APPLICATIONS: GAMMA RAYS

INTRODUCTION

In the preceding chapters, electrical devices are seen to respond primarily to the fluid content of earth formations. They are not used, therefore, to obtain information about the predominant constituent of formations, the rock matrix. Nuclear measurements used in logging respond to properties of both the formation and the contained fluids. These measurements employ gamma rays and neutrons. These two types of penetrating radiation are the only ones which are able to traverse the pressure housings of the logging tools and the formation of interest and still return a measurable signal.

One input to a more complete description of an earth formation is an analysis of its chemical composition. Knowledge of its major elemental constituents would be indicative of the dominant mineralogy. Instead of the obvious but time-consuming and expensive laboratory chemical analysis of formation samples, in-situ gamma ray spectroscopy can be used. This is based on the fact that the nucleus of any atom, after having been put into an excited state by a previous nuclear reaction, can emit gamma rays of characteristic energies which uniquely identify the atom in question. Gamma ray spectroscopy refers to the detection and identification of these characteristic gamma rays.

Another important property of an earth formation is bulk density. Its use in seismic interpretation (see Chapter 22) is well-known, but more important it is linearly dependent on formation porosity, a key ingredient for the interpretation of electrical measurements. Gamma rays are used to measure bulk density since this property of a material has a significant influence on the scattering and transmission of gamma rays through it. At very low energies the transmission of gamma rays is influenced additionally by the chemical composition. This additional absorption is related to the atomic number, Z, of the absorber and thus provides a third application of gamma rays.

Neutrons are used in well logging because of several different properties of their interaction with matter. First, the transmission and moderation of neutrons are influenced by the bulk properties of the medium, in particular, by the amount of hydrogen present. The scattering of neutrons by hydrogen is very efficient in reducing the neutron energy. Second, the interaction of high-energy neutrons can excite nuclei to emit characteristic gamma rays. The elemental identification can be achieved by gamma ray spectroscopy. At very low energies, neutrons can be absorbed, causing the emission of another set of characteristic gamma rays. Some of these capture gamma rays are emitted after considerable delay and are referred to as activation gamma rays. In summary, there are two types of measurements which can be based on the use of neutrons: the scattering or slowing-down properties of formations, and neutron production of gamma rays (either by absorption or inelastic high energy reactions with elements) of characteristic energies for use in spectroscopic identification. These are discussed in Chapters 11 through 13.

In this chapter,* a basic vocabulary for the description of nuclear radiation is developed which can be applied to the discussion of gamma rays or neutrons. This includes the quantification of intensity and energy, and the notions of cross sections, reactions, and counting statistics. The principal interaction of gamma rays with matter are described in relation to the physical parameters of the formation and methods for their detection. Unlike most of the resistivity devices considered, which consist of complicated arrays of very simple sensors, logging tools which use gamma rays have somewhat more sophisticated sensors. Understanding the detectors allows a better understanding of the limitations of the measurements.

NUCLEAR RADIATION

In the earliest investigation of radioactive materials, three types of radiation were identified and named, quite unimaginatively, α, β, and γ radiation. It was subsequently discovered that α radiation consisted of fast-moving He atoms stripped of their electrons, and that β radiation consisted of energetic electrons. The gamma rays were found to be packets of electromagnetic radiation also referred to as photons.

---

* Much of this chapter appears in slightly different form in the *SPE Petroleum Production Handbook*.[1]

The discovery of this radiation was followed by its quantification, namely the measurement of the amount of energy transported. The unit chosen is known as the electron-Volt (eV), which is equal to the kinetic energy acquired by an electron accelerated through an electric potential of 1 Volt. For the types of radiation discussed in the following sections, the range of energies is between fractions of an eV and millions of electrons Volts (MeV). Another convenient multiple for discussing gamma ray energies is the kilo electron-Volt (keV).

Since α and β radiation consist of energetic charged particles, their interaction with matter is primarily Coulomb in nature. This leads to atomic excitation or ionization; that is, the interactions are with the electrons of the medium. The α and β particles rapidly lose energy as they transfer it to electrons in their passage through the medium. Their ranges are rather limited and in most materials are a function of the material properties (Z, the number of electrons per atom, and density) and the energy of the particle. Consequently they have not been of any practical importance for well logging applications. Gamma rays, on the other hand, are extremely penetrating, which makes them of great importance for well logging applications.

RADIOACTIVE DECAY AND STATISTICS

Radioactive decay is a time-varying property of nuclei, in which a transition from one nuclear energy state to another lower one is made spontaneously. The excess energy is shed by the nucleus, by means of one or more of the types of radiation previously mentioned. The basic experimental fact associated with radioactivity is that the probability of any one nucleus decaying, within an interval of time Δt, is proportional to Δt; i.e., it is independent of external influences, including the decay of another nucleus. So for a *single* radioactive atom, the probability P(dt) of decaying in the interval of time dt is expressed as:

$$P(dt) = \lambda dt ,$$

where λ is the decay constant. For a *collection* of N_p identical radioactive particles, the number decaying dN is just:

$$dN = -\lambda dt N_p,$$

resulting in the expression for radioactive decay:

$$N_p = N_i e^{-\lambda t} ,$$

where N_p is now the number of particles remaining in the collection at time t, out of the initial number of particles N_i present at time zero. The constant of proportionality λ is related to the better-known parameter, the half-life $t_{1/2}$, by:

$$t_{1/2} = \frac{0.693}{\lambda} .$$

One can never measure any physical quantity exactly, but in the case of nuclear processes, where the number of events observed is small, randomness is important. The practical complication of this statistical process of nuclear decay is that only the bulk or average properties can be predicted with any certainty. We can only talk about the measurement of a group of particles together and the distribution of the measured value about some mean.

In order to understand an important property of nuclear radiation, it is necessary to digress a moment for a short review of the binomial distribution which was discovered in the eighteenth century by Bernoulli. It describes the probability, P_x, that a discrete event which has a probability P of occurring in a single observation will occur x times when the observation is repeated z times. The probability thus specified was identified with the binomial expansion of $(P+q)^z$, where:

$$q = 1-P,$$

so that the general term of the expansion is:

$$P_x = \frac{z!}{x!(z-x)!} P^x(1-P)^{z-x}, \tag{1}$$

which gives the probability of x occurrences in z trials.

This expression can be applied to radioactive decay, in which P_x represents the probability of having x nuclei decay in time dt when there are z atoms present. For this case, the probability P of observing the decay of a single nucleus in a unit time is very small, but the number of particles observed (z) is very large. This condition allows simplification of Eq. (1): to

$$P_x = \mu^x \frac{e^{-\mu}}{x!},$$

which is known as the Poisson distribution. It gives the probability of observing x decays in a given time in which an average of μ decays is to be expected. Fig. 8–1 shows the general form of the Poisson distribution with the maximum probability at the mean value μ, which was chosen as 100 for this example. For this case, with $\mu \gg 1$, the distribution is nearly symmetric about the mean value. It resembles the usual bell-shaped distribution curve whose width is specified by an independent parameter σ, the standard deviation.

An important property of the Poisson distribution is that the appropriate σ for the Poisson distribution which characterizes the statistics of counting random nuclear events is not an independent parameter (as is the case for most measurements) but is related to the mean value μ by:

$$\sigma = \sqrt{\mu}.$$

Hence, if N_r counts from a radiation detector are expected per time interval, then in repeated observations about 32% of the measurements will exhibit deviations beyond values of $N_r \pm \sqrt{N_r}$. The only sure approach to reduce this

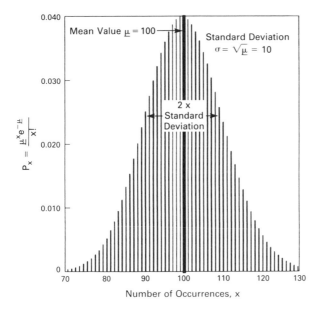

Figure 8-1. The Poisson distribution evaluated for an expected number of counts (100) in a given interval. The standard deviation is 10; 62% of all observations will deviate from the mean by a value greater than this. From *SPE Petroleum Production Handbook*.[2]

type of statistical fluctuation is to increase the average number measured, either by using higher output sources, more efficient counters, or longer counting times per sample.

This important implication of Poisson statistics is a consequence of the fact that the mean specifies the distribution about the mean; μ specifies the distribution completely. The error reduction noted above can be seen by consideration of the fractional uncertainty on a single measurement. The fractional uncertainty f can be expressed as:

$$f = \frac{\sigma}{N},$$

where N is the average number expected in the time interval considered. Since σ is given by the square root of the mean, f can be written as:

$$f = \frac{\sqrt{N}}{N} = \frac{1}{\sqrt{N}}.$$

Since N increases linearly with both t, the time of observation, and the source strength Q:

$$f \propto \frac{1}{\sqrt{Q\,t}}.$$

Thus if the source strength or observation time is increased by a factor of four, then the fractional uncertainty decreases by a factor of two.

RADIATION INTERACTIONS

There are certain interactions between radiation and materials that are of special interest in well logging. Before discussing these, a few mathematical definitions are presented to help describe the mechanisms of the interactions.

Using Fig. 8–2, consider the question of how readily these reactions will take place. A beam of radiation (e.g., gamma rays or neutrons) of intensity Ψ_i is seen to enter the slab of material and exit with an intensity Ψ_o. The intensity of the radiation Ψ is referred to as the flux and has dimensions of numbers of particles per unit surface area per unit time.

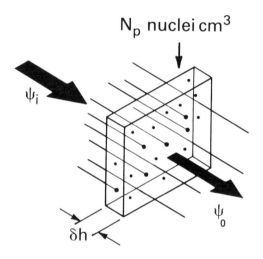

Figure 8–2. A flux of gamma rays of intensity Ψ_i impinging on a thin slab of material characterized by N interacting particles per cubic centimeter. The flux reduction in traversing the material is found to be proportional to the thickness of the slab δh and the number density of interacting particles N_p. From *SPE Petroleum Production Handbook.*[2]

The slab of material is characterized by N_p, the number of particles per unit volume with which the flux of radiation may interact. The experimental observation is that after passing through a slab of material of thickness δh, a certain fraction of the incident particles have undergone interactions, and that number is proportional to the thickness and the number of target nuclei, and the incident flux. This is expressed mathematically as:

$$\delta\Psi = \Psi_i - \Psi_o = \sigma\Psi N_p \, \delta h \,, \tag{2}$$

where the constant of proportionality σ is called the total cross section for the interaction. The units of this microscopic cross section σ are area/interacting target nucleus. Cross section is used, because in a classical sense it is the apparent area each target nucleus presents to the incoming beam. In effect, it collects all the nuclear interaction details into one useful number. The practical unit of cross section is called the "barn" and is equal to 10^{-24} cm^2. The so-called macroscopic cross section, Σ, is the product of σ and N_p, and has the dimensions of inverse length and is the reciprocal of the interaction mean free path. From σ, Σ can be calculated easily because N_p is related to Avogadro's number, N_{Av}, and the material density, ρ_b, by:

$$N_p = \frac{N_{Av}}{M} \times \rho_b,$$

where M is the molecular weight of the target material for a single particle per molecule.

In general, the cross sections for most reactions have to be determined experimentally. They depend on the type of interaction and the material. Each of these subsets has an additional dependence on the energy of the radiation and the angle between the incoming radiation direction and the resultant radiation. For this reason the data are often available in graphical or tabular form.

The quantity $\sigma \Psi N_p$ in Eq. (2) has dimensions of (cm^3 sec)$^{-1}$ and represents the reaction rate per unit volume between the incident flux and the target material.

FUNDAMENTALS OF GAMMA RAY INTERACTIONS

For our purpose, there are three types of gamma ray interactions which are of interest: the photoelectric effect, Compton scattering, and pair production. The probability of a specific gamma ray interaction occurring will depend on the atomic number of the material and the energy of the gamma ray. The ordering of these three interactions in the following discussion reflects the change of the dominant process as the gamma ray energy increases.

The photoelectric effect results from interaction of a gamma ray with an atom in the material. In this process the incident gamma ray disappears and transfers its energy to a bound electron. If the incident gamma ray energy is large enough, the electron is ejected from the atom and begins interacting with the adjacent material. Normally the ejected electron is replaced by another less tightly bound electron with the accompanying emission of a characteristic fluorescence X ray with an energy (generally below 100 keV) which is dependent on the atomic number of the material.

The cross section for the photoelectric effect σ_{pe} varies strongly with energy, falling off as nearly the cube of the gamma ray energy (E_{GR}). It is also highly dependent on the atomic number (Z) of the absorbing medium. In the energy range of 40 to 80 keV, the cross section per atom of atomic number Z is given by:

$$\sigma_{pe} \propto \frac{Z^{4.6}}{E_{GR}^{3.15}}.$$

For most earth formations, the photoelectric effect becomes the dominant process for gamma ray energies below about 100 keV.

The photoelectric effect is an important process in the operation of conventional gamma ray detection devices. Also it is the mechanism by which one type of well logging tool is made sensitive to the lithology of the formation. This tool measures the so-called photoelectric absorption factor, P_e, which is proportional to the photoelectric cross section per electron (i.e., $\frac{\sigma_{pe}}{Z}$). Since P_e is very sensitive to the average atomic number of the medium (Z), it can be used to obtain a direct measurement of lithology. This is due to the facts that the principal rock matrices (sandstone, limestone, and dolomite) have considerably different photoelectric absorption characteristics and that the pore fluids play only a minor role because of their low average atomic numbers.

Moving up the gamma ray energy scale, the dominant process becomes Compton scattering, which involves interactions of gamma rays and individual electrons. It is a process in which only part of the gamma ray energy is imparted to the electron; the remaining gamma ray is of reduced energy. Unlike the photoelectric effect, the cross section for Compton scattering changes relatively slowly with energy.

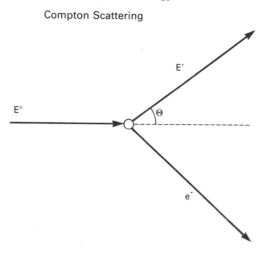

Figure 8–3.   Schematic representation of the Compton interaction. A gamma ray of energy E° transfers a portion of its energy to an electron, and a gamma ray of reduced energy E' leaves the site of the collision at an angle Θ with respect to the direction of the incident gamma ray.

Compton scattering is of great importance both as a measurement technique and as an interaction mechanism for gamma rays in detector materials. Therefore we will consider it in more detail. Fig. 8–3 illustrates the process: A gamma ray of incident energy E^0 interacts with an electron of the material, scatters at an angle Θ, and leaves with an energy E'. The difference between the incident gamma ray energy and the scattered gamma ray energy is imparted to the electron.

Fig. 8–4 illustrates the Compton energy-angle relationship for an incident gamma ray energy of 660 keV. This relationship gives the gamma ray energy as a function of the initial energy and scattering angle:

$$E' = \frac{E^0}{1 + \frac{E^0}{m_o c^2}(1-\cos\Theta)},$$

where m_o is the rest mass of the electron, and c the velocity of light. The quantity $m_o c^2$ is numerically equivalent to 511 keV.

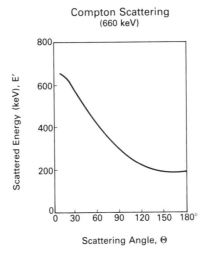

Figure 8–4. The energy-scattering angle relationship for Compton scattered gamma rays of initial energy 660 keV.

Since in all gamma ray detectors it is the energetic secondary electron which is used to produce a measurable signal, it is of some interest to look at the distribution of electron energies which result from Compton scattered gamma rays. Because the initial gamma ray energy is shared between the outgoing gamma ray and the scattered electron, it is relatively simple to derive a curve for the energy of the resultant electron energy as a function of the gamma ray scattering angle. This is is shown in Fig. 8–5 where it is seen that minimum electron energy is zero and that the maximum is around 450 keV (in this example, for 660 keV incident gamma rays) when the gamma ray is back-scattered at 180°.

The results of Fig. 8–5 show the energy of the electron as a function of all possible scattering angles. To determine the distribution of electron energies, something must be known about the distribution of scattering angles. In general, scattering is not isotropic; there is a probability distribution

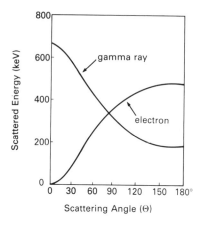

Figure 8–5. The energy-scattering angle relationship for the Compton electron. The curve for the electron energy is derived from energy conservation and the gamma ray-scattering angle relationship from the previous figure. Note the relative insensitivity of the electron energy to the gamma ray scattering angle above about 120° scattering.

associated with the scattering angle Θ. For Compton scattering, the preferred scattering angle at very high energy is close to zero. At low energies, however, Compton scattering is not too far from being isotropic, so that the curve of electron energies in Fig. 8–5 can be transformed into the electron distribution curve of Fig. 8–6 by considering the solid angle available for the gamma ray scattering, which varies as sinΘ. The resulting electron distribution seen in Fig. 8–6 has a close connection with observed gamma ray spectra. In the figure, the full energy of the gamma ray is indicated. This is the energy which would be registered by a detector if all of the incident gamma ray energy were absorbed in the detector material. However, if the gamma rays incident on the detector interact by a single Compton scattering, then they will register a distribution of energies as indicated. In gamma ray spectroscopy this degradation, or induced feature, is referred to as the Compton tail. The uppermost portion of this distribution is called the Compton edge (see Fig. 8–12).

To appreciate the bulk effect of Compton scattering in a material consisting of atoms of mass A, and atomic number Z, one can examine the so-called linear absorption coefficient. This macroscopic cross section is just the Compton cross section σ_{Co}, multiplied by the number of electrons per cubic centimeter:

$$\Sigma_{Co} = \sigma_{Co} \frac{N_{Av}}{A} \rho_b Z.$$

The final factor, Z, in the equation above takes into account that there are Z electrons per atom. Consequently the attenuation of gamma rays due to Compton scattering will be a function of the bulk density ρ_b and the ratio Z/A. The fact that Z/A is constant ($\approx$ ½) for most elements of interest is the basis for the determination of bulk density from gamma ray scattering devices.

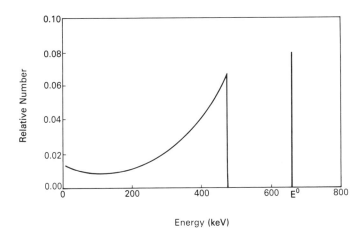

Figure 8–6. The distribution of Compton electron energy based on the assumption of isotropic scattering of the gamma rays.

The third and final gamma ray interaction is pair production. It, like the photoelectric process, is one of absorption rather than scattering. In this case the gamma ray interacts with the electric field of the nucleus, and if the gamma ray energy is above the threshold value of 1.022 MeV, it disappears and an electron-positron pair is formed. The onset of this interaction corresponds to the rest mass energy of the electron and positron. The subsequent annihilation of the positron (positively charged electron) results in the emission of two gamma rays of 511 keV each. The nuclear cross section of this process is zero below the threshold energy of 1.022 MeV and rises quite rapidly with increasing energy. It is also dependent on the charge of the nucleus, increasing approximately as Z^2.

In order to observe the regions of dominance of the three types of interactions, refer to Fig. 8–7. It shows, as a function of gamma ray energy and atomic number of the absorber, the boundaries at which the linear absorption coefficients for the adjacent processes are equal. The horizontal line, corresponding to an atomic number of 16, indicates the upper limit of Z for common minerals encountered in logging.

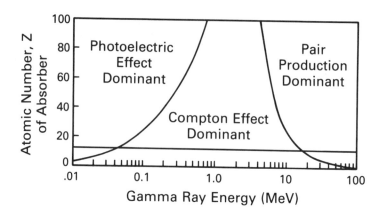

Figure 8–7. Regions of dominance of the three principal gamma ray scattering mechanisms as a function of energy and the atomic number, Z, of the scattering material. Adapted from Evans.[1]

ATTENUATION OF GAMMA RAYS

From the earlier definition of cross section, the fundamental law of gamma ray attenuation can be stated as:

$$\Psi = \Psi_i \, e^{-n\sigma h},$$

where Ψ_i is the flux incident on a scatterer of thickness h, n is the number of scatterers per unit volume, σ is the cross section for scattering per scatterer, and Ψ is the flux leaving the scatterer.

For gamma rays in the energy range of hundreds of keV, the primary interaction is Compton scattering. In this case, the scatterers are electrons and σ refers to the Compton cross section/electron. This results in the following expression for the attenuation of the source energy gamma rays:

$$\Psi = \Psi_i \, e^{-\rho_b \frac{Z}{A} N_{Av} \sigma_{Co} h}. \tag{3}$$

The attenuation of the gamma rays is seen to be proportional to the thickness of material h, the bulk density, and a property of the scattering material, Z/A. For most sedimentary rocks the ratio of Z/A is nearly 1/2, as has been noted earlier.

Another unit for measuring the gamma ray attenuation properties of a material is the mass absorption coefficient μ, which regroups the constants in

Eq. (3), i.e.:

$$\mu = \frac{Z}{A} N_{Av} \sigma,$$

so that the gamma ray attenuation equation can be written as:

$$\Psi = \Psi_i e^{-\mu \rho_b h}.$$

The mass absorption coefficient has units of cm^{2}g^{-1}. The convenience of using the mass absorption coefficient for Compton scattering stems from the fact that it is the same for all materials to within the approximation that Z/A = ½. Fig. 8–8 shows the mass attenuation coefficients (in cm^2/g) for aluminum.

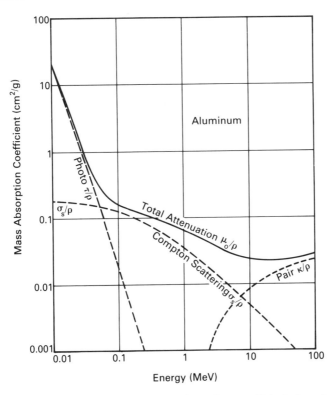

Figure 8–8. The gamma ray mass absorption coefficient for aluminum as a function of gamma ray energy. Adapted from Evans.[1]

GAMMA RAY DETECTORS

All common gamma ray detectors exploit one or more of the three modes of gamma ray interactions with matter described earlier. Three general types of gamma ray detectors in current use will be described next. The first variety, the gas ionization counter, is a direct descendant of one of the earliest efforts at nuclear radiation detection. The second and most common present-day gamma ray detector used in well logging is the scintillation detector. The third type of device, the solid state detector, is just beginning to be used in logging applications.

The common form of the ionized gas or gas-discharge counter consists of a metal cylinder with an axial wire passing through it (Fig. 8–9) and insulated from it. The cylinder is filled with a gas which is normally nonconductive, and a moderate (several hundreds of Volts) electrical potential is maintained between the central wire and the cylinder. For gamma rays to be detected with such a device, the gas must somehow be initially ionized. As the gas density, even at the rather high pressure available in commercial tubes, is moderate and the atomic number of useful gases is relatively low, there is little possibility of the gamma rays interacting directly with the gas. The main detection mechanism is photoelectric absorption or recoil electron ejection from Compton scattering in the metal shield. For the gamma rays absorbed near the inner radius of the cylinder, there is some probability of the ejected electron escaping into the gas and providing the initial ionization of detector gas molecules. The electrons freed in this process are accelerated by the radial electric field and, in collisions with gas molecules, produce additional free electrons. A fraction of the electrons are collected at the central wire, producing a voltage pulse.

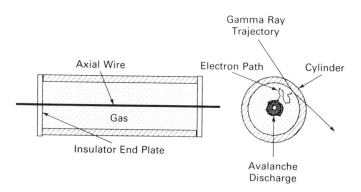

Figure 8–9. Schematic diagram of a gas-discharge counter, a simple but inefficient gamma ray detector.

The detection efficiency of such detectors is not high. It can be improved somewhat by the incorporation of conductive high atomic number gamma ray absorbers, such as silver, as an inner lining of the cylinder. Although they can be operated in a proportional mode*, the energy resolution of these detectors is not exploited in logging due to the poor efficiency. The most positive aspects of gas-discharge counters are their simplicity, ruggedness, and reliability for functioning in the hostile environment of well logging.

A more common type of gamma ray detector employs a scintillation crystal. Here too the active detector element is sensitive to ionizing radiation such as energetic electrons. When these particles travel within the crystal lattice, they impart their energy to a cascade of secondary electrons which are finally trapped by impurity atoms. As the electrons are trapped, visible or near-visible light is emitted. The light flashes are then detected by a photomultiplier tube optically coupled to the crystal and transformed into an electrical pulse. This is indicated schematically in Fig. 8–10. The output pulse height can be related to the total energy deposited in the crystal by the initial energetic electron. The great advantage of such a detection scheme is the possibility of doing gamma ray spectroscopy, that is, deducing the actual energy of the incident gamma ray. This, in some cases, permits identification of the source of the emitted gamma ray, as in the case of induced gamma ray logging.

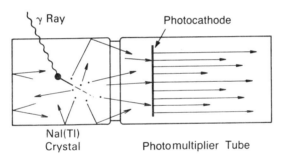

Figure 8–10. A scintillation detector with its associated photomultiplier. The photo cathode responds to a flash of light in the crystal by releasing electrons. The release of electrons is amplified by the rest of the photomultiplier structure into a detectable electrical pulse.

A scintillation detector is a detector of gamma rays only to the extent that an electron is produced in the crystal through one or more of the three basic gamma ray interaction mechanisms. Thus the gamma ray detection efficiency of a scintillator depends upon its size, density, and average atomic number (for photoelectric absorption). The most commonly used scintillator is sodium iodide doped with a thallium impurity, NaI(Tl). It has good gamma

---

* This mode can be attained by a reduction of the voltage between the outer cylinder and the central wire. In this case, the size of the output pulse will be proportional to the initial ionization produced by the ejected electron and thus to the absorbed gamma ray energy.

ray absorption properties and a fairly rapid scintillation decay time (≈ 0.23μs). The latter permits spectroscopy at high counting rates.

The use of such a device for gamma ray spectroscopy implies that the output light pulse is related uniquely to the incident gamma ray energy. However this is only possible for the case of total absorption of the gamma ray. Some of the difficulties that can complicate the detected spectrum are shown in Fig. 8–11. This schematically depicts a tool designed to look for the unique gamma rays emitted by excited states of carbon and oxygen. The

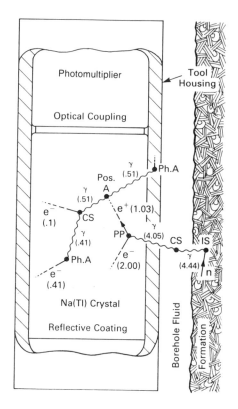

Figure 8–11. The life of a single gamma ray of 4.44 Mev which is emitted in the formation and ultimately detected by an NaI detector in the borehole.

figure illustrates what might happen to a 4.44 MeV gamma ray from carbon which is produced at the site marked (IS). It first makes a Compton scattering in the borehole fluid (CS) and loses 390 keV of energy before traversing the tool housing and entering the NaI detector with an energy of 4.05 MeV. At the point marked (PP) it suffers a pair production interaction, producing one electron and one positron with energies of 2.00 and 1.03 MeV the missing 1.02 MeV having gone into the creation of the electron-positron pair. Both particles impart their energy to the scintillation process. When the positron has given up all its kinetic energy, it annihilates with an electron to produce two gamma rays, each of 0.51 MeV energy. One of the gamma rays undergoes Compton scattering at (CS) and the reduced-energy gamma ray (0.41 MeV) is finally absorbed photoelectrically within the crystal at point (Ph.A). The other 0.51 Mev gamma ray is shown escaping the crystal, to the right, and being absorbed in the tool housing without contributing to the total energy transferred to the crystal. The light flash produced in the crystal for the sequence of events depicted corresponds to 3.54 Mev (4.05 Mev − 1.02 Mev pair-production + 0.511 Mev annihilation) instead of the 4.44 Mev which we would like to be measuring.

Thus the structure of the measured gamma ray spectrum is seen to be complicated by the physics of the many processes involved in the detection. Only if the energy of the gamma ray is totally absorbed by the detector is the light output of the scintillator proportional to the incident gamma ray energy. This would be the case for photoelectric absorption, for example. Fig. 8–12 shows the energy deposited in this case as the single line to the right marked E_γ. If a Compton interaction occurs, and the gamma ray then leaves the crystal, only a fraction of the energy will be registered. The possible range of energy deposition in this case follows the distribution shown earlier in Fig. 8–6. It extends from zero to the Compton edge, which corresponds to maximum energy being transferred from the gamma ray to the electron. Additionally, if the gamma ray is of sufficiently high energy, there may be a

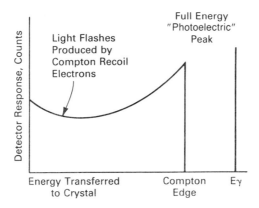

Figure 8–12. A representation of the detection of a monoenergetic gamma ray by an NaI detector. It shows the full energy peak, where gamma rays have been absorbed by the photoelectric process, and a broad Compton tail for gamma rays which have undergone Compton collisions in the crystal, transferring only part of their energy to electrons which provide the detection.

pair production reaction. In this case, if one or more of the 511 keV photons escapes the detector without interaction there will be produced in the detected spectrum the so-called first and second escape peaks. Fig. 8–13 indicates the additional complication introduced by this process; instead of a single line, three are present, and they are far from looking like "lines."

The broadening of the line spectra, represented in Fig. 8–13, is a distortion produced by the detector. A measure of this broadening is referred to as the detector resolution. The width of the observed gamma ray lines is, in the case of an NaI detector, primarily a function of the gamma ray energy, the size of the crystal, and the optical coupling between the crystal and photomultiplier, as well as the characteristics of the photomultiplier.

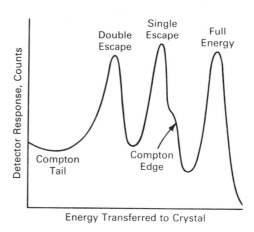

Figure 8–13. An example of further complication which takes place in the detector. Shown are the two escape peaks and the full energy peak. The two escape peaks, present only for gamma rays above the pair production threshold, correspond to the escape of one or both of the 511 keV annihilation gamma rays. The resolution of the detector is seen to have broadened each of the otherwise sharp peaks.

One of the major drawbacks of scintillation detectors is their poor energy resolution.* In this type of device, detection requires a number of inefficient steps. The result is that the energy required to produce one information "carrier" (a photoelectron in the photomultiplier) is about 1000 eV. Thus the number of carriers for a typical radiation detection is rather small. The statistical fluctuations on such a small number place an inherent limitation on the energy resolution.

---

* The resolution is usually quoted in percentage. It compares the observed width (ΔE) to the energy (E) of the gamma ray line at which it is measured. The width is determined at half the maximum of the peak. For an NaI detector at 600 keV, a typical value is about 10%.

The use of semiconductor materials as radiation detectors can produce many more information carriers per detected event and thus achieve a very high energy resolution. In a solid state device such as the germanium detector, the semiconductor properties are used to transfer the charged particle energy into a usable electrical pulse in a much more direct manner. The energetic charged particles transfer energy to electrons bound (by only 0.7 eV for Ge) in the crystal lattice, enabling many of them to become free. Each free electron leaves a positive hole in the electron structure of the crystal. Under a strong electrical field applied to the detector crystal, the free electrons and holes migrate quickly to electrodes and create an electrical impulse.

The excellent resolution arises because the band gap is so small. About 3.5×10^5 electrons are freed by the detection of a 1 MeV gamma ray. These contribute to the resulting pulse with no intervening inefficient steps. The result is sharp energy resolution. Another requirement, however, is that the detector must be operated at extremely low temperatures. This is because at room temperature (not to mention borehole temperatures) some electrons have sufficient thermal energy to cross the 0.7 eV band gap and camouflage those freed by gamma ray interactions. Although the gamma ray spectra obtained with Ge detectors is superb, their overall efficiency is poorer than that obtained by NaI detectors. This latter disadvantage is due to the small detector volumes available for this type of device. Applications of solid state detectors are limited to devices concerned with precise spectroscopic elemental definition or in situ chemical analysis.

References

1. Evans, R.D., *The Atomic Nucleus*, McGraw, New York 1967, pp. 426-438.
2. Ellis, D.V., "Nuclear Logging Techniques," in *SPE Petroleum Production Handbook*, edited by H. Bradley, SPE, Dallas (in press).

Problems

1. Show that the maximum of the Poisson distribution occurs for $x = \mu$ when $\mu \gg 1$.
2. As will be seen in Chapter 9, one method of measuring the density of a formation is to use the attenuation of gamma rays. The gamma-gamma density tool is basically measuring the attenuation of 662 keV γ-rays over a path length of about 40 cm.

a. Using the simple exponential flux attenuation law ($N = N_o e^{-\mu\rho x}$), derive an expression which relates the uncertainty in density ρ to the uncertainty in the measured counting rate dN.

b. Using the results from part a, what fractional uncertainty in the counting rate would permit the determination of porosity to within 1 PU (i.e., 1% porosity) in a sandstone of porosity 10 PU?

c. What counting rate does this imply for a 1-second measurement?

3. The ^{137}Cs γ-ray source which is used in the density logging sonde emits 10^{10} γ/sec and has a half-life of approximately 30 years. The actual form of the source material is microspheres of CsCl, which has a density of 4 g/cm^3. Assuming cubic packing for the microspheres, what volume in cm^3 does the source occupy?

4. Show that the efficiency of a γ-ray detector should vary as $(1-e^{-\mu\rho x})$, where x is the thickness of the detector.

5. Using the data of Fig. 8–8, sketch the efficiency curves of a detector of density 3 g/cm^3 and $Z = 13$ for thickness of 1" and 2" between 10 keV and 100 keV.

6. Radioactive decay is responsible for a portion of the thermal gradient observed in the earth. Show that the heat generation of the Cs^{137} density logging source is on the order of a milliwatt (1eV = 1.6×10^{-19} Joule).

9
GAMMA RAY DEVICES

INTRODUCTION

Previous log examples demonstrate that two logging measurements are reputed to respond to the difference between clean and shaly formations. One of the measurements, the SP, has been analyzed in some detail, and it is known to have a marked response to clean permeable zones. This is evident in the right track of Fig. 9-1 where the more negative potential of the zone between 8510'-8540' is due to the simultaneous absence of clay and free communication between the borehole fluid and the formation waters. For the same well, on the left side of the figure, the gamma ray log has a structure similar to the SP trace: a low reading in the clean zone and a high reading in an apparently shaly zone. Fig. 2-18 shows another example, in track 1, where the gamma ray log and SP correlate to a high degree.

As is evident from its name, the gamma ray (or GR) responds to the natural gamma radiation in the formation. We shall first address the question of the origin of this natural radiation. The few isotopes which are responsible for it can be attributed to a small list of common minerals which are rarely encountered in isolated occurrences in logging. The association of measurable quantities of radioactive isotopes in shales is primarily due to the presence of clay minerals, some of which are naturally radioactive or have radioactive ions associated with them.

Gamma ray logging was introduced in the late 1930s as the first non electrical logging measurement. It was immediately useful for distinguishing shaly from clean formations, among other applications to be discussed. Two types of devices are routinely used for determining formation radioactivity. The *GR tool*, in a form nearly indistinguishable from its 1930s predecessor, uses a simple gamma ray detector to measure the total radioactivity of the formation. *Spectral gamma tools* additionally quantify the concentrations of the radioisotopes present. The two types of devices have similar depths of investigation and suffer from minor environmental effects. The calibration of both types of devices is made with respect to artificial "shale" formations.

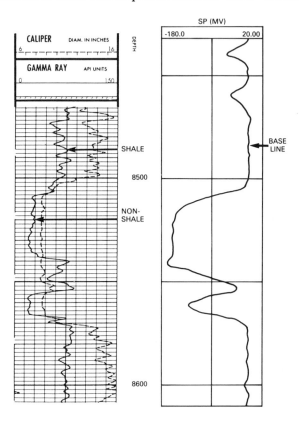

Figure 9–1. Comparison of the gamma ray curve with the SP and caliper over clean and shale zones.

Although the GR log is an important component of the traditional analysis of shaly formations, the interpretation of this measurement is somewhat imprecise. A few examples are shown to illustrate the usefulness of spectral gamma ray measurements. Numerous interpretation schemes utilizing the additional spectral information have been proposed and enjoy varying degrees of success. However, the full potential of this information is just beginning to be realized as a component of geochemical logging, discussed in Chapter 19.

SOURCES OF NATURAL RADIOACTIVITY

In order to suggest which naturally occurring isotopes might be responsible for the gamma ray activity of formations, it is instructive to compare half-lives with the estimated age of the earth, which is about 4×10^9 years. There are only three isotopes with half-lives of that magnitude or greater: ^{40}K: 1.3×10^9 years, ^{232}Th: 1.4×10^{10} years, and ^{238}U: 4.4×10^9 years. The decay of K is accompanied by the emission of a single characteristic gamma ray at an energy of 1.46 MeV. Thorium and uranium both decay through two different series of a dozen or more intermediate isotopes to a stable isotope of lead. This gives rise to complicated gamma ray spectra with emissions at many different energies, as shown in Fig. 9–2. The prominent gamma ray emission from the uranium series is due to an isotope of bismuth, while that of the thorium series is from thallium. In addition to potassium, the source of all significant gamma ray activity in sedimentary rocks is attributed to isotopes of the thorium or uranium series.

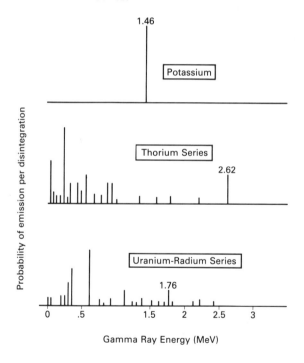

Figure 9–2. The distribution of gamma rays from the three naturally occurring radioactive isotopes.

The largest source of formation radioactivity is potassium, a fairly common element in the earth's crust. Fig. 9–3 shows the crustal abundances of the common elements. Only eight elements are found at concentrations of 1% or greater. Potassium and magnesium both occur at the 1% level. The minerals containing potassium in sedimentary formations are numerous. Table 9–1 lists a number of evaporites which are potassium-rich, the most commonly known being sylvite. Feldspars, which, after quartz, are the most

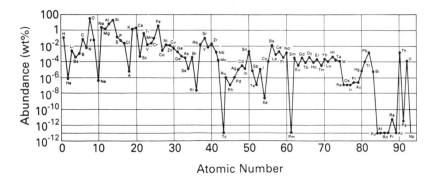

Figure 9–3. Concentration of the elements in the earth's crust in weight percent. From Garrels and MacKenzie.[1]

abundant minerals found in sandstones, have a family of potassium-rich members. One of the five groups of clay minerals (discussed in Chapter 19) prevalent in sedimentary formations contains potassium as part of the lattice structures. Commonly encountered minerals of this group are are illite, mica, and glauconite.

The potassium-bearing minerals of evaporites

Name	Composition	K (% weight)
Sylvite	KCl	52.44
Langbeinite	$K_2SO_4, (MgSO_4)_2$	18.84
Kainite	$MgSO_4, KCl, (H_2O)_3$	15.7
Carnallite	$MgCl, KCl, (H_2O)_6$	14.07
Polyhalite	$K_2SO_4, MgSO_4, (CaSO_4)_2, (H_2O)_2$	13.37
Glaserite	$(K,Na)_2, SO_4$	24.7

Table 9–1. Potassium-bearing evaporitic minerals. From Serra.[2]

By contrast, thorium and uranium-bearing minerals are rare. In logging applications, the uranium may be due to the odd rare mineral, but frequently it is from the precipitation of uranium salts. The solubility of uranium compounds accounts for its transport and its frequent occurrence in organic shales. In the latter case, the presence of uranium results from the absorption of uranium by plant or animal substances which later make up the shale. Thorium is frequently associated with heavy minerals, such as monazite or zircon, which are also known as resistates. Fig. 9–4 summarizes the distribution of thorium and uranium for some sediments. Unlike potassium, which we can expect to find at mass concentration at the level of a few percent, thorium and uranium may be expected, at most, at the level of tens of parts per million. The largest concentrations are seen to be associated with shales, which are considered next.

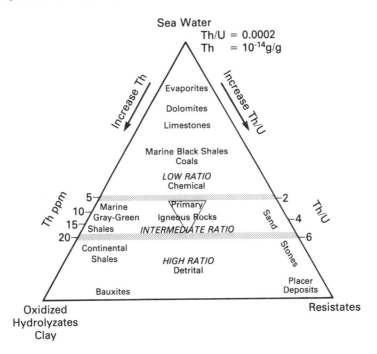

Figure 9-4. Classification of sediments from thorium and Th/U ratio. Distribution of Th/U in sediments. From Adams and Weaver.[3]

For this discussion, shale is considered to be a fine-grained rock composed of silt and clay minerals. The silt is predominantly quartz but may contain feldspars and organic matter. The clay minerals are primarily responsible for two sources of radioactivity associated with most shales. We have already seen that the illite group contains potassium. As Fig. 9-5 demonstrates, potassium shows up in association with other types of clay as well. Clay minerals, which are formed during the decomposition of igneous rocks, in general have a very high cation exchange capacity. As a result of this property, they are able to retain trace amounts of radioactive minerals which may have originally been components of the feldspars and micas which go into their production. This property may be responsible for the retention of trace amounts of thorium, which occurs in relatively insoluble minerals. Uranium, because of its solubility, is easily transported from the site of clay mineral formation. It is associated with the organic matter in the shales rather than the clay minerals.

186 Well Logging for Earth Scientists

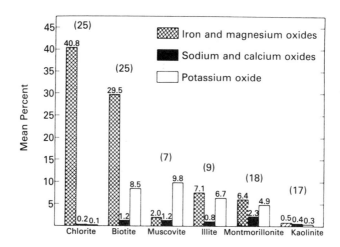

Figure 9-5. Concentration of potassium associated with several clay minerals. From Serra.[2]

A statistical study of the geochemistry of more than 500 core samples, conduced by Hassan et al., supports the view presented above.[4] The correlation between clay minerals and elemental concentrations was found to be largest for thorium and potassium, while that for uranium was negligible. The correlation between clay minerals and thorium was largest, because potassium is also associated with other components of the shale, such as feldspars. The only significant correlation for uranium was found to be the organic carbon content of the samples.

GAMMA RAY DEVICES

One of the principal uses of the gamma ray log is to distinguish between the shales and the nonshales. The first gamma ray devices measured only the total gamma ray flux emanating from the formation. These older gamma ray devices use Geiger counters or NaI detectors, measuring the gamma rays above some practical lower limit (on the order of 100 keV). This total counting rate is a function of the distribution and quantity of radioactive material in the formation. It will be influenced by the size and efficiency of the detector used. For this reason, some calibration standards have been established by the American Petroleum Institute, and all total intensity gamma ray logs are now recorded in API units.

The definition of the API unit of radioactivity comes from an artificially radioactive formation, constructed at the University of Houston facility to simulate about twice the radioactivity of a typical shale. This formation, containing approximately 4% K, 24 ppm Th and 12 ppm U, was defined to be

200 API units. The details of this calibration facility can be found in another work.[5]

The response of a gamma ray device GR_{API} is given by:

$$GR_{API} = \alpha\ ^{238}U_{ppm} + \beta\ ^{232}Th_{ppm} + \gamma\ ^{39}K_{\%},$$

where the subscripts refer to the mass concentration units of the isotope. Note that although ^{40}K is the radioactive isotope, the reference concentration is for the much more commonly occurring ^{39}K. The relative natural abundance of ^{40}K is only 0.012%. The coefficients α, β, and γ depend on the actual detector used and the sonde design details. It was for this type of variability that the API calibration standard was proposed.

However, different types of shale have different total gamma ray activity, depending on the Th, U, and K concentrations associated with them. Fig. 9–2 shows the various gamma ray lines associated with each radioactive isotope. This indicates that, by determining the intensities of particular gamma rays, it is possible to identify the quantity of each radioactive emitter in the formation. With the development of improved spectroscopic-quality gamma ray detectors, it was natural to refine the gamma ray tool into a device capable of determining the actual concentrations of the three components.

Spectral gamma ray devices employ the same basic type of detection system as the total gamma ray devices, but instead of using one broad energy region for detection, the gamma rays are analyzed into a number of different energy bins. Calibration in standard formations, where the concentrations of K, U, and Th are known, then permits determination of the mass concentrations present in a measured formation as well as the total activity.

To first order, the gamma ray intensity from a uniformly distributed source of constant mass concentration, is independent of the formation density, even though the attenuation is a direct function of the formation density. This can be seen from the following argument.

Consider an infinite homogeneous medium containing n gamma ray emitters per unit volume, each with an emission rate of one gamma ray per second. To calculate the total gamma ray flux which would be detected at any given point in this medium, refer to Fig. 9–6. The contribution to the total counting rate from a spherical shell of thickness dr, at a distance r from the detector, is proportional to the flux $d\Psi$ from this volume. It is given by the number of emitters contained in this shell multiplied by the attenuation over the path length r to the detector:

$$d\Psi = n\ 4\pi r^2 dr\ \frac{e^{-\mu \rho_b r}}{4\pi r^2},$$

and the total flux is the integral:

$$\Psi = n \int_0^\infty e^{-\mu \rho_b r} dr = n\ \frac{1}{\mu \rho_b}.$$

Since μ is independent of density, the total counting rate is a direct measure of n/ρ_b, which can be expressed as the weight percent of the isotope which is radioactive. Consequently, the GR log responds directly to the mass concentration of radioactive elements.

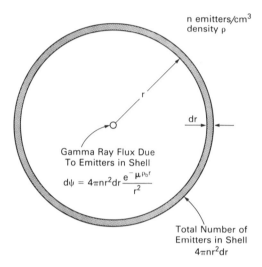

Figure 9–6. Geometry for calculating the total gamma ray intensity at a point detector due to a uniform distribution of gamma ray emitters. From *SPE Petroleum Production Handbook.*[6]

Gamma ray measurements suffer to some degree from the borehole environment. Because of mud in the borehole and varying hole diameters, the gamma rays emitted from the formation have to pass through different amounts of gamma ray absorber in order to reach the detector. Additional complications can arise from mud additives such as barite or KCl. Barium in the mud is a very efficient absorber of low energy gamma rays emanating from the formation. The potassium in KCl makes the mud an unwanted source of radioactivity. Reference 7 discusses a method for correcting spectral gamma ray measurements for these effects.

USES OF THE GAMMA RAY MEASUREMENT

The gamma ray log has traditionally been used for correlating zones from well to well, for crude identification of lithology, and for rough estimation of the volume of shale present in the formation. Continuous shale beds can be readily identified in wells separated by large distances from their characteristic gamma ray "signature." Due to the simplicity of the gamma ray tool, it is present as an auxiliary sensor on most other logging services to provide routine depth control. With the current state of knowledge of clay composition and with other more refined lithology determinations available, it is apparent that the GR log will be used in the future only for correlation and depth control.

The use of the GR measurement to estimate the shaliness of a formation has been the subject of some confusion, which arises from two sources. First, log analysts have used the terms *clay* and *shale* interchangeably. Second, the GR log responds to neither clay nor shale, but rather to associated radioactive isotope concentrations. For estimating the volume fraction of shale in a formation V_{sh}, the traditional approach is to scan the log for minimum and maximum gamma ray readings, γ_{min} and γ_{max}. The minimum reading is then assumed to be the *clean point* (0% shale), and the maximum reading is taken as the *shale point* (100% shale). Then the gamma ray reading in API units at any other point in the well (γ_{log}) may be converted to the gamma ray index I_{GR} by linear scaling:

$$I_{GR} = \frac{\gamma_{log} - \gamma_{min}}{\gamma_{max} - \gamma_{min}}.$$

This index can be scaled into percent of shaliness according to charts (see Fig. 9–7) depending on rock type. The linear transformation of I_{GR} to shale fraction is frequently used. The two non linear conversions of Fig. 9–7,

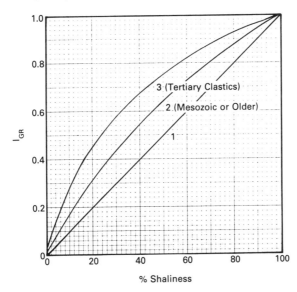

Figure 9–7. Conversion of the GR index to shaliness depending on rock types, based on work of Larinov. Heslop proposed a modification based on a distinction between "shale" and "clay."[8] From Dresser.[9]

which result in small shale volume estimates for a given gamma ray index, attempt to compensate for the clay mineral proportions of different shales. At best, this is seen to be a somewhat fuzzy approach to determining the nature of a shaly sand. Depending upon the application, sometimes the shale fraction is desired, and sometimes the clay mineral volume would be more appropriate. The linear interpolation for shale volume is appropriate for shaly zones which contain the same proportions of clay minerals as those zones used for the determination of the GR end points. Further discussion of this technique, which attempts to determine too much from too little information, may be found in Heslop.[8]

SPECTRAL GAMMA RAY LOGGING

One of the difficulties in the interpretation of the gamma ray measurements is a lack of uniqueness. There are nonradioactive clays, and there are "hot" dolomites. The use of spectral gamma ray devices can point out anomalies such as a "hot" dolomite or other formations with some unusual excess of U, K, or Th. They permit recording the individual mass concentrations of the three radioactive components of the total gamma ray signal. For one type of tool, the relationship between the concentration of the three radioactive components and the total gamma ray signal in API units is given approximately by:

$$\gamma_{API} = 4\ Th + 8\ U + 16\ K,$$

when thorium and uranium are measured in ppm and potassium in percent by weight. This expansion shows, for example, that a shaly sand containing a mineral rich in potassium, such as mica, could be interpreted erroneously. A false indication of the percentage of shale would be reported as a result of the additional radioactivity produced by the mica. Another use of this decomposition is to provide a total gamma ray signal minus the uranium contribution. This can provide a uranium-free gamma ray index which is more representative of the clay minerals in the shale, by eliminating effects of organic shales or the deposit of uranium salts in fractures.[10] Examples of the use of the spectral gamma ray device to detect such anomalies are presented next.

Fig. 9-8 shows a log example in a micaceous sand. At 10,612'-10,620', a shale is indicated having a total gamma ray signal of about 90 API units. If only the total gamma ray is used as an indicator, it appears that the zone between, 10,568' and 10,522' contains about half the amount of shale estimated for the lower zone. However, the decomposition of the gamma ray signal shows quite clearly that the amounts of U, Th, and K in these two zones are quite different. In fact, the upper zone is a mixture of sand and mica, whereas the lower zone is indeed shale.

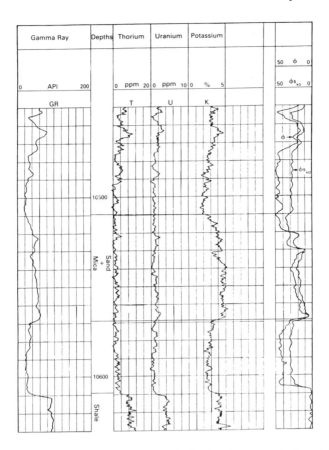

Figure 9–8. A spectral gamma ray log indicating the concentration of Th, U, and K. The zone indicated as containing mica shows an abnormally high K content. In the zone, the GR curve would incorrectly imply the presence of a nonnegligible amount of clay. From *SPE Petroleum Production Handbook*.[6]

In the next example, Fig. 9–9, the gamma ray log alone would indicate that below the lower boundary of the shale bed, at 12,836', there is a relatively clean sand. It can be seen, however, from the K trace that the high level of potassium in the shale zone persists several feet below 12,836'. This excess potassium was found from subsequent core analysis to result from the presence of feldspar. This is an important piece of knowledge because feldspar affects the choice of grain density to be used in the interpretation of density logs.

The third example, Fig. 9–10, shows how a uranium-rich formation would be misinterpreted (in a simple gamma ray analysis) as being shale. The sudden increase in uranium content signals that this is not a simple shale like those at nearby depths. Core analysis showed this zone to be rich in organic material. This is consistent with the notion that U is often trapped in organic complexes.

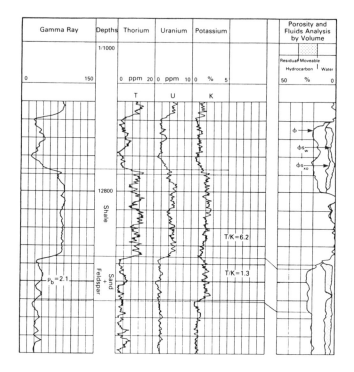

Figure 9–9. A log showing the effect of feldspar on the spectral and total gamma ray logs. From *SPE Petroleum Production Handbook*.[6]

There are two important reasons for using a spectral gamma ray measurement rather than the standard gamma ray, which is reliable only for correlation. The first is for the resolution of radioactive anomalies like those described above. The second is to help identify the clay types by classifying them in terms of the relative contributions of the three radioactive components. It will become apparent in Chapter 19 that clay mineral identification is a task much too complicated to be attempted solely from a knowledge of the associated radioactive elements. However, several schemes for extracting some additional information from spectral gamma ray logs, discussed below, enjoy a certain amount of success.

As indicated earlier, the distinguishing of mica from shale has been an important application for the spectral gamma ray tools. Resistivity logs need to be corrected for the presence of conductive clay minerals, but mica is an insulator. Hodsen et al., in the tradition of well log data analysis, used a cross plot of measured thorium and potassium values to distinguish shaly sands from those containing mica, as shown in Fig. 9–11.[11] The end members

Gamma Ray Devices   193

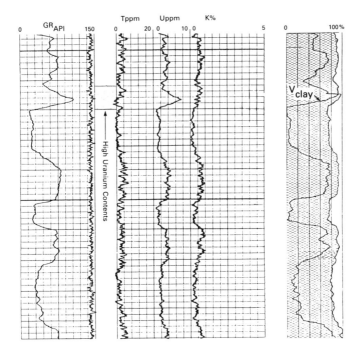

Figure 9–10.   A log showing the result of an anomaly of U. If undetected, the volume of clay in the entire zone is compromised. From *SPE Petroleum Production Handbook.*[6]

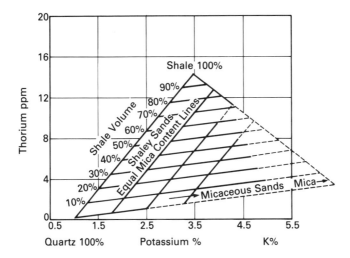

Figure 9–11.   A cross plot of Th and K for determining shale volume and distinguishing micaceous sands from shaly sands. From Hodsen et al.[11]

in this plot correspond to clean sand, shale (appropriate to certain zones of a particular well), and mica. The plot can be scaled to give the fractional volumes of the three components. An improved clay indicator was proposed by Ruhovets and Fertl.[12] It is based on an approximate knowledge of potassium and thorium contents of some common clay minerals, as shown in

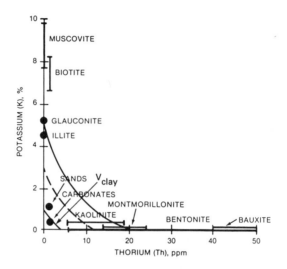

Figure 9–12. Determination of V_{clay} for thorium and potassium distributions. After Ruhovets and Fertl.[12]

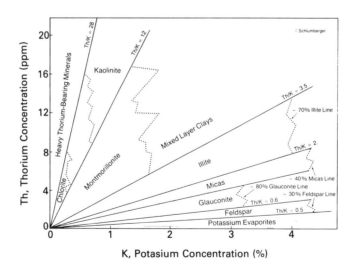

Figure 9–13. Identification of clay minerals from thorium and potassium. From Schlumberger.[14]

Fig. 9–12. Lines of constant clay volume should follow a parabolic shape, as indicated in the figure. The scaling is adopted to the clay minerals present in the zones under investigation and determined from the cross plot. Quirein et al. have suggested a means to determine the presence of major clay minerals, such as illite and kaolinite, and to separate them from feldspars.[13] The method is summarized in Fig. 9–13 and consists of an ambitious reinterpretation of the potassium-thorium cross plot.

Spectral Stripping

To measure the concentration of the three radioactive isotopes responsible for the total gamma ray signal, a spectral analysis of the detected gamma radiation is performed. Fig. 9–14 shows how the line emission spectrum of the three isotopes are distorted in an NaI detector. Of the roughly 20 lines, only 3 are seen clearly.

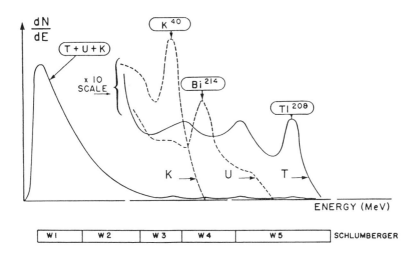

Figure 9–14. A schematic representation of the distortion of the spectrum of natural radiation as registered by an NaI detector.

One data-reduction technique divides the observed spectrum into a number of windows.[15] In the illustration of Fig. 9–14, the number of windows is five. By a series of measurements in specially constructed formations containing known concentrations of the three radioactive isotopes, it is possible to construct a *response matrix*. This matrix relates the counting rates $(W_1, \cdots, W_5)$ in the five windows to the concentrations of U, Th, and K.

This is shown symbolically as:

$$\begin{bmatrix} W_1 \\ W_2 \\ W_3 \\ W_4 \\ W_5 \end{bmatrix} = \overline{A} \times \begin{bmatrix} Th \\ U \\ K \end{bmatrix}, \qquad (1)$$

where the response matrix $\overline{A}$ is a 5×3 matrix. The entries of $\overline{A}$ correspond to the counting rate contributed by each radioactive material to each window. Normally, if the number of equations were equal to the number of unknowns, the solution of Eq. (1) for the concentrations would be simple. In this overdetermined system, the method of least squares can be used to determine the coefficients of the inverse matrix. The procedure begins by considering that there is statistical noise present in the five window counting rates. Consequently, in a formation with precisely known Th, U, and K concentrations, there will be a residual counting rate r_i for each window measurement, when compared to the noise-free standard response of Eq. (1). The residual can be expressed by:

$$\sum_{i=1}^{5} W_i - A_i Th - B_i U - C_i K = \sum_i r_i \,,$$

where A_i, B_i, C_i are the appropriate elements of $\overline{A}$. The least-square procedure is to minimize the square of the residuals:

$$\sum_{i=1}^{5} (W_i - A_i Th - B_i U - C_i K)^2 = \sum_i r_i^2 \,. \qquad (2)$$

This can be done by taking the derivative of Eq. (2) with respect to Th, U, and K. This results in three equations in three unknowns. The solution is of the form:

$$\begin{bmatrix} Th \\ U \\ K \end{bmatrix} = \overline{m} \begin{bmatrix} W_1 \\ W_2 \\ W_3 \\ W_4 \\ W_5 \end{bmatrix},$$

where the elements of $\overline{m}$ are a combination of the elements of the original response matrix. The optimized solution takes into account the variance of the counting rates of each window (the square root of the counting rate), and the set of equations to be solved are determined from differentiating the following expression:

$$\sum_{i=1}^{5} \frac{1}{W_i} (W_i - A_i Th - B_i U - C_i K)^2 = \sum_i r_i^2 \,.$$

The set of coefficients $\dfrac{1}{W_i}$ is referred to as the weighting matrix and may represent the average counting rates expected in shales, for example, or can be determined, on a measurement-by-measurement basis, using the actual counting rates. Bevington treats this data-reduction problem in terms of matrix operations and provides programs for their implementation.[16]

A NOTE ON DEPTH OF INVESTIGATION

How deeply does the GR tool see into the formation? The depth of investigation of such a device is difficult to measure experimentally; however, it can be determined from Monte Carlo simulations. For simplicity, the depth of investigation for a single component of the GR signal is considered. Fig. 9–15 shows the depth of investigation of detected radiation from the decay of ^{40}K computed in this manner.[17] For this simulation, the tool was taken to be eccentered in an 8" diameter borehole. The figure shows the normalized integrated signal J(r) produced by coaxial cylinders of K-bearing formation around the borehole. What is seen from this figure, computed for a density of 2.5 g/cm^3, is that 90% of the signal of unscattered gamma rays comes from an annulus which is about 15 cm thick. For multiply scattered gamma rays, the depth of investigation increases by only a few centimeters. This is to be contrasted, for example, with the integrated radial geometric factor of the deep induction, which shows a total insensitivity for such a shallow zone.

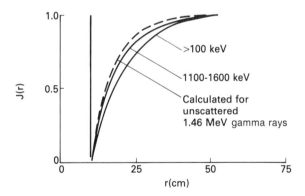

Figure 9–15. The computed radial geometric factor for the detection of K gamma rays by a borehole sonde in an 8" borehole. From Wahl.[17]

The result of the sophisticated calculation can be approximated rather easily by considering the mean free path of gamma rays. The general gamma ray attenuation relationship (from Chapter 8) can be written as:

$$N = N_o e^{-\mu \rho x},$$

where μ is the mass absorption coefficient, ρ is the material bulk density, and x is the distance over which the attenuation is taking place. From this expression, the mean free path λ is taken to be:

$$\lambda = \frac{1}{\mu\rho}.$$

It is the distance over which the flux is reduced by a factor of 1/e. The mass absorption of a 1.46 MeV gamma ray in nearly any substance is found to be 0.05 cm^2/g (see Fig. 8–8) so the mean free path, in centimeters, is given by:

$$\lambda = \frac{20}{\rho},$$

where ρ is in g/cm^3. This means that the mean free path for 1.46 MeV gamma rays varies between 7 and 10 cm for formations with densities between 2.0 and 3.0 g/cm^3. Using the method developed in the appendix of Chapter 12, an expression for the integrated radial geometric factor J can be developed. It shows the signal to grow as:

$$J \propto \left(1 - e^{-\frac{r}{\lambda}}\right).$$

To compare this prediction with the Monte Carlo calculation, λ is taken to be 8 cm, corresponding to a density of 2.5 g/cm^3. The 90% point will be attained for a shell (in this spherical approximation) of total thickness of about 18 cm., which is in close agreement with the Monte Carlo calculation.

References

1. Garrels, R. M., and MacKenzie, F. T., *Evolution of Sedimentary Rocks*, W. W. Norton, New York, 1971.

2. Serra, O., *Fundamentals of Well-Log Interpretation*, Elsevier, Amsterdam, 1984.

3. Adams, J. S., and Weaver, C. E., "Thorium to Uranium Ratio as Indicator of Sedimentary Processes: Examples of Concept of Geochemical Facies," *Bull. AAPG*, Vol. 42, 1958.

4. Hassan, M., Hossin, A., and Combaz, A., "Fundamentals of the Differential Gamma Ray Log-Interpretation Technique," Paper H, SPWLA Transactions, 1976.

5. Belknap, W. B., Dewan, J. T., Kirkpatrick, C. V., Mott, W. E., Pearson, A. J., and Rabson, W. R., "API Calibration Facility for Nuclear Logs," *Drill. and Prod. Prac.*, API, Houston, 1959.

6. Ellis, D. V., "Nuclear Logging Techniques," in *SPE Petroleum Production Handbook*, edited by H. Bradley, SPE, Dallas (in press).

7. Ellis, D. V., "Correction of NGT Logs for the Presence of KCl and Barite Muds," Presented at the SPWLA Twenty-Third Annual Logging Symposium, 1982.
8. Heslop, A., "Gamma-Ray Log Response of Shaly Sandstones," Trans. SPWLA, McAllen, Texas, 1974.
9. *Well Logging and Interpretation Techniques: The Course for Home Study,* Dresser Atlas, Dresser Industries, 1983.
10. Serra, O., Baldwin, J., and Quirein, J., "Theory, Interpretation and Practical Applications of Natural Gamma Ray Spectroscopy," Paper Q, SPWLA Transactions, 1980.
11. Hodsen, G. W., Fertl, W. H., and Hammack, G. W., "Formation Evaluation in Jurasic Sandstones in the Northern North Sea Area," *The Log Analyst,* Vol. 17, No. 1, 1976.
12. Ruhovets, N., and Fertl, W. H., "Digital Shaly Sand Analysis Based on Waxman-Smits Model and Log-Derived Clay Typing," Paris SPWLA Symposium, 1981.
13. Quirein, J. A., Gardner, J. S., and Watson, J. T., "Combined Natural Gamma Spectral/Litho-Density Measurement Applied to Complex Lithologies," Paper SPE 11143, Fifty-seventh Annual Fall Technical Conference, New Orleans, 1982.
14. *Schlumberger Interpretation Charts,* Schlumberger, New York, 1985.
15. Marett, G., Chevalier, P. Souhaite, P., and Suau, J., "Shaly Sand Evaluation Using Gamma Ray Spectrometry Applied to the North Sea Jurassic," Paper DD, SPWLA Symposium, 1976.
16. Bevington, P. R., *Data Reduction and Error Analysis for the Physical Sciences,* McGraw-Hill, New York, 1969, pp. 164–176.
17. Wahl, J. S., "Gamma-Ray Logging", *Geophysics,* Vol. 48, No. 11, 1983.

Problems

1. It was stated in the derivation of the counting rate from a uniformly distributed gamma ray emitter (with n emitters per cubic centimeter) that $\frac{n}{\rho}$ is proportional to the percent by weight of the isotope. What is the constant of proportionality?

2. Estimate the counting rate (in counts/sec) for a perfectly efficient γ-ray detector with surface area of 1 cm^2 which is placed in the 200-API gamma ray standard formation. Recall that the material in this standard

consists of 12 ppm U, 24 ppm Th, and 4% ^{39}K. For this calculation assume a reasonable average value of the mass absorption coefficient μ for the energy range of emissions, and further assume that only 1 γ-ray is emitted per decay of the radioactive material.

3. On the log of Fig. 9–1, estimate the amount of clay, V_{clay}, at the following depths: 8540', 8549', and 8560'. Do the same using the SP curve and compare. Why does the GR yield higher values for V_{clay} in the bottom zone?

4. Using the total GR curve of Fig. 9–8, compute the value of V_{clay} at the top and bottom of the zone indicated as sand + mica.

5. Using the response matrix of Eq. (1), derive an expression for an appropriate two-window estimate of Th and U. Write it in terms of elements of the response matrix; for example, a_{13} corresponds to the contribution in window 1, due to the concentration of K, and a_{24} corresponds to the counting rate in window 4 from U.

6. For the spectral gamma device described in Fig. 9–14 where the gamma rays from Th and U are predominantly in windows 4 and 5, and where the K gamma ray is contained only in window 3 and below, determine the response equations for U and Th. This can be done from a single measurement where the window counting rates are known, as well as the Th and U concentrations. Suppose window 5 and window 4 have 40 and 100 counts, respectively, in a formation containing 5 ppm Th and 20 ppm U.

10
GAMMA RAY SCATTERING AND ABSORPTION MEASUREMENTS

INTRODUCTION

As noted in Chapter 8, the transmission of gamma rays through matter can be related to the electron density if the predominant interaction is Compton scattering. In the borehole environment, a transmission measurement is not possible. However, a gamma ray transport measurement through a formation can be used to determine its density. With some information on the material composition (lithology and pore fluids), the porosity can be determined.

The motivation for the measurement of formation bulk density comes from its direct relationship with the formation porosity and from geophysical applications. As seen earlier, porosity is an essential petrophysical descriptor and an important ingredient in the interpretation of resistivity measurements in terms of water saturation, S_w. Bulk density is used to compute the acoustic impedances of adjacent layers for seismic interpretation and for estimating overburden pressure.

The basic equation which relates the bulk density of the formation to the porosity ϕ is:

$$\rho_B = \phi \, \rho_f + (1-\phi)\rho_{ma} , \qquad (1)$$

where ρ_f is the density of the fluid filling the pores and ρ_{ma} is the density of the rock matrix. Although this equation is exact, it presents several problems for the determination of porosity. What value is to be used for the matrix density? For normally encountered formations, it is generally between 2.65 and 2.87 g/cm^3, depending on the lithology. For values of fluid density, it is necessary to know the type of fluid in the pores. The fluid density for hydrocarbon ranges from 0.2 to 0.8 g/cm^3. Salt-saturated water (NaCl) density may be as high as 1.2 g/cm^3, and with the presence of CaCl$_2$, values even as great as 1.4 g/cm^3 may occur. It is fortunate that the uncertainty that can be tolerated in ρ_f is much greater than that for ρ_{ma}.

For the moment, this problem of interpretation is overlooked while we discuss, instead, the measurement technique for density determination and how it naturally leads to an auxiliary measurement of the photoelectric factor P_e, which is closely related to the formation lithology.

DENSITY AND GAMMA RAY ATTENUATION

In Chapter 8 it is shown that the interaction of gamma rays by Compton scattering is dependent only upon the number density of the scattering electrons. This in turn is directly proportional to the bulk density of the formation. The reduction of the flux Φ_o in traversing a thickness of material x is given by:

$$\Phi = \Phi_o e^{-\rho_b \frac{Z}{A} N_o \sigma x}, \qquad (2)$$

where the term $\rho_b \frac{Z}{A} N_o$ is the number density of electrons in a material of mass density ρ_b, and σ is the cross section for Compton scattering.

It is natural, therefore, to exploit the attenuation of gamma rays for the determination of bulk density. An idealized device would consist of a detector and a source of gamma rays whose primary mode of interaction is Compton scattering. Finding such a source would be difficult for any arbitrary group of materials. However, for the types of earth formation generally encountered in hydrocarbon logging, the average atomic number rarely exceeds 13 or 14. It was seen from Fig. 8-7 that, for logging applications, there is a large range of gamma ray energies which will be predominantly governed by Compton interaction.

It is worth noting that the basic gamma ray flux attenuation law, Eq. (2), indicates that there is a slight difficulty in the interpretation of a flux attenuation measurement. The attenuation will be strictly related to the bulk density ρ_b only if the ratio of Z/A remains constant. For most elements the value of Z/A is about ½, but there are several significant departures; hydrogen, for example, has a Z/A ratio of nearly 1. For this reason, it is convenient to define a new quantity, ρ_e, the *electron density index*, to be:

$$\rho_e \equiv 2\frac{Z}{A} \rho_b . \qquad (3)$$

In this manner the tool response (or measured flux, Φ) can be specified as:

$$\Phi \propto e^{-\rho_e x},$$

where x corresponds to the source-detector spacing.

NAME	FORMULA	MOLECULAR WEIGHT	Z	P_e	ρ_b	ρ_e(3)	U(4)
A. ELEMENTS							
	H	1.008	1	0.00025			
	C	12.011	6	0.15898			
	O	16.000	8	0.44784			
	Na	22.991	11	1.4093			
	Mg	24.32	12	1.9277			
	Al	26.98	13	2.5715	2.700	2.602	
	Si	28.09	14	3.3579			
	S	32.066	16	5.4304	2.070	2.066	
	Cl	35.457	17	6.7549			
	K	39.100	19	10.081			
	Ca	40.08	20	12.126			
	Ti	47.90	22	17.089			
	Fe	55.85	26	31.181			
	Sr	87.63	38	122.24			
	Zr	91.22	40	147.03			
	Ba	137.36	56	493.72			
B. MINERALS							
Anhydrite	$CaSO_3$	136.146		5.055	2.960	2.957	14.95
Barite	$BaSO_4$	233.366		266.8	4.500	4.011	1070.
Calcite	$CaCO_3$	100.09		5.084	2.710	2.708	13.77
Carnallite	$KCl \cdot MgCl_2 \cdot 6H_2O$	277.88		4.089	1.61	1.645	6.73
Celestite	$SrSO_4$	183.696		55.13	3.960	3.708	204.
Corundum	Al_2O_3	101.90		1.552	3.970	3.894	6.04
Dolomite	$CaCO_3 \cdot MgCO_3$	184.42		3.142	2.870	2.864	9.00
Gypsum	$CaSO_4 \cdot 2H_2O$	172.18		3.420	2.320	2.372	8.11
Halite	NaCl	58.45		4.65	2.165	2.074	9.65
Hematite	Fe_2O_3	159.70		21.48	5.210	4.987	107.
Ilmenite	$FeO \cdot TiO_2$	151.75		16.63	4.70	4.46	74.2
Magnesite	$MgCO_3$	84.33		0.829	3.037	3.025	2.51
Magnetite	Fe_3O_4	231.55		22.08	5.180	4.922	109.
Marcasite	FeS_2	119.98		16.97	4.870	4.708	79.9
Pyrite	FeS_2	119.98		16.97	5.000	4.834	82.0
Quartz	SiO_2	60.09		1.806	2.654	2.650	4.79
Rutile	TiO_2	79.90		10.08	4.260	4.052	40.8
Sylvite	KCl	74.557		8.510	1.984	1.916	16.3
Zircon	$ZrSiO_4$	183.31		69.10	4.560	4.279	296.
C. LIQUIDS							
Water	H_2O	18.016		0.358	1.000	1.110	0.40
Salt Water	(120,000 ppm)			0.807	1.086	1.185	0.96
Oil	$CH_{1.6}$			0.119	0.850(1)	0.948(1)	0.11
	CH_2			0.125	0.850(1)	0.970(1)	0.12
D. MISCELLANEOUS							
"Clean" Sandstone	(1)			1.745	2.308	2.330	4.07
"Dirty" Sandstone	(1)			2.70	2.394	2.414	6.52
Average Shale	(2)			3.42	2.650(1)	2.645(1)	9.05
Anthracite	C:H:O = 93:3:4			0.161	1.700(1)	1.749(1)	0.28
Bituminous Coal	C:H:O = 82:5:13			0.180	1.400(1)	1.468(1)	0.26

(1) Variable; values shown are illustrative.
(2) Pettijohn, 1949: p. 271.
(3) ρ_e is electron density = $\rho_b \times 2Z/A$.
(4) $U = P_e \rho_e$.

TABLE 10–1. DENSITY AND PHOTOELECTRIC PARAMETERS FOR VARIOUS MATERIALS

Table 10–1 lists the density and photoelectric parameters of a number of common elements, minerals and liquids. Of interest for this discussion are the two columns labeled ρ_b and ρ_e. It can be seen by comparison that the bulk density and electron density index for the three major minerals (calcite, dolomite, and quartz) are practically identical in these three cases. However, for the case of water there is an 11% discrepancy between the two (due to the anomalous Z/A value for H). Thus there will be an increasing discrepancy between ρ_b and the density tool response parameter ρ_e, for increasing porosity.

For this reason, the density reading on logs has been adjusted slightly to give precisely the bulk density of water-filled limestone. The relationship between the log reading ρ_{log} and the electron density index ρ_e is:

$$\rho_{log} = 1.0704\rho_e - 0.188 . \tag{4}$$

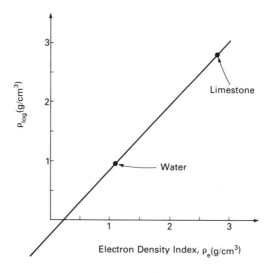

Figure 10-1. The transform between the measured parameter, electron density (ρ_e), and the density value presented on logs.

This relationship is illustrated in Fig. 10–1 which shows the log density as a function of ρ_e for calcite and for water. The straight line joining the two points is described by the equation above and provides the exact bulk density for water-filled limestone. Further reference to the data of Table 10–1 will indicate that Eq. (4) is nearly true (to within a few 1/1000s g/cm³) for many other matrices.

DENSITY MEASUREMENT TECHNIQUE

Fig. 10–2 shows schematically the evolution of the density device from the notion of the simple transmission of monoenergetic gamma rays through a thin sample, to a borehole device consisting of a source, shield, and gamma ray detector. In the upper panel, the gamma rays are transmitted without much scattering, as indicated by the line spectrum detected. As the sample thickness is increased, the gamma ray line intensity decreases, due to exponential attenuation. At the same time, it is accompanied by a buildup of low energy Compton-scattered gamma rays. In the final example, the detector is well-shielded from the source. No source-energy gamma rays reach it. However, the level of multiply scattered gamma rays will still vary exponentially with the scattering material density.

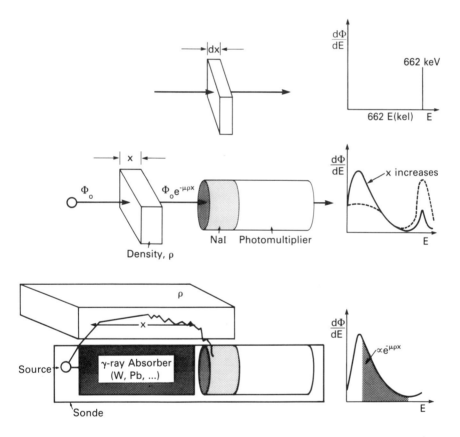

Figure 10-2. A schematic view of the effect of geometry on the determination of density by the use of the attenuation and scattering of gamma rays. In the top panel, the idealized experiment is shown for a very thin piece of material. In the second panel, the thickness has increased so that there is considerable scattering of gamma rays. In the third panel, an approximation to the logging situation is shown.

One explanation for this behavior considers the multiply scattered gamma rays, detected far from the source, to have undergone most of their scattering in the formation close to the detector. These multiply scattered gamma rays are fed by a virtual source. It consists primarily of unscattered source-energy gamma rays which travel from the source, nearly parallel to the borehole wall, to reach the site of their last few collisions before detection. Their intensity will depend on the probability of source gamma rays arriving at this site unscattered, and thus varies exponentially with formation density. Tittman prefers to treat this as a diffusion problem and also concludes that the multiply scattered spectrum will vary exponentially with the formation density.[1,2]

The gamma ray source usually used in density logging is ^{137}Cs, which emits gamma rays at 662 keV, well below the threshold for pair production. This isotope has a half-life of about 30 years, thus providing a stable intensity during a reasonable period of time. Some devices use ^{60}Co, which emits two gamma rays at 1332 and 1173 keV.

The earliest devices consisted of the gamma ray source and a single detector, as indicated in Fig. 10–2.[2] However, to compensate for the frequent occurrence of intervening mudcake, modern devices (Fig. 10–3) incorporate

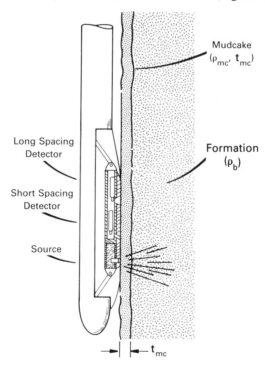

Figure 10-3.   A formation density device in the borehole situation applied to the borehole wall and separated from it by the thickness of the mudcake, t_{mc}. From Ellis et al.[4]

two detectors (generally both NaI) in a housing that shields them from direct radiation from the source.[3,4] The complete device is forced up against the formation with a hydraulically operated arm. This arm also provides a measurement of the diameter (along one axis) of the borehole.

Two detectors at different spacings from the gamma ray source are used in order to compensate for the possible intervening presence of mudcake or drilling fluid. Normally, in addition to the density curve, the log will also show a trace of the compensation, generally referred to as the $\Delta\rho$ curve. This curve represents the correction made to the apparent density seen by the long spacing detector (ρ_{ls}). It is based on the discrepancy between the long and short spacing measurements.

The measurement principle derives from the fact that the counting rate of a detector varies exponentially with the density of the formation, as expected from the general attenuation relation:

$$N = N_o e^{-\mu \rho x}, \qquad (4a)$$

where N is the counting rate of a detector at a distance x from the source. This is illustrated in Fig. 10–4 which shows an exponential relationship between the counting rate and the bulk density. Note that the shorter spacing detector has less density resolution or sensitivity than the farther detector.

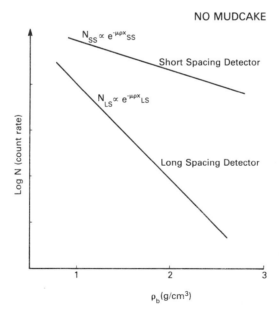

Figure 10-4.   Idealized counting rate response of two detectors for variation of formation density in the presence of no mudcake.

For a given density variation, its counting rate exhibits a smaller fractional change than that of the far detector. This can be seen from the preceding equation, where the product μx corresponds to the slope of the logarithm of the counting rate as a function of density. The spacing x of the nearer detector produces a slope in its response curve that is less than that of the long-spacing detector. The formation density could be determined simply from an observed counting rate from either detector. However, when mudcake of unknown density and/or thickness intervenes, there will be a perturbation of each counting rate. Each counting rate can be translated into an apparent density using Eq. (4a) or Fig. 10-4. Because of the different density sensitivities of the two detectors, the apparent densities for the two detectors will differ.

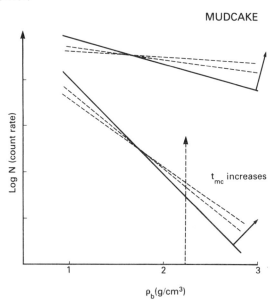

Figure 10-5. Idealized response of the two detectors for two different thicknesses of mudcake. Its density is approximately 1.8 g/cm^3. For a given thickness of mudcake and formation density/mudcake density contrast, both detectors experience approximately the same percentage of change in counting rate.

Fig. 10-3 shows the usual logging condition with the intervening mudcake. Fig. 10-5 indicates the counting-rate behavior for the two detectors in this situation. At a fixed formation bulk density, if the mudcake density is less than the value of ρ_b, then the counting rates of the two detectors increase by approximately the same percentage. This behavior is sketched in Fig. 10-5, where the mudcake density shown is about 1.8 g/cm^3 (this is the point about which the counting rate/mudcake thickness curves pivot). For a given mudcake density, it is seen from the figure that, depending on the formation density, the counting rates either increase or decrease in comparison to the counting rate expected for a clean borehole wall. If the mudcake is more dense than the formation, then the counting rates decrease, and if the mudcake density is less than the formation density, then the counting rates increase.

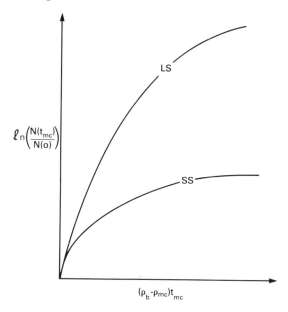

Figure 10-6. Effect of mudcake on the normalized counting rate of near and far detectors. The controlling parameter is the contrast in mudcake and formation density multiplied by the mudcake thickness.

This behavior is further elaborated in Fig. 10–6 which shows the behavior of the normalized counting rates. Each detector's counting rate is normalized to its counting rate with no mudcake present. The parameter that controls the counting rate is the density contrast $\rho_b - \rho_{mc}$ between formation and mudcake, multiplied by the mudcake thickness. This is equivalent to thinking of the last layer of formation, through which the gamma rays have to pass in order to reach the detector, as a filter of variable density.

Fig. 10–7 is the traditional way of presenting the tool response to mudcake. Although this "spine and ribs" plot is illustrative, it is not as convenient as some other representations for making a correction for the presence of mudcake. The "spine" is the locus of the two counting rates without mudcake. The "ribs" trace out the counting rates, at a fixed formation density, for the presence of intervening mudcake.

As an alternative for correcting the intervening mudcake, one can use apparent densities of the long- (ρ_{LS}) and short-spacing (ρ_{SS}) detectors (from the counting-rate calibration determined with no mudcake present) from a series of laboratory measurements to define the correction $\Delta\rho$. This is the correction to be applied to the apparent density from the long-spacing detector in order to arrive at the true value of the formation density behind the mudcake. Data of this type are shown in Fig. 10–8 as a function of the apparent-density differences between the long- and short-spacing detectors.

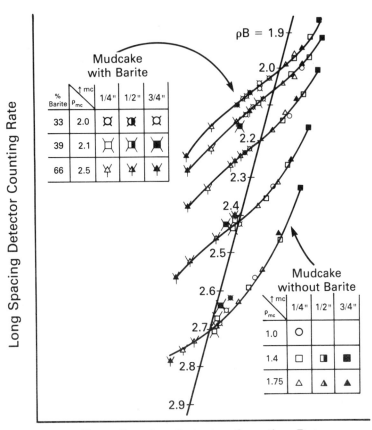

Figure 10–7. A "spine and ribs" representation of the response of a two-detector density device to formation density and mudcakes. Because of its characteristic outline, it is known as the "spine and ribs" chart. From Tittman et al.[3]

If there is not any intervening material between the tool surface and the formation being measured, then the two apparent-density values will be equal. As the mudcake thickness increases, the two density values diverge for some reasonable value of thickness (generally less than 1″). (Eventually, in principle, the two values must again converge to the mudcake density as the thickness becomes very large.) The quantity $\Delta\rho$ is determined, from data such as those shown in Fig. 10–8, to be the value to be added to the apparent

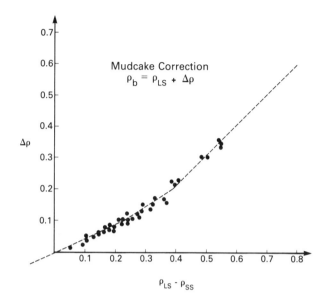

Figure 10–8. A normalized "spine and ribs" chart showing the correction necessary for the case of light mudcakes. Adapted from Ellis et al.[4]

long-spacing density ρ_{LS}, in order to match the known bulk density of the formation, i.e.:

$$\rho_b = \rho_{LS} + \Delta\rho . \qquad (5)$$

A common misconception is that $\Delta\rho$ is a measure of the mudcake thickness t_{mc}. It is, in fact, proportional to the product of mudcake thickness and the density contrast between the mudcake, ρ_{mc}, and the formation density, i.e.:

$$\Delta\rho \propto t_{mc} (\rho_b - \rho_{mc}) . \qquad (6)$$

Beyond some thickness ($\approx 1''$), depending on the tool design details, the compensation scheme breaks down and the estimate of ρ_b will be in doubt. However, this point cannot be identified by use of a simple, single cutoff value of $\Delta\rho$. A very small gap of water ($\rho_{mc} = 1$) in front of a low porosity formation (high density) would yield a large value of $\Delta\rho$ and yet be perfectly compensated for, whereas a $1''$-thick mudcake of medium density in front of a high porosity zone may yield a small $\Delta\rho$ with some residual error in the compensation.

An example of the use of the $\Delta\rho$ curve as a quality control is shown in the log of Fig. 10–9 taken from the shaly sand portion of the simulated reservoir model. Portions of the $\Delta\rho$ curve are highlighted, as well as the corresponding smooth portions of the caliper. In these cases the $\Delta\rho$ curve shows negligible correction, which is probably indicative of little or no mudcake. In the very rough sections of the borehole, the value of $\Delta\rho$ is seen to be quite large, because of poor pad contact with the borehole wall.

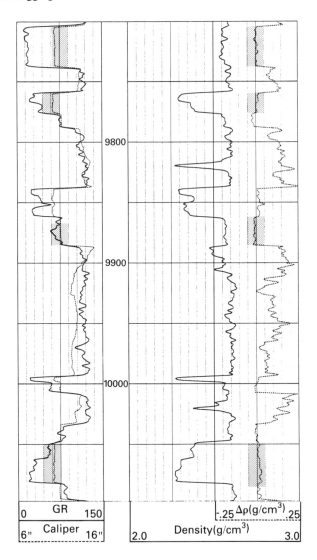

Figure 10–9. A sample density log showing the qualitative nature of the Δρ curve for identifying smooth sections of borehole.

LITHOLOGY LOGGING

In modern density tools, the shape of the low energy portion of the scattered gamma ray spectrum is measured also.[4,5] This is correlated with the photoelectric absorption parameters of the formation, which are in turn tightly linked with the lithology of the formation. Fig. 10–10 indicates the utility of measuring the average atomic number Z of the formation. It allows one to distinguish among the three major types of rock matrix. The spread associated with each lithology in Fig. 10–10 corresponds to variations in average atomic number with porosity (from 0 to 40 PU). In all three cases the average atomic number of the mixture decreases as porosity increases, since the average atomic number of water is about 7.5.

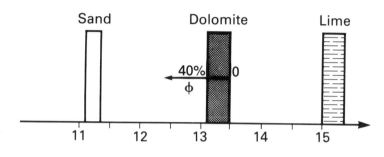

Figure 10–10. The average atomic number of three common lithologies. The width of each corresponds to the range of P_e for the porosity range 0–40%.

In order to appreciate the relationship between average atomic number (determined essentially by lithology) and the photoelectric effect, recall that the cross section for photoelectric absorption τ, is given by:

$$\tau = 12.1 E^{-3.15} Z^{4.6}, \qquad (7)$$

where τ is in barns (10^{-24} cm^2) per atom, E is the energy of the gamma ray in keV, and Z is the atomic number of the absorber. The attenuation of a flux Φ_o, of gamma rays by photoelectric absorption alone can be written as:

$$\Phi = \Phi_o e^{-n\tau x},$$

where n is the number of atoms per cm^3 and x is the attenuation depth. In Fig. 10–11, the behavior of τ is shown as a function of Z for several energies.

It is convenient to define a new parameter P_e, the photoelectric index, as:

$$P_e \equiv \left[\frac{Z}{10}\right]^{3.6}, \qquad (8)$$

where we see from Eq. (7) that P_e is proportional to the photoelectric cross section per electron with the energy dependence suppressed. Then the attenuation equation can be rewritten as:

$$\Phi \propto \Phi_o e^{-n_e P_e x}, \qquad (9)$$

where n_e is the number density of electrons, as in the case of Compton scattering.

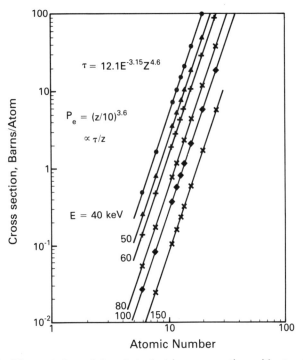

Figure 10–11. The variation of the photoelectric cross section with atomic number. From Bertozzi et al.[6]

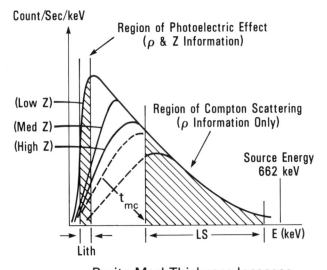

Figure 10–12. A schematic representation of the detected spectral variation of gamma rays in three formations of increasing average atomic number. Additional low energy absorption is noted for the presence and increasing thickness of absorbing barite mudcake. From Ellis et al.[4]

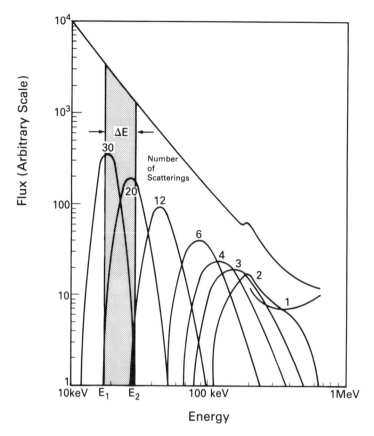

Figure 10–13. The theoretical behavior of the multiply scattered gamma ray spectrum produced by an infinite homogeneous formation containing uniformly distributed sources. It is for the special case of no photoelectric absorption. The spectrum is composed of the contribution of individual spectra of scattering order K. The total spectrum between E_1 and E_2 can be computed from the contributions of the individual spectra for gamma rays that have scattered between 20 and 30 times. From Bertozzi et al.[6]

Qualitatively, the gamma ray spectra observed with a logging device equipped with a window nearly transparent to low energy gamma rays (such as Be) is shown in Fig. 10–12. As the average atomic number, Z, of the formation increases, the lower energy portion of the spectrum is progressively reduced. Thus a measurement of this spectral shape at low gamma ray energies should yield the photoelectric absorption properties, and thereby the Z of the formation.

But first one may ask why the spectrum of gamma rays looks the way it does in Fig. 10–12. In fact, the first question is: "How would the spectrum of gamma rays, emitted in an infinite homogeneous medium containing uniformly distributed Cs sources, look without the presence of any photoelectric absorption?" The answer is contained in Fig. 10–13, which shows the results of a Monte Carlo calculation for this situation. It indicates that the steady-state multiply scattered spectrum falls off with energy as 1/E. The components of this spectrum (grouped by multiplicity of scattering) are shown individually. The spectra formed by gamma rays which have scattered, once, twice, three times, etc., are shown beneath the nearly straight line that results. Each one of these spectra can be imagined to be a snapshot view of the energy distribution of gamma rays that have experienced the same number of scatterings. As shown in the figure, the value of the total flux in an interval ΔE can be computed by summing the individual fluxes over the scattering order K.

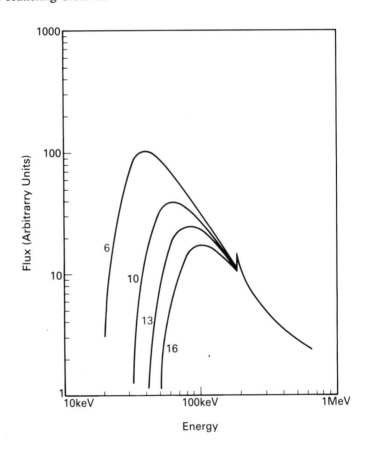

Figure 10–14. A Monte Carlo calculation of the spectrum due to uniformly distributed sources, including the effect of photoelectric absorption. The evolution of the spectra is seen for materials of four different atomic numbers. As the atomic number increases, the low energy portion of the spectrum is attenuated. From Bertozzi et al.[6]

When photoelectric absorption is included in the calculation, it is necessary to take into account the fact that at each collision there is a probability that the gamma ray will be absorbed rather than scattered. The lower the gamma ray energy and the higher the Z of the scattering material, the greater is this probability. The spectra which result are shown in Fig. 10–14, for the range of Z values of interest to us, assuming all the formations have the same density.

Experimental curves over a much narrower range of Z values are shown in Fig. 10–15. Although the curves are not exactly the same shape as Fig. 10–14, there is considerable similarity in the behavior of the low energy region as the Z of the scattering material changes. Furthermore the curves of Fig. 10–14 are calculations of the actual gamma ray flux and do not include the effects of the NaI detector, which considerably modify the spectra.

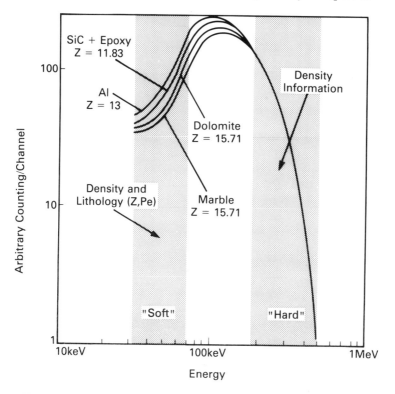

Figure 10–15. Experimental gamma ray spectrum taken with a density device which also measures photoelectric absorption properties of the formation. The spectra have been normalized in the high energy region to emphasize the variations due to atomic number and to eliminate overall amplitude levels due to differences in formation density. A measure of the formation atomic number can be obtained from the ratio of counting rates in the two windows indicated. From Bertozzi et al.[6]

As indicated in Fig. 10–15, two bands of gamma ray energy can be used to determine the density and the lithology effects. The higher energy window contains density information, and the lower window, a combination of density and lithology information. The ratio of the two is primarily dependent on Z. The experimental data of Fig. 10–16 illustrate this.

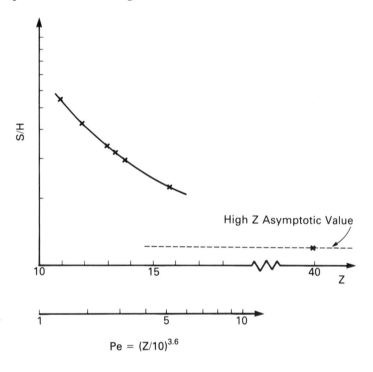

Figure 10-16. The ratio of the "soft" to "hard" windows of Fig. 10–15, as a function of the formation average atomic number, Z. Also shown is a scale corresponding to the photoelectric factor P_e, which is normally presented on logs. From Bertozzi et al.[6]

The calibration data for the photoelectric portion of one type of tool is shown in Fig. 10–17. Once this calibration curve has been established for a given tool, the counting-rate ratio can be used to provide a value of P_e at any depth. The soft/hard ratio dependence seen in Fig. 10–17 agrees with the results of theoretical analysis.[6]

In normal logging circumstances, the P_e log readings should range between 1 and 6. This can be seen from Table 10–1, which also lists the value of P_e for the three principal matrices: 1.83 for sand, 3.1 for dolomite, and 5.1 for limestone. Fig. 10–18 contains logs in a shaly sand portion of the simulated reservoir; the value 1.8 has been indicated in the track corresponding to P_e. It can be seen that only a few portions of the log correspond to this value. However, where the value of P_e indicates sandstone,

Gamma Ray Scattering and Absorption   219

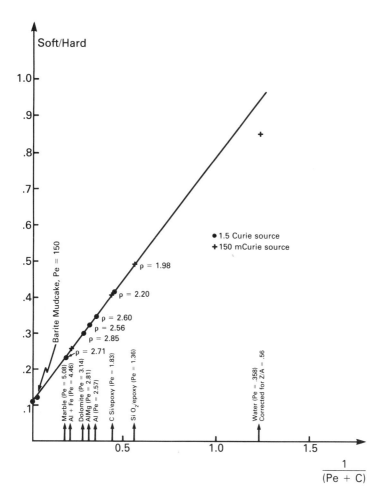

Figure 10-17. The response of an experimental lithology-sensitive device to the photoelectric factor of various laboratory formations.

the gamma ray reading is also at a minimum, indicating that higher P_e readings, in this case, may be associated with the shaliness.

Fig. 10–19 shows a cross plot of P_e versus ρ_b, with lines indicating expected P_e values for porosity variation from 0 to 50 PU for the three principal matrices. Data from a 400′ interval have also been plotted. It is relatively straightforward to separate the sands from the shaly sands and from the shales on this presentation. One anomaly, appearing at a P_e value of about 4.25, corresponds to the presence of an unidentified heavy (high Z) mineral.

After having seen an example of the variation of P_e with lithology, it is natural to wonder how to scale intermediate values of P_e in the case of a mixture. For example, what value of P_e would be expected for a sandstone with a large amount of calcite cement?

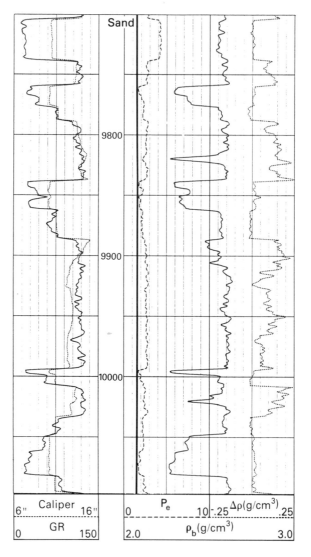

Figure 10-18. The P_e and density response of a logging tool in the shaly sand section of the simulated reservoir.

Since P_e is proportional to the photoelectric cross section per electron, the mixing law must be formulated in terms of the electron density. Eq. (9) makes clear that the calculation of P_e for a mixture will involve weighting the electron density of each atomic species by the Z for that species. It is the mass fraction that enters, rather than the usually desired volume fraction.

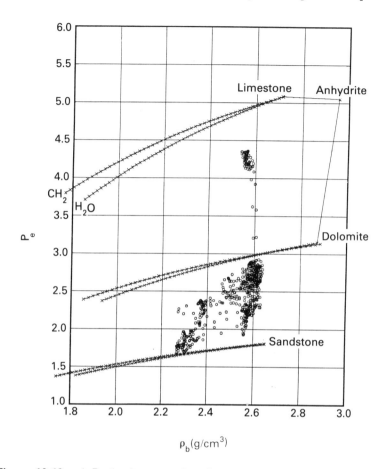

Figure 10-19. A P_e–density cross plot of the data of Fig. 10–18. The three groups of lines correspond to the three major lithologies: sand, dolomite, and limestone. Porosity increases to the left on each of the mineral lines and is marked in 1 porosity unit intervals.

Since the P_e of mixtures does not combine volumetrically, for interpretation purposes a new parameter, U, was developed which has the property of combining volumetrically for mixtures. This parameter is suggested by Eq. (9). Since the electron density n_e is proportional to the electron density index ρ_e, the attenuation is proportional to $e^{-P_e \rho_e x}$. From this we can see that the product of P_e and electron density index, or U, appears to be a macroscopic linear cross section and specifies the absorption of a given thickness of material. The dimensions of U are cross section/cm^3, which indicates that it combines volumetrically.

From this definition of U, the recipe for obtaining the P_e of any mixture first involves computing the U value:

$$U_{total} = P_{e,1}\rho_{e,1}V_1 + P_{e,2}\rho_{e,2}V_2 + \cdots,$$

where $\rho_{e,i}$ is the electron density of material i, $P_{e,i}$ is the photoelectric factor of material i, and V_i is the volume fraction of that material. The final value of the average, $\overline{P_e}$, is obtained from:

$$\overline{P_e} = \frac{U_{total}}{\overline{\rho_e}},$$

where the average electron density index, $\overline{\rho_e}$, is given by:

$$\overline{\rho_e} = \rho_{e,1}V_1 + \rho_{e,2}V_2 + \cdots.$$

Table 10–1 lists some useful values of U and ρ_e for a number of commonly encountered minerals. The use of U in quantitative evaluation of mineral content of a formation will be covered in Chapters 18 and 19.

Table 10–1 shows the enormous sensitivity of the parameter U or P_e to elements with large atomic number. In particular, note the values of P_e for the several iron compounds and for barium. In the case of iron, this sensitivity can be exploited to make a determination of the shale content of the formation if there is iron associated with the clay mineral. This application is also discussed in Chapter 19. However, the sensitivity to barite makes the P_e measurement difficult in heavily weighted barite muds. If there is a substantial thickness of barite mudcake between the tool skid and the formation, or if there is invasion of $BaSO_4$ particles into the formation, the resultant photoelectric absorption can seriously influence the measurement.

ESTIMATING POROSITY FROM DENSITY MEASUREMENTS

The basic output of the gamma-gamma density device, bulk density, is conceptually the simplest measured parameter to interpret in terms of porosity. The basic relation is:

$$\rho_b = \phi\rho_f + (1-\phi)\rho_{ma},$$

which volumetrically links the density of the pore fluid ρ_f and the rock matrix density ρ_{ma} to the bulk density ρ_b. The solution of this equation for porosity yields:

$$\phi = \frac{\rho_{ma} - \rho_b}{\rho_{ma} - \rho_f}. \tag{10}$$

Thus the problem rests on knowing the values to insert for fluid and matrix density. Before we examine the means for determining these values, it is of some interest to know to what precision these two parameters must be known.

First, let us look at the uncertainty which can be tolerated in the value used for the matrix density ρ_{ma}. From the previous equation we can write:

$$\partial\phi = \left[\frac{\rho_b - \rho_{ma}}{(\rho_f - \rho_{ma})^2} - \frac{1}{\rho_f - \rho_{ma}}\right]\partial\rho_{ma}.$$

If this is evaluated for the case of a sand of about 30% porosity, we can use the following values:

- $\rho_b = 2.16$ g/cm^3
- $\rho_f = 1.00$ g/cm^3
- $\rho_{ma} = 2.65$ g/cm^3

and obtain

$$\partial \phi = 0.43 \, \partial \rho_{ma}.$$

Thus for the uncertainty in ϕ to be less than 0.02 g/cm^3, the uncertainty in ρ_{ma} must be less than 0.05 g/cm^3.

For values of fluid density, it is necessary to know the type of fluid in the pores. However, the uncertainty which can be tolerated in ρ_f is much greater than that for ρ_{ma}. An error analysis similar to the one above, for the values chosen, shows that $\partial \phi = 0.18 \partial \rho_f$, which allows about double the margin of error.

The value for matrix density can, in simple cases, be taken from Table 10–1, which shows a rather narrow variation between 2.65 g/cm^3 for quartz to 2.96 g/cm^3 for anhydrite. Grain densities for shales are a considerably more complex issue and will not be covered here. The obvious remaining problem is assigning a matrix density, which clearly requires some knowledge of the lithology. In essence, all of density interpretation aimed at porosity determination revolves on this. Various approaches to this problem are presented in Chapter 18.

References

1. Tittman, J., *Geophysical Well Logging,*, Academic Press, Orlando, 1986.

2. Tittman, J., and Wahl, J. S., "The Physical Foundations of Formation Density Logging (Gamma-Gamma)," *Geophysics,* Vol. 30, 1965.

3. Wahl, J. S., Tittman, J., Johnstone, C. W., and Alger, R. P., "The Dual Spacing Formation Density Log," Presented at the Thirty-ninth SPE Annual Meeting, 1964.

4. Ellis, D., Flaum, C., Roulet, C., Marienbach, E., and Seeman, B., "The Litho-Density Tool Calibration," Paper SPE 12048, SPE Annual Technical Conference and Exhibition, 1983.

5. Minette, D. C., Hubner, B. G., Koudelka, J. C., and Schmidt, M., "The Application of Full Spectrum Gamma-Gamma Techniques to Density/Photoelectric Cross Section Logging," Paper DDD, SPWLA Twenty-seventh Annual Symposium, 1986.

6. Bertozzi, W., Ellis, D. V., and Wahl, J. S., "The Physical Foundation of Formation Lithology Logging with Gamma Rays," *Geophysics*, Vol. 46, No. 10, 1981.

Problems

1. In the lowest portion (which is water-bearing) of a clean sandstone reservoir known to be of constant porosity, the density tool reads 2.21 g/cm^3. Further up in the same reservoir, above the oil-water contact (where the formation is fully hydrocarbon-saturated), the density tool reads 2.04 g/cm^3. What is the density of the hydrocarbon?

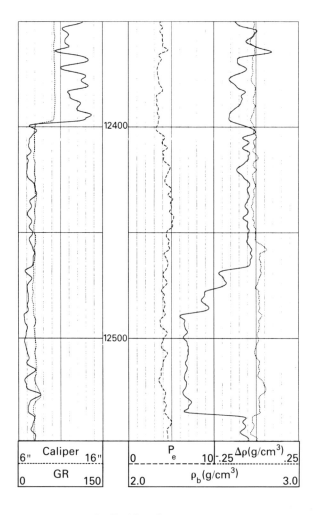

Figure 10-20. Log example for Problem 3.

2. The equation:

$$\rho_{log} = 1.0704\, \rho_e - 0.188$$

relates the electron density ρ_e to the tool reading ρ_{log}, which is closely related to the bulk density ρ_b. This equation has been defined so that the tool reading corresponds to ρ_b in water-limestone mixtures. What transform would be used for the log reading to coincide with the bulk density of mixtures of 120 Kppm salt-water and sand (SiO_2)?

3. Fig. 10–20 is a short section of the LDT log in the carbonate section of the simulated reservoir model.

   a. From a knowledge of the density curve alone, what ranges of porosity would you ascribe to the seemingly uniform layer from about 12490′ to 12540′?

   b. By including the P_e measurement, could you refine your porosity estimate? What new average value would you estimate it to be?

   c. What proportion of dolomite and limestone does the matrix seem to be? If not constant, what is the range of mixtures?

4. Fig. 10–19 shows the overlay of P_e versus ρ_b for three different matrices as a function of porosity for lithology determination.

   a. Frequently salt plugging occurs in dolomite formations which contain very saline formation waters. In this case the preexisting porosity of the dolomite can be replaced with depositions of NaCl. Plot the trend line on this cross plot for the case of a 20-PU water-filled dolomite in which the porosity progressively becomes salt-filled. Make use of the mixing law for U.

   b. From this cross plot alone, with what might you confuse the fully plugged case?

5. A density log has been run in a sandstone reservoir where core analysis has determined porosity to be 23% In this zone the density reading is 2.40 g/cm³.

   a. What do you estimate the grain density of the sandstone to be?

   b. What is the error in porosity if you assume the matrix to be pure sandstone (SiO_2)?

   c. From the core analysis the formation is known to consist of SiO_2 and pyrite (FeS_2). What volume fraction of the matrix does the apparent grain density correspond to?

   d. The actual grain density from core analysis gives a value of 2.76 g/cm³. What does this imply as a value for ρ_{fl}?

11
BASIC NEUTRON PHYSICS FOR LOGGING APPLICATIONS

INTRODUCTION

The use of neutrons to probe formations has had a long history in well logging. The first neutron device appeared shortly after World War II. The initial application was to determine formation porosity. Currently, in addition to logging tools that detect neutrons of several energy ranges in order to determine formation porosity, there are tools which use pulsed neutrons to analyze the absorption rate of the emitted neutrons, and gamma ray spectroscopy tools which detect neutron-induced gamma rays to produce a limited chemical analysis of the formation. The key to understanding the responses of these tools is the interactions that are exploited. The purpose of this chapter is to describe these interactions in order to provide a basis for succeeding chapters.

As in the case of gamma rays, neutrons can interact with materials in a number of different ways, each with an appropriate cross section to describe its probability of occurrence. The interactions of neutrons with matter are much more varied and complex than those of the gamma rays. For simplicity, we will confine ourselves to two groups of those interactions: scattering and absorption. Four types of cross sections to describe these interactions will be taken up, after a review of some useful terminology and the kinematics of neutron elastic scattering.

Unlike gamma ray sources, which come from naturally occurring or easily produced isotopes, neutron sources used in logging are the result of deliberate nuclear reactions. Several of these reactions will be discussed, along with the techniques for the detection of neutrons.

FUNDAMENTALS OF NEUTRON INTERACTIONS

The reaction rate of neutrons with matter depends on four parameters. The first two are the density (number/volume) of neutrons, n, and their velocity, v. The product of these two quantities is called the flux (identical with Ψ, used earlier to describe gamma ray intensities), and the units are number of neutrons per cm^2–sec. The reaction rate also depends on the nuclear density, N_i of the particles with which they will interact, and finally upon the cross section σ_i for the particular reaction. Thus an expression for R, the reaction rate (number of neutron reactions of type i / cm^3), is given by:

$$R = n \, v \, \sigma_i \, N_i \, .$$

The density of particles of type i in a material of molecular weight M and bulk density ρ is:

$$N_i = \frac{6.02 \times 10^{23}}{M} \, \rho,$$

if there is a single nucleus of type i per molecule.

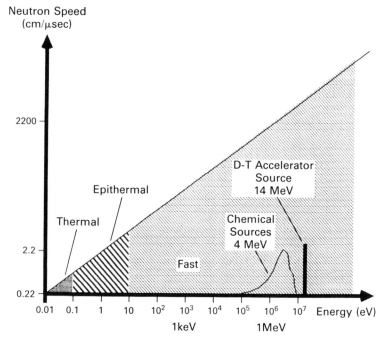

Figure 11–1. The classification of neutrons according to broad energy ranges and their corresponding velocities. From Ellis.[1]

Fig. 11–1 defines, in broad terms, the energy range of interest for neutrons. For logging applications, this range is over about nine decades: from source neutrons of 5 to 15 MeV, in the broad fast neutron range above 10 eV, to epithermal neutrons in the range of 0.2–10 eV, and thermal neutrons which are distributed around 0.025 eV at room temperature.

For later discussions of the time scale associated with the slowing-down process, it is useful to note the relationship between neutron energy and its associated velocity. To evaluate the velocity of a neutron, we can use, at low energies, the classical relationship between kinetic energy E, velocity v, and mass m:

$$E = \frac{1}{2}mv^2, \tag{1}$$

so that the velocity v is given by:

$$v = \sqrt{\frac{2E}{m}}. \tag{2}$$

If this expression for velocity is evaluated for thermal energies (0.025 eV), the result is 2200 m/sec or 0.22 cm/μsec. Thus the velocity at any energy E (in eV) is given by:

$$v = 0.22\sqrt{\frac{E}{0.025}}, \tag{3}$$

where v is in cm/μsec. Therefore the speed of an epithermal neutron of 2.5 eV is 2.2 cm/μsec, and for a near-source-energy neutron of 2.5 MeV, its velocity is 2200 cm/μsec. These velocities are also noted on Fig. 11–1.

Of the four principal types of interaction, the first two are generally referred to as moderating interactions, or interactions in which the energy (or speed) of the neutron is reduced. One of these is known as elastic scattering, and the other as inelastic scattering. Let us consider elastic scattering first. Classical mechanics (elastic billiard ball analysis) can be used to describe the moderating power of the struck nucleus. The energy of the neutron is reduced more efficiently in collisions with nuclei of mass not too different from the mass of the neutron. Thus hydrogen and other low atomic mass elements are quite effective in reducing fast neutron energy.

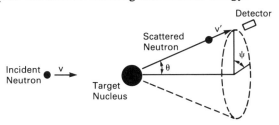

Figure 11–2. The idealized scattering of a neutron with a target nucleus. From Rydin.[3]

The physical variables for describing elastic scattering can be obtained by a consideration of the concept of center of mass. Fig. 11–2 shows the laboratory view of a collision between a stationary nucleus and a neutron moving with velocity v. After collision, the neutron has deviated from its initial direction by an angle Θ and has some reduced velocity, v'.

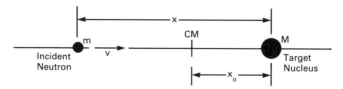

Figure 11–3. The scattering reaction drawn to suggest the center of mass (CM) system. From Rydin.[3]

Another approach is to define the center of mass as shown in Fig. 11–3. This new coordinate system is defined by:

$$Mx_0 = m(x-x_o),$$

where M is the mass of the target, and m is the mass of the neutron. The coordinate x_o is given by:

$$x_o = \frac{mx}{m + M} = \frac{x}{1 + A},$$

after substituting 1 for the mass of the neutron and A for the mass of the nucleus. The velocity of the center of mass v_{cm}, as seen in the laboratory system, can be found from:

$$v_{cm} = \frac{dx_o}{dt} = \frac{1}{1 + A} \frac{dx}{dt} = \frac{1}{1 + A} v.$$

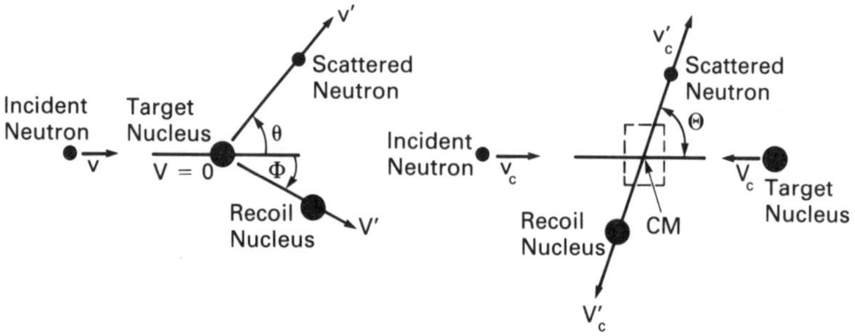

Figure 11–4. The scattering reaction drawn from the perspective of the laboratory and center of mass systems. From Rydin.[3]

Two views of the reaction are shown in Fig. 11–4: the laboratory view on the left and the center of mass system on the right. In the center of mass system, the two particles are seen to be approaching each other with velocities v_c and V_c. These two velocities are given by:

$$v_c = v - v_{cm} = (1 - \frac{1}{1+A})v = (\frac{A}{1+A})v,$$

and

$$V_c = -v_{cm} = -\frac{1}{1+A}v.$$

The total momentum in the center of mass system is given by:

$$mv_c + MV_c = \frac{A}{A+1} 1 \cdot v - \frac{1}{1+A} A \cdot v,$$

which is seen to be zero. This unique result, for an elastic collision viewed in the center of mass system, means that the neutron and nucleus enter and leave the reaction with the same velocities and are oppositely directed.

An analysis of conservation of energy shows that the neutron energy E', after scattering through an angle Θ in the center of mass system, can be related to the energy E_o before the collision by the following :[3]

$$\frac{E'}{E_o} = \frac{A^2 + 2A\cos\Theta + 1}{(A+1)^2}.$$

From this expression, it is seen that the minimum energy after collision is a fraction α of the initial energy, where α is related to the mass A of the scattering nucleus by:

$$\alpha = \left[\frac{A-1}{A+1}\right]^2.$$

Fig. 11–5 illustrates, for most of the elements of interest, the permitted ranges of reduction in neutron energy on a single collision. It is seen that for the most common earth formation elements the maximum energy reduction per collision for the heavy elements is about 10–25%. However, for the case of hydrogen, the entire neutron energy can be lost in a single collision. This sensitivity of elastic scattering energy loss to hydrogen is exploited in neutron porosity devices.

In the case of inelastic scattering, a portion of the energy of the incident neutron goes into exciting the target nucleus. This reduces the energy of the incident neutron. The target nucleus will usually produce one or more characteristic gamma rays upon de-excitation. This type of reaction always has a threshold energy (below which it will not happen). The de-excitation gamma ray is exploited in the measurement of the carbon-to-oxygen ratio in earth formations.

Another general category of neutron interaction is known as *absorption*. It also is divided into two types: radiative capture and reactions which produce nuclear particles. In radiative capture, unlike the moderating

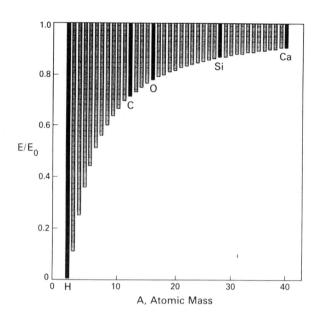

Figure 11–5. The allowed distribution of neutron energy after a single elastic scattering with nuclei ranging in mass from H to Ca. The energy scale is normalized to the incident neutron energy. From Ellis.[1]

interactions considered above, the neutron (usually near thermal energies) is absorbed by the target nucleus, producing a compound nucleus. This nucleus de-excites instantly with the emission of characteristic gamma rays. This type of reaction is exploited in pulsed neutron logging tools or in gamma ray spectroscopy of the induced gamma rays for chemical analysis.

The category of particle reactions is quite broad; it is sufficient to say that the interaction of neutrons with some nuclei can provoke the emission of particles such as alphas, protons, βs, or even additional neutrons. These reactions, although common, have a very small probability for occurring relative to the other interactions of interest in logging described above. Usually they are possible only above a relatively high neutron energy.

The complexity of the cross sections for neutron interactions is illustrated in Fig. 11-6, which schematically indicates the variations with energy. The top figure refers to the total cross section as a function of neutron energy E, and the four following figures indicate how this can be decomposed. The first, (n,n), refers to elastic scattering, which is shown to be rather constant with energy except for some resonances at low energies. The next sketch shows inelastic interactions, (n,n'), showing some characteristic threshold below which this reaction is not possible; the third sketch is one of the many particle reactions possible, (n,α); and the final (although there could be others) is the radiative capture, (n,γ), which is seen to increase in probability at low energies.

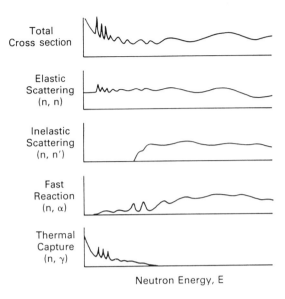

Figure 11–6. A schematic illustration of the energy variation of the total neutron cross section and four of its components. From Ellis.[1]

NUCLEAR REACTIONS AND NEUTRON SOURCES

Since neutron sources are almost never found in nature, it is appropriate to briefly discuss the techniques for creating them. There are two types in use in logging: so-called chemical, or encapsulated, sources, and accelerator sources.

The classic reaction, which resulted in the discovery of the neutron, was the bombardment of beryllium by alpha particles. It can be written as:

$$^4Be + {}^2He \rightarrow {}^6C + n + 5.76 \text{ MeV} .$$

This forms the basis for the cheapest, easiest, and most reliable method for neutron production. The physical explanation of this reaction is beyond the scope of interest of the present work and may be found in References 2 and 4. The practical construction of this kind of chemical neutron source consists of mixing a naturally occurring α-emitter with an appropriate light element having a large (α,n)* cross section. Some α-emitters which have been used for this purpose are Pu, Ra, Am, and Po. Three common target elements are Be, B, and Li. The actual spectrum (energy distribution) of emitted neutrons is quite complicated. It depends somewhat on the geometric details of the α-emitter and target, but the peak of the neutron distribution is generally around 4 MeV.

---

* This shorthand, (α,n), indicates a reaction of an α particle with an unspecified nucleus, resulting in the production of a neutron and another unspecified nucleus.

Another method of exploiting particle-induced reactions in the production of neutrons is by the use of charged particle accelerators. In one realization, currently in use in well logging, deuterium and tritium ions are accelerated toward a target impregnated with the hydrogen isotopes deuterium (D) and tritium (T). The reaction is written as:

$$^2D + {}^3T \rightarrow {}^4He + n + 17.6 \text{ MeV} .$$

The cross section for this reaction has a maximum at about 100 keV of ^2D projectile energy. This dictates the required accelerating voltages in such a device.

Despite the engineering difficulties of constructing such a device, the advantages for logging are many. One is the relative high energy of the produced neutrons. They are emitted at 14.1 MeV (not 17.6 MeV, because some of the energy of this reaction is given up to the alpha particle). These high energy neutrons are useful for producing other interesting nuclear reactions in the formation, as is discussed later. Another advantage is that a source of this type can be controlled, i.e., switched off and on at will. This provides a degree of safety unparalleled for radioactive sources as well as permitting measurements involving timing as a means of determining some interesting nuclear properties of the formation, a topic covered in Chapter 12.

USEFUL BULK PARAMETERS

Despite the complexities of the cross sections shown in Fig. 11–6, which govern the details of the interactions, some gross properties can be specified for neutron interactions with materials. The first is the macroscopic cross section, which is defined as the product of the cross section (σ_i) in question times the number of atoms/cm³, N, i.e.:

$$\Sigma_i = N \sigma_i = \frac{N_{Av} \rho_b}{A} \sigma_i , \qquad (4)$$

where N_{Av} is Avogadro's number, ρ_b is the bulk density, and A is the atomic weight. The dimensions of the macroscopic cross section Σ_i are inverse centimeters. Its reciprocal is the mean free path length between interactions of type i. Frequently, in logging, special use is made of the macroscopic absorption cross section evaluated at thermal energies. It is convenient to define special units for it. These so-called capture units (cu) are 1000 times the Σ_i as defined above. In this case, σ_i refers to the thermal absorption cross section which dominates at thermal energies for most elements. Fig. 11–7 shows the total mean free path in limestones of 0, 20, 40, and 100 PU (porosity units) as a function of energy for fast neutrons. At the energy of chemical source emission (2–4 MeV), it is seen that there is very little porosity dependence. It is only as the neutrons are slowed down that the mean free path becomes strongly dependent on the hydrogen concentration of the formation.

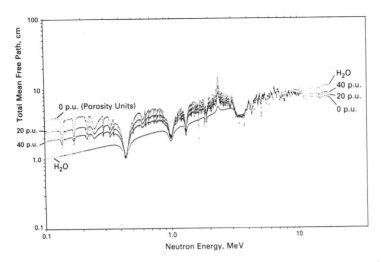

Figure 11–7. The mean free path of neutrons in limestones of various porosity and water. They are given as a function of the neutron energy. From Ellis.[1]

As mentioned earlier in the discussion of elastic scattering, low mass nuclei are very effective in reducing the energy of the scattered neutron. As can be inferred from Fig. 11–5, the result of a collision can be considered, on average, as a percentage decrease of the neutron energy. This is usually expressed as the average logarithmic energy decrement ξ, which is defined by:

$$\xi \equiv \overline{\ln(E_i) - \ln(E)} = -\overline{\ln(E/E_i)}, \qquad (5)$$

where E_i is the initial energy and E is the energy of the neutron after collision. It can be shown from classical mechanics that the average log energy decrement is simply related to the atomic mass, A, of the struck nucleus by:[3]

$$\xi \approx \frac{2}{A + 2/3}, \qquad (6)$$

for large values of atomic mass A. The average log energy decrement allows an estimation of the average number of collisions, n, to reduce the neutron from an initial energy E_i to some lower energy E, from the following reasoning. If the sequence $E_1, E_2, \cdots, E_n$ represents the average energy after each collision, then we can write:

$$\overline{\ln(\frac{E_i}{E_n})} = \overline{\ln(\frac{E_i}{E_1} \frac{E_1}{E_2} \cdots \frac{E_{n-1}}{E_n})} \qquad (7)$$

$$= \overline{\ln(\frac{E_i}{E_1})^n} = n \overline{\ln(\frac{E_i}{E_1})} \qquad (8)$$

$$= n\xi . \qquad (9)$$

Thus the average number of collisions is given by:

$$\overline{n} = \frac{1}{\xi} \overline{\ln\left(\frac{E_i}{E_n}\right)}. \qquad (10)$$

The constant ξ can be computed for a mixture of elements by weighting the value of each ξ_i for element i with the appropriate total scattering cross section $N_i \sigma_i$. Table 11–1 shows some typical values for the average logarithmic energy decrement and the number of collisions necessary to reduce source energy neutrons (4.2 MeV) to 1 eV.

Neutron Slowing-down Parameters		
Moderator	ξ	$\overline{n}$*
H	1.0	14.5
C	0.158	91.3
O	0.12	121
Ca	0.05	305
H_2O	0.92	15.8
20 PU Limestone	0.514	29.7
0 PU Limestone	0.115	132

*(avg. number of collisions from 4.2 MeV to 1 eV)

Table 11–1. Average logarithmic energy decrement and average number of collisions for reducing neutron energy from 4.2 MeV to 1 eV for selected moderators.

There are two more parameters which help to characterize neutron interactions with bulk material. One parameter is known as the slowing-down length, L_s and the other as the thermal neutron diffusion length, L_d. The slowing-down length is proportional to the root-mean-square distance from the point of emission of a high energy neutron to the point at which it arrives at the lower edge of the epithermal energy region. This distance can be calculated from a detailed knowledge of the cross sections of the constituent elements.[5] Fig. 11–8 shows the slowing-down length as a function of water-filled porosity for limestone, sandstone, and dolomite.

In order to understand the variations of the slowing-down length seen in Fig. 11–8, it is interesting to compare it with a random walk. The random walk in one dimension is shown in Fig. 11–9, which plots for three trials (three different neutrons) the distance from the starting point as a function of the number of equal-length steps taken. At each step the probabilities for a forward or backward displacement are equal. It is obvious that the average displacement from the starting point for a large number of trials is zero. However, there is a distribution of terminal points around the origin. A measure of the width, or spread, of the distribution is the root-mean-square displacement, which can be shown to be equal to $\sqrt{n}$ times the length of the step. Fig. 11–10 shows the probability distribution for three series of random walks, each containing a different number of steps. They all center about zero (no displacement from the origin), but the width increases as the number of steps taken increases.

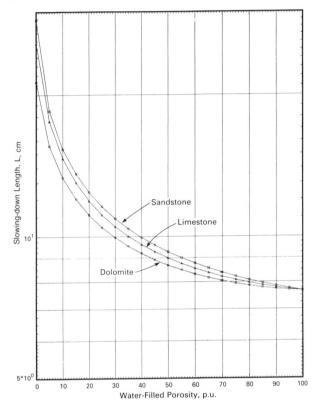

Figure 11-8. The slowing-down length of sandstone, limestone, and dolomite as a function of porosity. From Ellis.[1]

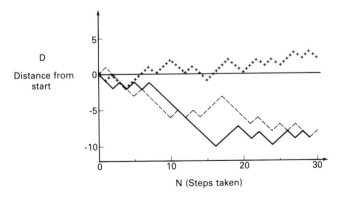

Figure 11-9. Three trials of a random walk. Adapted from Feynman.[6]

Although the slowing-down of neutrons is a three-dimensional process, and the free path between collisions varies somewhat, it can still be thought of as a random walk. One important feature which distinguishes the random walk in a zero porosity limestone from one in water is the number of collisions (the number of steps taken in the random walk). Fig. 11–11 illustrates this idea, along with a few useful parameters. From Table 11–1, the number of collisions in the slowing-down process in limestone is about nine times that in water. Consequently, if we associate the slowing-down length with the root-mean-square displacement for a random walk, we expect

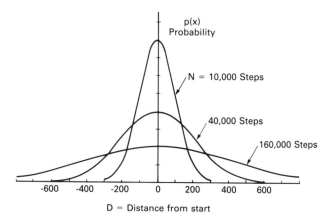

Figure 11–10. The probability distribution of terminal points for random walks with three large step numbers. Adapted from Feynman.[6]

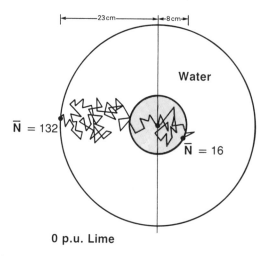

Figure 11–11. A schematic of the slowing-down trajectories of neutrons in water and limestone, suggesting the connection with the random walk. The ratio of the slowing-down length in limestone to that in water follows the expectation from the average number of collisions. From Ellis.[1]

the value of L_s in 0-PU limestone to be about three times larger than that in water, which is illustrated in Fig. 11-11 and which is in agreement with the calculations displayed in Fig. 11-8.

The diffusion length L_d can be thought of as the rectified distance a thermal energy neutron travels between the point at which it became thermal until its final capture. This distance is given by:

$$L_d = \sqrt{(D/\Sigma)}, \qquad (11)$$

where D is the thermal diffusion coefficient (discussed in Chapter 12) and Σ is the macroscopic thermal absorption cross section of the material. The diffusion coefficient D can also be calculated from knowledge of the cross sections of the material. It is shown in Fig. 11-12 as a function of porosity for the three principal matrices.

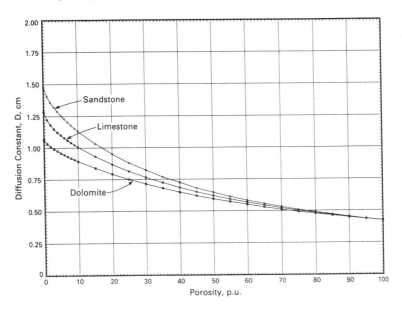

Figure 11-12. The diffusion coefficient of thermal neutrons as a function of porosity for the three principal rock types.

Since thermal neutrons are strongly affected by the presence of thermal absorbers, it is interesting to look at an abbreviated list of elements frequently found in formations which have large macroscopic thermal absorption cross sections. This is found in Table 11-2 where the units are capture cross section (cu) per gram of material per cm^3. Chlorine is of particular interest since saltwater will have a measurable effect on the thermal neutron population. Iron and boron, which are frequently associated with clays, may dominate the capture cross section of the formation if present in sufficient concentration.

Macroscopic Thermal Absorption X-sections	
ELEMENT	$\Sigma(cu/g/cm^3)$
Boron	42300
Chlorine	564
Hydrogen	198
Manganese	146
Iron	27.5

Table 11–2. Mass-normalized macroscopic thermal absorption cross section for selected elements.

Another sometimes useful parameter, the migration length (L_m), has been defined as:

$$L_m^2 = L_s^2 + L_d^2 . \tag{12}$$

It can be viewed as a distance which represents the combination of the path traveled during the slowing-down phase (L_s) and the distance traveled in the thermal phase before being captured (L_d). This parameter provides a convenient way of predicting the response of a thermal neutron porosity device, which is discussed in more detail in the next chapter.

NEUTRON DETECTORS

Neutrons are detected in a two-step process. First, the neutrons react with a material in which energetic charged particles are produced. Then the charged particles are detected through their ionizing ability. Thus most neutron detectors consist of a target material for this conversion, coupled with a conventional detector, such as a proportional counter or scintillator, to achieve the measurement. Since the cross section for neutron interactions in most materials is a strong function of neutron energy, different techniques have been developed for different energy regions. For well logging applications, at present, it is the detection of thermal and epithermal neutrons which is of interest. The detection schemes considered in this section are appropriate for these low energy neutrons.

Nuclear reactions useful for neutron detectors must satisfy several criteria: the cross section for reaction must be very large, the target nuclide should be of high isotopic abundance, and the energy liberated in the reaction following the neutron absorption should be high enough for ease of detection by conventional means. Three target nuclei have been found to generally satisfy these conditions: ^{10}B, ^6Li, and ^3He. In the case of the first two targets, the (n,α) reaction is utilized, and for ^3He, it is the (n,p) reaction.

The boron reaction is widely exploited in the form of BF_3 in a proportional counter. In this case the boron trifluoride serves as both the target and the ionization medium. For this application, the gas is enriched in ^{10}B, to attain a high detection efficiency. Another approach is to use a boron coating on the inner wall of a proportional counter, which may use some other proportional gas more suitable than BF$_3$ for applications involving fast timing, for example.

Since a suitable lithium compound gas does not exist, the lithium reaction is not exploited in proportional counters. However LiI scintillators, similar to sodium iodide for gamma ray detection, are available. Due to the large energy released by the (n, α) reaction, neutrons are registered at an energy of about 4.1 MeV, which provides a means of discriminating against the gamma rays, which are also readily detected by the LiI crystal.

The most common neutron detector in well logging, however, is based on the ^{3}He (n,p) reaction. In this case, ^{3}He is used as the target and proportional gas in a counter. It is preferred since it has a higher cross section than the boron reaction and the gas pressure can be made much higher than for BF$_3$ without degradation of its proportional operation. The simplicity of a proportional counter is also preferred to the complications associated with a scintillator.

For the three reactions discussed above, the cross sections vary inversely with the square root of the neutron energy, so that the detection efficiency for neutrons will vary in the same manner. The detectors employing these reactions respond primarily to thermal neutrons. For some logging applications it is desirable to measure the epithermal neutron flux, while being insensitive to thermal neutrons. This can be achieved by making a minor modification on any of the three types of detectors previously mentioned. It consists of using a shield of thermal-neutron absorbing material with a large cross section, such as cadmium, around the detector. Thermal neutrons will be absorbed in the shield, but the reaction particles, whose range is small (on the order of tenths of mm), do not reach the counter. The higher energy epithermal neutrons which manage to penetrate the shield are detected with somewhat reduced efficiency.

References

1. Ellis, D. V., "Nuclear Logging Techniques" in *SPE Petroleum Production Handbook,* edited by H. Bradley, SPE, Dallas (in press).

2. Weidner, R. T., and Sells, R. L., *Elementary Modern Physics,* Allyn and Bacon, Boston, 1960, pp. 371–392.

3. Rydin, R. A., *Nuclear Reactor Theory and Design,* University Publications, Blacksburg, 1977.

4. Evans, R. D., *The Atomic Nucleus,* McGraw-Hill, New York 1967, pp. 426–438.

5. Kreft, A., "Calculation of the Neutron Slowing Down Length in Rocks and Soils," *Nukleonika*, Vol. 19, 1974.

6. Feynman, R. P., Leighton, R. B., and Sands, M. L., *Feynman Lectures on Physics*, Vol. 1, Addison-Wesley, Reading, Mass., 1965.

Problems

1. A neutron generator used in logging applications employs the D–T reaction illustrated in Fig. 11–13. The result of the reaction is two particles (a neutron and a ^{4}He) which share 17.6 MeV of energy.

    a. Applying conservation of energy and momentum to the reaction products, calculate the neutron energy.

    b. If the reaction products were instead a neutron and a ^{3}He, sharing the same 17.6 MeV, what would the resultant neutron energy be?

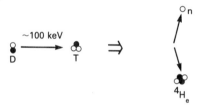

Figure 11–13. The D–T reaction used for producing 14 MeV neutrons.

2. Using the data of Figs. 11–8 and 11–12

    a. Compute the diffusion length in water.

    b. Compute the diffusion length in 0-PU and 20-PU limestone.

3. From the data in Table 11–2, estimate the macroscopic thermal absorption cross section of water. Assume that the oxygen can be neglected. Express the answer in capture units. How does it compare with the standard value of 22 cu?

4. What is the mean free path of a thermal neutron in water? What is the mean free path of a 4 MeV neutron in water?

5. Analysis of a shale core sample whose density is 2.60 g/cm^3 indicates that the concentration of boron is 400 ppm. What is its contribution, in cu, to the total Σ of the sample?

12
NEUTRON POROSITY DEVICES

INTRODUCTION

Historically, the neutron logging tool was the first nuclear device to be used to obtain an estimate of formation porosity. The measurement principle is based on the fact that hydrogen, with its relatively large scattering cross section and small mass, is very efficient in the slowing-down of fast neutrons. A measurement of the spatial distribution of epithermal neutrons resulting from the interaction of high energy source neutrons with a formation can be related to its hydrogen content. If the hydrogen (in the form of water or hydrocarbons) is contained within the formation pore space, then the measurement yields porosity. The simplest version of the device, illustrated in Fig. 12–1, consists of a source of fast neutrons, such as Pu-Be or Am-Be, with average energy of several MeV, and a detector, sensitive to much lower energy neutrons, at some distance from the source. Two general categories of neutron porosity tools will be considered in the following discussion. These are distinguished by the energy range of neutrons detected, epithermal or thermal.

Although neutron interactions are complicated, a simple theory useful for predicting the response trends of neutron porosity devices will be reviewed. The results of this theory will be compared to laboratory measurements with logging devices. The general design of neutron porosity logging tools will be shown to be a consequence of severe environmental perturbations to the measurement. One goal of this design is to minimize the influence of the hydrogen-rich borehole on the estimation of formation hydrogen content.

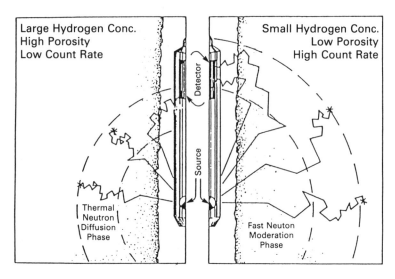

Figure 12-1. A schematic representation of a single-detector neutron porosity device. Adapted from Arnold and Smith.[1]

Several important effects that complicate the porosity interpretation of neutron logs will be considered. They are related to the response of the neutron porosity devices to rock type, shale, and the presence of gas in the invaded zone. For this latter effect, the sensitivity of neutron porosity tools is largely dependent on the depth of investigation of the measurement; gas displaced by mud filtrate beyond this depth of investigation will not be sensed. To give an idea of the limits of gas detection, the experimental determination of the depth of investigation of neutron porosity tools is compared to that given by a simple geometric theory which is discussed in the appendix to this chapter.

BASIS OF MEASUREMENT

How can the detection rate of neutrons in the device illustrated in Fig. 12-1 be related to the properties of the formation? Formally this is a problem of neutron transport: Neutrons are emitted at one point and transported, through a material, to another point (where they are detected, in our case). The properties of the material affect the transport process. They influence the spatial and energy distribution of the neutron population and, consequently, the counting rate at any detector location.

A formal description of the transport of neutrons is given by the Boltzmann transport equation (BTE). In its time-independent form, appropriate for neutron porosity logging, it may be written as:[2]

$$\overline{\Omega} \cdot \overline{\nabla \Phi} + \Sigma_t \overline{\Phi} = \int dE' \int d\overline{\Omega}' \; \Sigma_s(E' \rightarrow E, \Omega' \rightarrow \Omega) \overline{\Phi} + S.$$

This formidable-looking equation is nothing more than an expression of the conservation of the total number of neutrons in an elemental cylindrical volume, shown in Fig. 12–2. It is written in terms of the angular flux $\overline{\Phi}$, a vector quantity which specifies the number of neutrons crossing a unit surface area per unit time, in a given direction ($\overline{\Omega}$) and energy interval at any point in space. The BTE relates the loss rate of neutrons in the volume by absorption and scattering to the rate of increase of neutrons through source production and scattering from other regions of space and from neutrons of higher energies (which may even be contained within the volume considered). The first term, $\overline{\Omega}\cdot\overline{\nabla}\Phi$, represents the net leakage rate of neutrons out of the volume in the direction $\overline{\Omega}$ due to streaming of the neutrons into and out of the volume. The loss rate of neutrons from the volume, energy region, and direction of interest is given by the reaction rate $\Sigma_t\overline{\Phi}$, where Σ_t represents the the total interaction cross section.

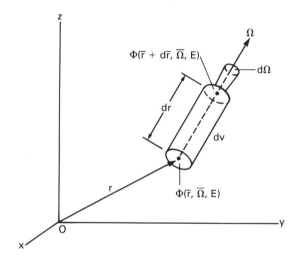

Figure 12–2. The geometry for deriving the neutron flux balance in a typical cell of material leading to the Boltzmann transport equation.

The two preceding loss rate terms are balanced by the rate at which neutrons from within the volume scatter into the energy and direction of interest, and by any sources (S) contained within the region. The scattering term is written as:

$$\int dE' \int d\overline{\Omega}' \, \Sigma_s(E'\to E, \overline{\Omega}'\to\overline{\Omega})\overline{\Phi},$$

where the integration is performed over neutrons of energy E' (greater than E), and over all other directions, $\overline{\Omega}'$. The cross section $\Sigma_s(E'\to E, \overline{\Omega}'\to\overline{\Omega})$ must take into account the angular dependence of scattering as well as the fact that there may be a limited energy range (E') over which neutrons can scatter to the specified energy E.

As it stands, the BTE is difficult if not impossible to solve analytically. However, a number of numerical approaches have been developed. The discrete ordinates method reduces the BTE to a set (usually large) of coupled equations by allowing the variables in the equation to take on only a limited set of discrete values. Cross sections corresponding to discretized energies and angles must be computed from continuous data (when it exists) for the sampling intervals chosen. The Monte Carlo method, in its simplest form, performs a numerical simulation of the neutron transport by using cross section data to compute probabilities of interactions, path lengths between interactions, and angles of scattering. Appropriately sampled random numbers, in conjunction with the calculated probabilities, are used to trace the events in a neutron's life from birth to capture. By sampling large numbers of these "typical" histories, we can predict the behavior of a specified system containing neutron sources.

Analytical solutions of the BTE all depend in one way or another on reducing the complexity of the problem by integrating or averaging over one or more of the variables. One of the more useful approaches is the diffusion approximation. The basic simplification employed is that the angular flux is only weakly dependent on direction. A further simplification which considers only a single energy group of neutrons allows the BTE to be written in terms of the scalar flux, $\Phi(r)$, which has only a spatial variable. Consideration of the weak angular dependence or direct application of Fick's law (the current of neutrons is proportional to the gradient of the flux) allows the net leakage portion of the BTE to be expressed as a term containing a diffusion coefficient. In this case the BTE reduces to :[2]

$$D\nabla^2\Phi(r) - \Sigma_a\Phi(r) + S = 0, \qquad (1)$$

where the diffusion coefficient D is given by:

$$D = \frac{1}{3(\Sigma_t - \bar{\mu}\Sigma_s)} .$$

It depends on the difference in the total cross section (Σ_t) and the scattering cross section (Σ_s) times the average cosine ($\bar{\mu}$) of the scattering angle. The absorption cross section Σ_a in Eq. (1) is related to the previous cross sections by:

$$\Sigma_t = \Sigma_a + \Sigma_s .$$

A slight refinement of the preceding approach, which reduces the nine decades or so of neutron energy variation into a single average quantity, is to admit two broad energy regions of interest: epithermal and thermal. In this case, two coupled diffusion equations can be written, one for each of the energy bands:[3]

$$D_1\nabla^2\Phi_1 - \Sigma_{r1} + S = 0$$

and

$$D_2 \nabla^2 \Phi_2 - \Sigma_{r2} \Phi_2 + \Sigma_{r1} \Phi_1 = 0 \; .$$

The first of these is for the broad epithermal (subscript 1) energy region, which contains the source of strength S, and the second is for the thermal (subscript 2) region, in which scattering from the epithermal region ($\Sigma_{r1}\Phi_1$) plays the role of the source. The cross section Σ_{r1} is called the removal cross section. It accounts for the portion of scattering which may reduce the neutron energy sufficiently for it to be counted in the lower energy group. The cross section Σ_{r2} is the familiar thermal absorption cross section.

In simple geometries, these coupled equations can be solved for the spatial distribution of the flux. For the case of a point source in an infinite medium, the epithermal flux is given by :

$$\Phi_1(r) = \frac{Q}{4\pi D_1} \frac{e^{-r/L_1}}{r} \; , \qquad (2)$$

where L_1, also known as the slowing-down length (L_s), is given by $\sqrt{D_1/\Sigma_{r1}}$, where D_1 is the diffusion coefficient associated with the high energy neutrons. Chapter 11 provides a less formal definition of this same quantity.

The thermal flux is given by the slightly more complicated expression:

$$\Phi_2(r) = \frac{QL_2^2}{4\pi D_2(L_1^2 - L_2^2)} \times \left[\frac{e^{-r/L_1}}{r} - \frac{e^{-r/L_2}}{r} \right] \; . \qquad (3)$$

In this expression $L_2 = \sqrt{D_2/\Sigma_{r2}}$, where D_2 is the thermal diffusion coefficient. L_2 is also known as the thermal diffusion length and is usually written as L_d.

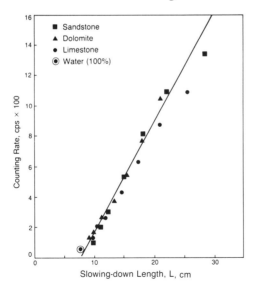

Figure 12–3. The counting rate of a single-detector epithermal device as a function of slowing-down length of calibration formations. Adapted from Edmundson and Raymer.[4]

In order to anticipate the response of an epithermal neutron porosity device we can use the results of Eq. (2). It shows that the flux of epithermal neutrons, in an infinite homogeneous medium containing a point source of fast neutrons, falls off exponentially with distance from the source r, with a characteristic length L_s, determined by the constituents of the medium. The implication for a borehole device such as that shown in Fig. 12–1 is that the epithermal neutron counting rates should vary nearly exponentially with the slowing-down length of the formation. An indication of this type of behavior can be seen in Fig. 12–3 which shows the counting rate of one of the early epithermal neutron devices as a function of slowing-down length in three types of formations. In Fig. 12–4 the same data is shown as a function of porosity. The separation of response trends according to formation lithology, seen in Fig. 12–4, is referred to as the matrix effect. The matrix effect is much reduced in the first presentation but not entirely eliminated. However, slowing-down length is the more representative parameter for describing counting rate variations, as expected from Eq. (2).

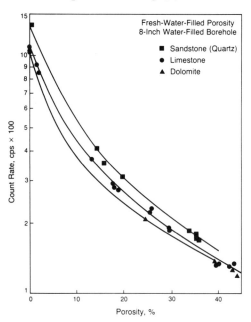

Figure 12–4. The same data as Fig. 12–3, but plotted as a function of porosity in formations of sandstone, limestone, and dolomite. Adapted from Edmundson and Raymer.[4]

Fig. 12–3 also shows how the counting rate in any other material, once its slowing-down length has been calculated, can then be estimated. Conversely, the slowing-down length of the formation can be determined from a measurement of the epithermal flux. Since the slowing-down length of a formation is strongly dependent on the amount of hydrogen present, the porosity can be determined using the data of Fig. 11–8. Referring back to this figure, which shows the slowing-down length as a function of porosity for three common lithologies, we see that if the rock matrix type is known, the formation porosity can be determined from the formation slowing-down length.

MEASUREMENT TECHNIQUE

Table 12–1 lists a few of the physical parameters of the formation and borehole which have an impact on the response of a neutron porosity device. Of the dozen or so mentioned, only one, porosity, is in fact, the desired quantity. All the others on the list can be related to a change in the local hydrogen density and thus will have an influence on a measuring device that is sensitive to the concentration of hydrogen. Of particular importance are the environmental effects such as borehole size and borehole fluid.

FORMATION
• Porosity
• Matrix
• Pore Fluid
• Fluid Salinity
• Temperature
• Pressure
BOREHOLE GEOMETRY
• Diameter
• Borehole Shape
• Mudcake Thickness
• Standoff
BOREHOLE MUD
• Density
• Solids
• Fluid Type
• Salinity
• Temperature
• Pressure

Table 12–1. Parameters affecting neutron porosity response.

One of the first quantitative neutron porosity devices employed a single epithermal detector in a skid applied mechanically against the borehole wall.[5] This side-wall epithermal neutron device had the advantage of reducing borehole size effects, but it could be disturbed by the presence of mudcake between the pad surface and the borehole wall.

To first order, compensation for these environmental effects was achieved by the addition of a second detector. The compensated device (see Fig. 12–5) uses a pair of thermal neutron detectors for increased counting rate, which decreases the statistical uncertainty of the derived porosity values at large formation porosity. The detector closest to the source is used to provide compensation for borehole effects on the farthest detector by a simple division of counting rates. Although thermal neutron detection is used, it can be seen from an examination of Eq. (3) that if the source-detector spacings are chosen to be large enough, the ratio of the two counting rates should vary exponentially as the inverse of the slowing-down length, just as in the case of the single epithermal detector.

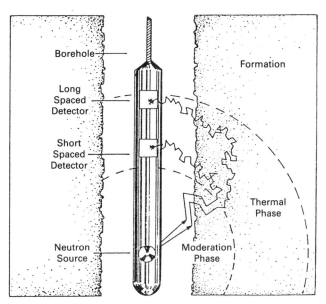

Figure 12–5. The representation of a two-detector neutron porosity logging device. Adapted from Arnold and Smith.[1]

In order to characterize the laboratory data of the thermal neutron tool, which are at some variance with the predictions of Eq. (3), use is made of another characteristic length, L_m. It is called the migration length and corresponds to the quadratic sum of the slowing-down length and the diffusion length:

$$L_m^2 \equiv L_s^2 + L_d^2 ,$$

where L_d is the diffusion length. In this manner, the migration length corresponds to $\frac{1}{\sqrt{6}}$ the distance traveled ($\sim L_s$) by the neutron, on average, during its slowing-down phase and additionally ($\sim L_d$) during its thermal phase until absorption. Use of the migration length (rather than L_s) accounts explicitly for the additional perturbation of absorption in the formation on the measurement of porosity by the thermal neutron tool. The thermal absorption of the formation is characterized by the macroscopic thermal absorption cross section, or Σ_{for}. Logging devices for the measurement of this important parameter are discussed in the next chapter.

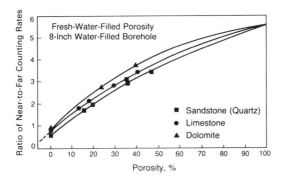

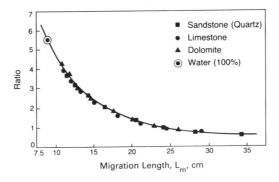

Figure 12–6. The measured response of a thermal neutron porosity device in formations of three different lithologies. In the upper figure, the data are plotted as a function of porosity. In the lower, they are plotted as a function of the appropriate migration length for the formation conditions used. Adapted from Edmundson and Raymer.[4]

The migration length, as defined above, provides a convenient way to characterize the response of the thermal neutron device. Fig. 12–6 shows, in the upper panel, the ratio of the near to far counting rate of such a device, for three types of lithologies, as a function of porosity. Notice the three trend lines, corresponding to the three lithologies present among the calibration formations. If the porosity values corresponding to the measurement points of this plot are converted through the use of Fig. 12–7, which shows the migration length L_m as a function of porosity, then the counting rates for the three lithologies are found to lie on a single line, as seen in the lower part of Fig. 12–6. This demonstrates that the response characteristics of the thermal neutron porosity tools can be described by a function of the slowing-down length and diffusion length, rather than by porosity.

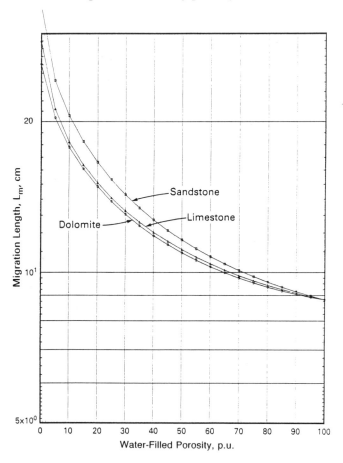

Figure 12–7. The calculated migration length as a function of formation porosity for the three principal lithologies. From *SPE Petroleum Production Handbook.*[6]

From the preceding discussion, conversion from the measured ratio, or migration length, to formation porosity would seem to be straightforward. Because of the dependence on the diffusion length, thermal absorbers in the formation and borehole can cause some deviation from the response derived from the limited set of calibration data shown in Fig. 12–6. In practice, it is found that some additional corrections have to be made to the inferred porosity if the thermal capture properties of the borehole and formation are significantly different from one another. These are generally provided by the service companies in the forms of charts or nomographs.[7,8,9] More recently they are provided as a part of computerized interpretation.

A recently developed neutron porosity device consists of a pair of thermal and a pair of epithermal detectors.[12,13] The epithermal portion allows the measurement of an apparent porosity unaffected by thermal absorbers. A comparison of thermal and epithermal measurements can be used to infer the thermal absorption properties of the formation. With respect to the pair of epithermal detectors, an analysis similar to that of the thermal tool, given above, can be made. The ratio of counting rates from the two epithermal detectors at two different spacings from the source, r_1 and r_2, can be shown by reference to Eq. (2) to result in an exponential dependence on slowing-down length:

$$R = \frac{N}{F} = e^{-(r_1 - r_2)/L_s},$$

where N and F are the near and far detector counting rates respectively. Fig. 12–8 shows the laboratory data for the ratio of the epithermal detector counting rates for a number of calibration formations. The data have been plotted versus $1/L_s$. Although they do not fall along a straight line as predicted, the important point to note is that the data, which represent measurements in three different matrices, fall along a single line. Thus the

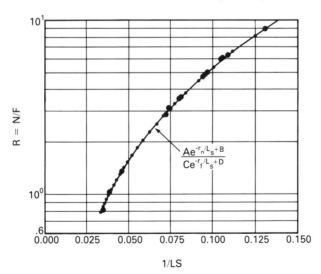

Figure 12–8. The ratio of the near and far epithermal detectors for a neutron porosity device as a function of the inverse slowing-down length for the formations measured. The behavior is slightly different than that anticipated for the case of an infinite homogeneous medium.

ratio of epithermal counting rates is, indeed, a measure of the formation slowing-down length. From the slowing-down length dependence on porosity, the porosity can be extracted.

Although the API committee that set up the gamma ray calibration standards also took some steps to standardize neutron log responses,* their recommendations for API units have not been implemented.[12] The conventional approach has been to calibrate the tool in limestone formations and to report all tool readings in apparent "limestone porosity." Conversion charts are then necessary to correct the apparent limestone porosity for the matrix in which the measurement was actually made. Based on the preceding discussion, some consideration should be given to using the slowing-down length and migration length as the quantities for reporting the log measurements. The conversion to porosity by use of charts similar to Fig. 12-7 would then lie entirely in the realm of interpretation.

THE NEUTRON POROSITY DEVICE: RESPONSE CHARACTERISTICS

Modern neutron porosity devices of either type, thermal or epithermal, require corrections for environmental effects which are specific for the type of tool. In this section perturbations which are common to both types of devices are discussed. These perturbations are fundamental to understanding neutron porosity tool response.

Shale Effect

As will be seen in the log examples to follow, one of the characteristics of neutron porosity logs is to indicate rather large values of apparent porosity in shale zones (and elevated values in porous shale-bearing zones). A common misconception is that the error in the thermal porosity readings in shales is caused by associated trace elements with large thermal capture cross sections. However, even without the additional effect of thermal absorbers, clays and shales present a problem for all neutron porosity interpretation because of the hydroxyls associated with the clay mineral structure.[13] The large apparent porosity values are due primarily to the hydrogen concentration associated with the shale matrix.

---

* At the time of the introduction of neutron porosity logging, the usual scale was neutron counting rate. This, of course, could vary significantly not only with porosity, but also from service company to service company, depending upon the source-to-detector spacings (usually marked on the log heading), the neutron source activity, and types of detection. Interpretation of porosity required comparison with core or cutting analysis, and makes today's interpretation problems look minimal.

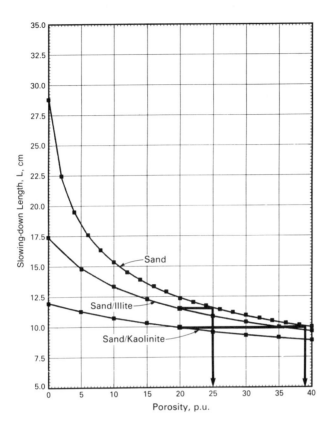

Figure 12-9. The effect of two types of clay minerals on the slowing-down length of sand. In the case of the two lower curves, the matrix is composed of equal volumes of sand and clay. The lowermost curve has a higher hydroxyl content than the intermediate curve. From *SPE Petroleum Production Handbook.*[6]

Fig. 12-9 illustrates this point by showing the variation of slowing-down length of sand, a sand-illite, and a sand-kaolinite mixture as a function of porosity. In the latter two cases, the sand and shale volumes are in equal proportions. It is clear that if the presence of clay, in addition to the sand, is not taken into account, large errors in porosity can result. Also note that the apparent porosity of kaolinite is much larger than that of illite. The examples shown in the figures indicate that a 20-PU sand/illite mixture appears to be about 25 PU, whereas the 20-PU sand/kaolinite mixture has an apparent porosity of about 39 PU. This effect, due to the differing hydroxyl contents of these two clay minerals, is shown in Chapter 19 to be useful in distinguishing clay types.

Matrix Effect

As in density logging, it is necessary to know the rock matrix in order to convert a measured apparent limestone porosity into an actual porosity estimate. Fig. 12–10 shows, for a thermal and an epithermal device, the matrix-dependent transform of measured units to porosity for a given rock type. It should be noted that these two charts are for two specific tools.

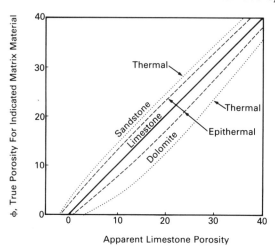

Figure 12–10. The matrix correction chart for two particular types of thermal neutron porosity tools. Adapted from Schlumberger.[8]

When performing such a transformation, we should use the chart for the appropriate tool since some of this so-called matrix effect is tool-design dependent. *Rather than elaborate on the use of such a correction chart, we will examine how it might be constructed. A large part of the matrix effect can be understood in terms of the two basic parameters used to describe the bulk parameters of the formation, i.e., the slowing-down length and the migration length.*

First, consider the case of epithermal detection. To demonstrate the construction of the matrix effect correction curves, there are four steps to consider:

1. The first step is the link between the measurement [in this case, the ratio of near (N_{Nn}) to far (N_{Nf}) detector counting rate] and porosity in primary laboratory calibration standards. This might be a fresh-water-filled limestone with an 8" borehole. Fig. 12–11 shows the behavior of the ratio $F = N_{Nn}/N_{Nf}$ as a function of limestone porosity, ϕ_{lime}. From this plot, a fit can be established to describe a functional relationship between the measured parameter F and the limestone porosity ϕ_{lime}:

$$F = f(\phi_{lime}) . \qquad (4)$$

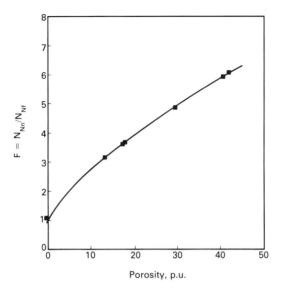

Figure 12–11. The measured ratio of near and far detector counting rates in limestone calibration formations. From *SPE Petroleum Production Handbook.*[6]

2. The relationship between the calculated parameter F and the slowing-down length L_s must be established for measurements in all three of the principal rock types. That this can be easily done is shown in Fig. 12–8, where measurements in quartz, dolomite, and limestone are shown for a range of porosities. From this plot, a new fit can be found:

$$F = f(L_s), \qquad (5)$$

which allows the prediction of the calculated ratio from the known slowing-down length of the formation.

3. The next step is to establish the connection between the slowing-down length L_s and porosity for limestone, as shown in Fig. 12–12. The middle curve now represents the limestone "transform" of Eq. (4) and the porosity axis represents true porosity.

4. The slowing-down lengths of sandstone and dolomite are now calculated as a function of porosity and are also shown in Fig. 12–12. They fall on either side of the limestone response because of their different chemical compositions (i.e., atomic weights), which influence their slowing-down lengths.

The apparent limestone porosity for either formation can be found by selecting a porosity, 10-PU sandstone, for example, and finding the corresponding slowing-down length (approximately 15.5 cm). The apparent

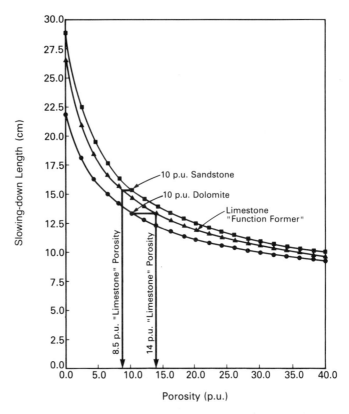

Figure 12–12. The origin of the lithology effect is demonstrated by considering the slowing-down length for the three principal matrices. If the limestone line is taken as the reference for converting slowing-down length to porosity, then the apparent "limestone" porosity of a 10% porosity sandstone and dolomite formation are illustrated. From *SPE Petroleum Production Handbook.*[6]

limestone porosity is then obtained by finding the porosity associated with the limestone formation of the same L_s value. In this case it is 8.5 PU. The same case for dolomite indicates 14 PU instead of 10 PU.

A similar procedure can be used for a thermal neutron porosity device. As shown earlier, it has been found useful to cast the results in terms of the migration length, L_m. As the quantity L_m contains some information concerning the macroscopic thermal absorption cross section, the results are qualitatively similar to the preceding analysis but differ slightly in magnitude. This can be seen by performing the preceding exercise on the plots of L_m versus porosity of Fig. 12–7.

The matrix effect of the neutron log can be exploited in conjunction with the density log for lithology determination. Fig. 12–13 shows the neutron-density cross plot for a dual detector thermal device. The bulk density is

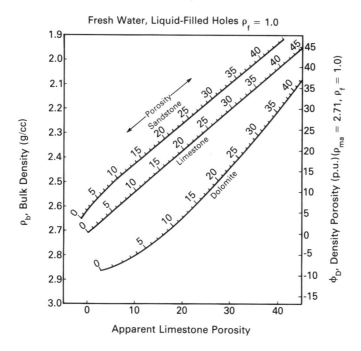

Figure 12–13. The neutron-density cross plot used for lithology identification. Adapted from Schlumberger.[8]

plotted as a function of the apparent limestone neutron porosity. The three indicated lines correspond to density and neutron porosity variation for values of porosity between 0 and 40 PU for the three principal rock types. (More recent versions of this chart show the line associated with dolomite to be straight.)[9,14] The matrix effect can be observed by comparing the equiporosity points on the individual lithology lines with the apparent limestone porosity scale on the abscissa. These discrepancies are the same as those indicated in the portion of Fig. 12–10 indicated as "thermal." The use of this cross plot of neutron porosity and density for lithology determination is discussed in Chapter 18.

Gas Effect

Neutron porosity devices are calibrated in formations containing liquid-filled porosity. In a gas-filled formation, with lower than expected hydrogen density, an error will result in the apparent porosity. This is because the replacement of the liquid in the pores by gas has considerable impact on the slowing-down length of the formation and thus on the apparent porosity. Partial replacement of the water component of the formation by a much less dense gas increases the neutron slowing-down length, and thus the apparent porosity decreases. The decrease in apparent porosity is a function primarily of the true porosity, the water saturation, the gas density, and to some extent the lithology. Replacement of fluid in the pores by a less dense gas also decreases the bulk density of the formation. These two effects have been exploited in well logging by making the density and neutron porosity measurements in a single measurement pass. On the log presentation, the density and neutron traces separate in a direction which makes the presence of gas easily visible.

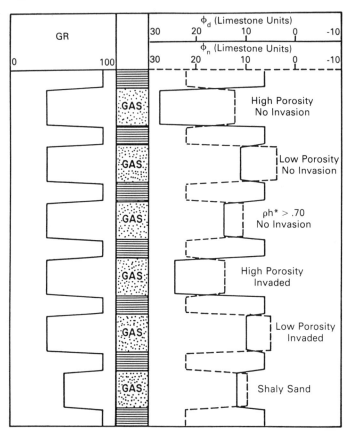

*ph Hydrocarbon Density

Figure 12-14. The idealized gas effect seen from a combination presentation of apparent formation porosity derived from a density tool and a neutron tool. From Asquith.[15]

This type of idealized behavior in a sequence of shale and gas sands is presented in Fig. 12-14. The apparent porosity from the density measurement and the neutron device are presented in limestone units. In the shale section, the characteristic separation is seen, with the neutron reading much higher than the density porosity. In the first gas sequence, where no invasion is presumed, the density porosity reads about 28 PU. (The actual porosity is much less, since this was derived from the assumption that the porosity was liquid filled, and thus ρ_{fl} is taken to be 1.00 g/cm^3.) The low hydrogen density increases the slowing-down length, and thus an abnormally low porosity reading is derived from the neutron device. Note that the neutron-density separation in the third gas sand is much reduced compared to the preceding two zones, since in this case the hydrocarbon density is 0.7 g/cm^3, and thus the density value is not far from the expected water-filled value. Reduced separation is also noted in other zones where there is invasion or shale.

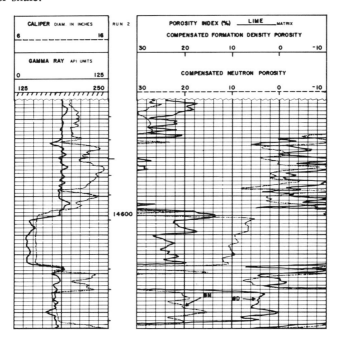

Figure 12-15. A log illustrating an actual case of neutron density separation indicating the presence of gas. From Asquith.[15]

Fig. 12-15 is an example in which the gas effect can be easily seen. In the zone below 14600', there are two sections of moderate porosity separated by a very washed-out zone, probably a shale. In fact, the lower 30' of the log appear, from the gamma ray, to be shaly. In the upper porous section, there is roughly a 10-PU difference indicated between the neutron and density porosity, the density porosity being higher. This indicates that the formation, presumably a limestone, contains gas. In the lower porous section, there is about a 15-PU separation between the two. This time the neutron porosity is greater than the density porosity, indicating the presence of shale.

In an attempt to quantify the traditional neutron-density gas separation and to illustrate the possibility of estimating gas saturation from an epithermal

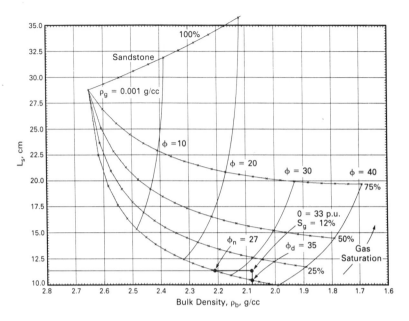

Figure 12–16. The variation of slowing-down length as a function of density. The third parameter is formation porosity. The several lines correspond to various values of gas saturation. The gas is assumed to be nearly zero density. From *SPE Petroleum Production Handbook.*[6]

neutron measurement and a density measurement, consider Figs. 12–16 and 12–17. In both, the slowing-down length of a sand formation has been computed to be between zero and 40 PU at 2-PU increments. These values of slowing-down length have been plotted as a function of the corresponding bulk density of the formation for the five gas saturations indicated on the figures. In the case of Fig. 12–16, the gas density ρ_g has been taken to be .001 g/cm^3, and in Fig. 12–17, the gas density has been taken to be about 0.25 g/cm^3, which covers the entire range possible under normal reservoir conditions. In either case, for the presence of gas at a fixed porosity, it is clear that the slowing-down length is larger than that associated with the water-filled porosity. The interpretation of this larger value of L_s is an apparent decrease in porosity. In the case of the total gas saturation curve of Fig. 12–16, L_s values greater than $\approx$ 28 cm would correspond to zero or negative apparent porosities on the log readings.

On the plots of L_s versus ρ_b, a pair of points (L_s, ρ_b) yields the saturation and porosity corresponding to the conditions of gas density specified. For both figures, the calculations correspond to a matrix of sandstone with no shale. The gas density difference for the two plots spans the range of expected gas densities. The saturation referred to on the figures refers to a gas/water mixture. For the lowest curve on both figures, the porosity is entirely filled with water.

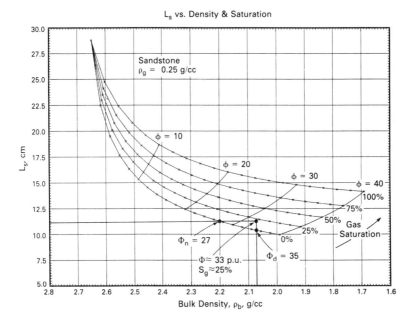

Figure 12–17. Slowing-down length as a function of density for gas saturations of a gas with density of 0.25 g/cm³. From *SPE Petroleum Production Handbook*.[6]

Since the slowing-down length is not normally presented on a neutron porosity log, an example taken from the log of Fig. 12–18 can serve as a guide for using this technique. In this example, the neutron- and density-derived porosity values are for a sandstone matrix. The 40' zone around 650' is a clean sand with two clear indications of gas. From the gas zone at 670', we can read the following values for use in the following example:

- ϕ_n, neutron porosity (sand)= 27 PU
- ϕ_d, density porosity (sand)= 35 PU

The lowermost curves on Fig. 12–16 and Fig. 12–17 serve to determine the correspondence between ϕ_n and the slowing-down length, L_s. It is indicated by the horizontal line to be about 11 cm for this example. The density porosity of 35 PU can be seen to correspond to a density of about 2.07 g/cm³, as indicated by the vertical line. The intersection of these two points at the coordinate (2.07, 11), shown in the figures, indicates about 25% gas saturation in the case of a gas density of 0.25 g/cm³ (Fig. 12–15) and a saturation of about 12.5% if the gas density is taken to be practically zero. The porosity of the formation can be found by following the slope of the equiporosity lines from the example coordinate to the lower liquid-saturated line. It is seen to indicate a value of about 33 PU for both cases.

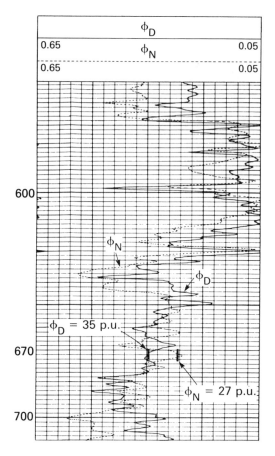

Figure 12-18. An example of neutron density crossover in which the gas saturation can be estimated. From *SPE Petroleum Production Handbook*.[6]

If the gas density and saturation were known from other means, the previous analysis could yield some information concerning the invasion of the drilling fluid into the zone of investigation of the neutron and density devices.

DEPTH OF INVESTIGATION

The depth of investigation for neutron devices is of interest when we are trying to evaluate gas saturation. However, unlike the depth of investigation of induction devices, it is a quantity which is rather difficult to compute exactly. It must be done by rather careful numerical simulation, using discrete ordinates or Monte Carlo codes, or it may be based on a series of experiments. The appendix to this chapter presents the results of a simple geometric factor approach to the determination of the depth of investigation which compares favorably with the experimental results presented next.

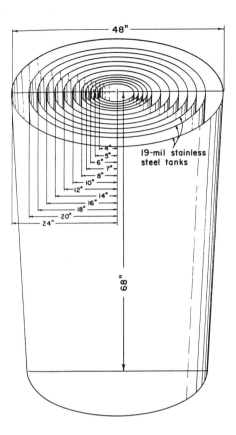

Figure 12–19. The laboratory setup for the determination of the depth of investigation or pseudogeometric factor of neutron porosity tools. Adapted from Sherman and Locke.[16]

Just as the calculation of the neutron tool depth of investigation is difficult, so is the measurement of it. For the determination of depth of investigation for a single formation porosity, the experimental setup is shown in Fig. 12–19. The test formation consisted of a number of coaxial cylinders of loose dry sand. The porosity of this formation was approximately 35 PU. With the tool in place, the cylindrical layers of sand nearest the borehole were saturated with water, one at a time. The observed change in apparent porosity as a function of this water-invasion depth is shown in Fig. 12–20. Data from an experimental epithermal neutron porosity device are shown. The three curves correspond to the values of porosity derived for the individual detectors and from the ratio of the two counting rates.

As we expect intuitively, the depth of investigation of the far detector is somewhat greater than that for the near detector. In practice neither of these detectors is used alone for the determination of porosity; rather their ratio is used. We see that the ratio has a depth of investigation larger than either of the two individual curves. This is somewhat analogous to the case of the induction tool described earlier. In that case, a second receiver coil was introduced to remove some of the signal originating close to the borehole from the farther receiver, thus weighting the signal contributions from deep in the formation more strongly than those from nearby. In the case of the neutron tool, taking the ratio of the near and far detector counting rates partially eliminates the common response of the two detectors to the shallow zone. This has the effect of increasing the relative sensitivity for hydrogen in the region beyond the near detector depth of investigation.

The curves of Fig 12–20 are similar in nature to the pseudogeometric factors for the resistivity electrode devices. The exact shapes of the response curves are dependent upon the actual experimental conditions. However, it is clear that the gas sensitivity in a 35-PU formation comes from roughly the first 8″ of formation.

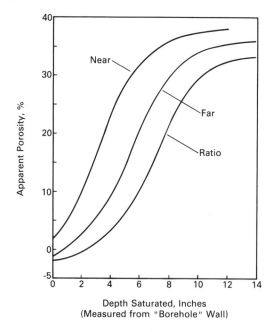

Figure 12–20. The results of the determination of the depth of investigation of a dual-detector epithermal neutron porosity device. Note that the combined measurement has a greater depth of investigation than either of the single detectors. Adapted from Sherman and Locke.[16]

LOG EXAMPLES

For the sake of completeness, the log from a thermal neutron porosity device in the simulated reservoir is shown in Fig. 12–21. It should be clear by now that a single log, without accompanying information, is insufficient for detailed interpretation. In the upper portion of the figure, the log is from the shaly sand portion of the reservoir. It is easy to spot the high neutron porosity values which correspond to the shale (they correlate with the increased gamma ray reading in each case). In the lower portion, the section shown is from a carbonate sequence. Once again the very high neutron

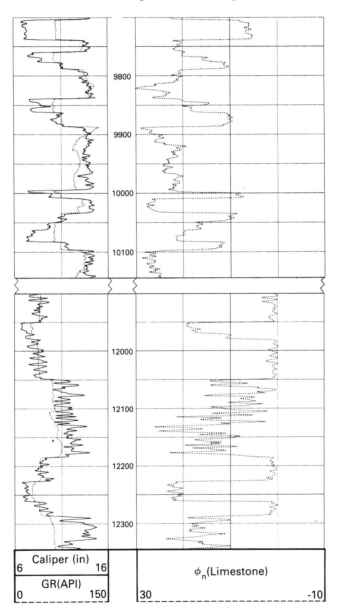

Figure 12–21. A log example of a thermal porosity device in the simulated reservoir.

porosity values correspond to shale sections, as can be confirmed from the comparison with the companion gamma ray. The matrix setting for the neutron log was limestone, which is known to be present in this lower section.

In Fig. 12–22, a comparison between the density and the neutron porosity logs has been made on an expanded depth scale. Note the unusual density scale. It was chosen so that, in the case of limestone formations, the density

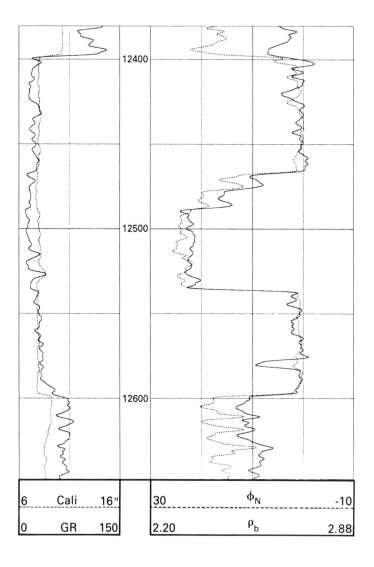

Figure 12–22. Comparison of the neutron porosity and density logs in the carbonate section of the artificial reservoir.

values would overlay the neutron porosity scale. From the match observed, it seems likely that the two low porosity sections (< 3 PU) are indeed limestone. The very high porosity section around 12500' shows a slight discrepancy between the density and neutron porosity. This may be due to a slight change in matrix or a fluid density which is not the 1.00 g/cm^3 assumed in the density to porosity conversion.

APPENDIX

Depth of Investigation of Nuclear Logging tools

For nuclear logging measurements, a principal controlling factor for depth of investigation is the mean free path of the radiation. For gamma rays of interest in probing the formation deeply, energies run from several MeV to several hundred keV. This is the region dominated by Compton scattering, and the mass absorption coefficient (see Fig. 8–8) is on the order of 0.025–0.08 cm^2/g. For typical formation densities ranging between 2.0 and 3.0 g/cm^3, the mean free path is thus between 4 and 20 cm. The mean free path of high energy neutrons (see Fig. 11–7) is of a similar order of magnitude.

To get some idea of the depth of investigation of a nuclear measurement, we return to the simplest case: a point detector of gamma rays in an infinite medium containing uniformly distributed gamma ray sources of a single energy (see Fig. 9–6). The density of material in which they are embedded determines the mean free path, λ, which is sufficient to characterize the gamma ray attenuation. The medium is further characterized by defining n gamma ray emissions/sec per unit volume. The contribution to the total unscattered flux $d\phi$ from a spherical shell at a distance r was shown to be:

$$d\phi = \frac{1}{4\pi r^2} e^{-r/\lambda} \, n \, 4\pi r^2 \, dr.$$

The total flux contributed by the material contained within a radius R_o is given by:

$$\phi(R_o) = \int_o^{R_o} \frac{1}{4\pi r^2} e^{-r/\lambda} \, n \, 4\pi r^2 \, dr = \frac{n}{\lambda}(1 - e^{-R_o/\lambda}).$$

This expression is very similar to an integrated radial geometric factor, and the form of the equation indicates that 90% of the total flux comes from a spherical region surrounding the detector with a radius of about 2.3λ.

After considering this simple case, which corresponds roughly to the natural gamma ray tool, let us extend the analysis to other nuclear logging devices which generally contain a source of radiation and a detector separated by some distance. In general, this complicated problem of radiation transport can be answered only by application of the Boltzmann transport equation. However, it is possible to get an idea of the depth of investigation from an analysis similar to that presented earlier for the induction tool.

The situation to be considered, which is a considerable simplification of a logging tool in its measurement environment, is sketched in Fig. 12–23. It shows the sonde containing a source of radiation, and a detector separated by a distance L. The distance* ρ_s denotes the distance from the source to the site of interaction contained in an elemental volume of the formation. The distance from this element of volume to the detector is denoted by ρ_d. In this type of device, the role of the embedded sources of the preceding case is now played by the density of interaction centers and the flux at the site, due to the source. For composite tools of this type, there is no longer a uniform source distribution, n, but one which has a complicated radial and axial dependence.

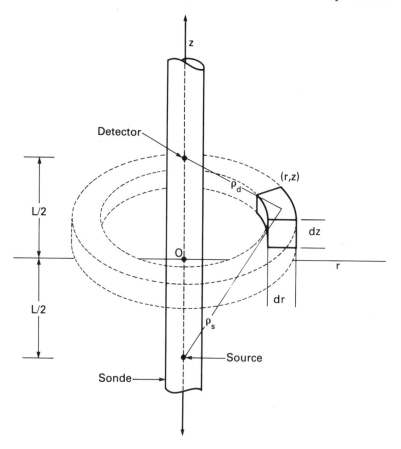

Figure 12–23. Geometry for describing the depth of investigation of logging tools.

---

* In this appendix, ρ denotes distance, not density.

The tool geometry determines, to some extent, the depth of investigation. If, for example, some sort of shielding is used at the source to exclude direct source flux from scattering regions close to the sonde and near the detector, then the depth of investigation will be somewhat deeper than in the case of the uniformly distributed emitters considered earlier. The detector, usually shielded from the direct source radiation, does not have a uniform isotropic efficiency for detection. This directionality may also affect the depth of investigation. For the following discussion, both of these practical details are ignored.

For the development of a radial geometric factor, the task is to determine the contribution to the signal of the detector from a typical element of volume (with cross-sectional area of dr dz) at position z and at a radial distance r. The first case to consider is that of a tool like the gamma-gamma density or the induced gamma ray spectroscopy device (see Chapter 13). In the case of the latter tool, the rate of gamma ray production in the volume element depends upon the flux of unattenuated high energy neutrons reaching it, as well as its volume, $2\pi r$ dr dz. Thus it is proportional to:

$$2\pi r \, dr \, dz \, \frac{1}{\rho_s^2} \, e^{-\rho_s/\lambda_s},$$

where ρ_s is the distance from the source to any position on the indicated ring volume, and λ_s is the mean free path for the source energy neutrons. For the density tool, the gamma ray scattering which takes place in the ring is given by the same expression as above, with λ_s corresponding to the mean free path of the source energy gamma rays. The flux of gamma rays reaching the detector, for either case, is proportional to:

$$\frac{1}{\rho_d^2} \, e^{-\rho_d/\lambda_d},$$

where ρ_d is the distance from the ring to the detector and λ_d is the mean free path which characterizes the attenuation of gamma rays in the material.

Thus a differential geometrical factor $g(r,z)$ can be defined for either of these devices as:

$$g(r,z) \propto \frac{r}{\rho_s^2 \rho_d^2} e^{-\rho_s/\lambda_s} e^{-\rho_d/\lambda_d} \, dr \, dz. \tag{A-1}$$

Although it involves a gross oversimplification, we can proceed further in this approach and consider the case of a neutron porosity tool. If we take the result shown earlier, in Eq. (2), the flux of epithermal neutrons at distance ρ_s from the source is given by an expression like:

$$\frac{1}{\rho_s} e^{-\rho_s/L_s},$$

where ρ_s is the distance from the source. By analogy we might expect the geometric factor (considering epithermal neutrons scattered in the unit volume to be detected without further diffusion) to be:

$$g(r,z) \propto \frac{r}{\rho_s \rho_d^2} e^{-\rho_s/L_s} e^{-\rho_d/\lambda_d} \, dr \, dz \, . \tag{A-2}$$

For a thermal neutron device, there is a characteristic slowing-down length on the source path and a diffusion length, L_d, on the final path. Its differential geometric factor should vary as:

$$g(r,z) \propto \frac{r}{\rho_s \rho_d} e^{-\rho_s/L_s} e^{-\rho_d/L_d} \, dr \, dz \, , \tag{A-3}$$

where the extra power of ρ_d has been removed, since the diffusing thermal flux from the scattering site to the detector is expected to fall off only as $1/\rho_d$.

Interesting consequences of these three expressions for differential geometric factors can be seen if an integration over the z-axis is performed in order to obtain the differential radial geometric factor $g(r)$:

$$g(r) = \int_{-\infty}^{\infty} g(r,z) \, dz \, .$$

In the upper portion of Fig. 12–24 this has been done numerically for the three cases, using a source-to-detector spacing, L, of 50 cm, and for convenience, characteristic mean free paths of 15 cm for all source and detector paths. The results show, for the cases corresponding to the density device, Eq. (A-1), and the epithermal neutron device, Eq. (A-2), that the portion of the formation closest to the sonde is the most important. However, for the case of the thermal neutron diffusion, Eq. (A-3), there is a peaking behavior in the radial sensitivity much like the case of an induction device with a 50-cm spacing, which is shown for comparison in the same figure.

To get an idea of the depth of investigation, the integral radial factors, $G(r)$, have been obtained from the radial geometric factors by integrating, starting at a radius of 1 cm:

$$G(r) = \int_{1cm}^{\infty} g(r) dr \, .$$

The results are shown in the lower portion of Fig. 12–24. The integrated results differ slightly from one another. If the 90% point is taken as a measure of the depth of investigation, it is found to be between one and two times the assumed 15-cm mean free path. How much of this depth of investigation is controlled by the spacing? In order to get a feeling for this, the same calculation was performed with a reduction of the source- to- detector spacing to 25 cm. These results are shown, in the same format, in Fig. 12–25. By comparison, the 90% points can be seen to have been shifted by only about 5 cm closer to the sonde for the three functions. This indicates a strong control due to the mean free paths in this case, where the spacing, L,

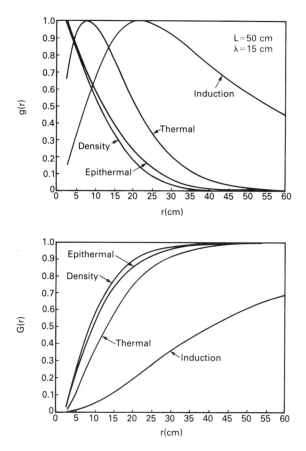

Figure 12-24. The differential and integrated radial geometric factors for hypothetical tools of 50-cm source-to-detector spacing. Compared are an induction device, a thermal neutron device, an epithermal device, and the density tool.

is much larger than the characteristic attenuation lengths. In the case of the induction device (with no attenuation considered), the 50% point falls at about one half the coil spacing, as was the case in the preceding figure.

The preceding results should only be taken as indicative of the relationship between depth of investigation, source-detector spacing, and mean free paths. As mentioned earlier, the depth of investigation can be modified by source and detector shielding. In fact, one type of shielding is imposed by the construction of the sonde: it is very unlikely that source radiation would travel directly to the detector because of the intervening material. This shielding will exclude source flux from zones of the formation close to the sonde and thus force the radial geometric factor to extend further into the formation. To simulate this type of collimation, an additional attenuation can be placed on the source terms of Eqs. (A-1), (A-2), and

(A–3), which vary with the angle between the vector ρ_s and the sonde axis. Choosing reasonable values for this attenuation and for the mean free paths allows the matching of the model results with measured tool response.

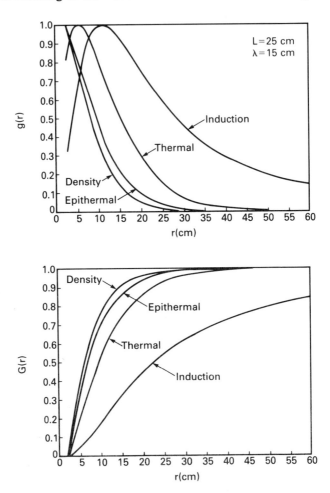

Figure 12–25. The differential and integral radial geometric factors for hypothetical tools of 25-cm source-to-detector spacing. Compared are an induction device, a thermal neutron device, an epithermal device, and the density tool.

The experimental data for a 35-PU sandstone are shown, in Fig. 12–26, for the long spacing detectors of the three devices. The J-factor of the graph is a renormalization of the data of the type seen earlier in Fig. 12–20, and corresponds to the integrated radial geometric factor G(r). Rather than attempt to use the actual source-to-detector spacing for each device, the model keeps a constant 50-cm spacing, which is close to the actual values.

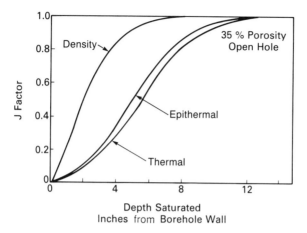

Figure 12–26. The integrated radial geometric factors derived from the data of Fig. 12-20. From Sherman and Locke.[16]

For comparison, note that the 50% points for the long spacing detectors of the three tools are: 10 cm for density, 12.7 cm for epithermal, and 15 cm for thermal. The model predictions of the radial geometric factor are shown in the upper portion of Fig. 12–27. Note that the addition of the source attenuation term has produced a peaking term in the epithermal result but not in the density case; see Eq. (A–1). The integrated radial factors are shown in the bottom portion of Fig. 12–27. The 50% values derived from these compare favorably with the experimental values.

To evaluate the effect of spacing on this slightly more realistic model, the same calculation performed with a source-to-detector spacing of 25 cm produced the results shown in Fig. 12–28. By comparison with the previous results, the integrated depth point for 50% of the signal is between 3 and 5 cm closer to the sonde. Although the case just considered was for a 35-PU sand formation, the numbers are representative of the order of magnitude of the depth of investigation for nuclear tools. There will be some variation depending on lithology and porosity, but they will still be in the range of tens of centimeters.

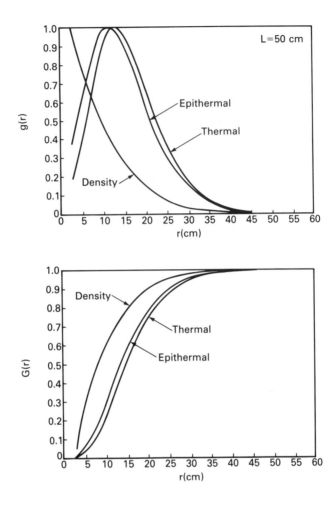

Figure 12-27. The differential and integrated radial geometric factors for hypothetical tools of 50-cm source-to-detector spacing. An attenuation term is included as a function of the angle of emission. Compared are a thermal neutron device, an epithermal device, and the density tool.

Neutron Porosity Devices 277

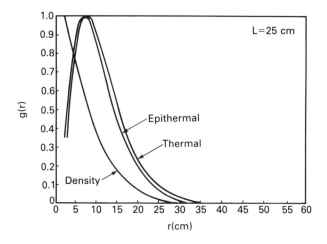

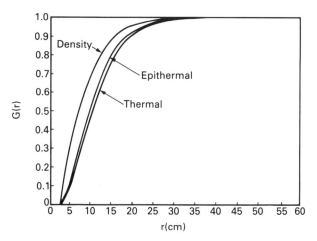

Figure 12-28. The differential and integrated radial geometric factors for hypothetical tools of 25-cm source-to-detector spacing. Compared to the results of Fig. 12-27, the 25-cm shortening of source-to-detector spacing has reduced the depth of investigation only by several centimeters.

References

1. Arnold, D. M., and Smith, H. D., Jr., "Experimental Determination of Environmental Corrections for a Dual-Spaced Neutron Porosity Log," Paper W, SPWLA Twenty-Second Annual Logging Symposium, 1981.

2. Duderstadt, J. J., and Hamilton, L. J., *Nuclear Reactor Analysis,* John Wiley, New York, 1976.
Henry, A. F., *Nuclear-Reactor Analysis,,* MIT Press, Cambridge, Mass., 1975.
Glasstone, S., and Sesonske, A., *Nuclear Reactor Engineering,* D. Van Nostrand Co., Princeton, 1967.

3. Allen, L. S., Tittle, C. W., Mills, W. R., and Caldwell, R. L., "Dual-Spaced Neutron Logging for Porosity," *Geophysics,* Vol. 32, No. 1 1967.

4. Edmundson, H. and Raymer, L. L., "Radioactive Logging Parameters for Common Minerals," Presented at the SPWLA Twentieth Annual Logging Symposium, 1979.

5. Tittman, J., Sherman, H., Nagel, W. A., and Alger, R. P., "The Sidewall Epithermal Neutron Porosity Log," *JPT* Vol. 18, 1966.

6. Ellis, D. V., "Nuclear Logging Techniques," in *SPE Petroleum Production Handbook,,* edited by H. Bradley, SPE, Dallas (in press).

7. *Well Logging and Interpretation Techniques: The Course for Home Study,* Dresser Atlas, Dresser Industries, 1983.

8. *Log Interpretation Charts,* Schlumberger, New York, 1982.

9. Gilchrist, W. A., Galford, J. E., Flaum, C., Soran, P. D., and Gardner, J. S., "Improved Environmental Corrections for Compensated Neutron Logs," Paper SPE 15540, SPE Annual Technical Conference, 1986.

10. Davis, R. R., Hall, J. E., and Boutemy, Y. L., Paper SPE 10296, SPE Fifty-sixth Annual Technical Conference, 1981.

11. Scott, H. D., Flaum, C., and Sherman, H., Paper SPE 11146, SPE Fifty-seventh Annual Technical Conference, 1982.

12. Belknap, W. B., Dewan, J. T., Kirkpatrick, C. V., Mott, W. E., Pearson, A. J., and Rabson, W. R., "API Calibration Facility for Nuclear Logs," *Drill. and Prod. Prac.,* American Petroleum Institute, Houston, 1959.

13. Ellis, D. V., "Neutron Porosity Logs: What do they Measure?" *First Break,* Vol. 4, No. 3, 1986.

14. Ellis, D. V., and Case, C. R., "CNT-A Dolomite Response," Paper S, SPWLA Twenty-Fourth Annual Logging Symposium, 1983.
15. Asquith, G. B., and Gibson, C. R., *Basic Well Log Analysis for Geologists*, AAPG, Tulsa, 1982.
16. Sherman, H., and Locke, S., Paper Q, SPWLA Sixteenth Annual Logging Symposium, 1975.

Problems

1. You are faced with interpreting a set of old epithermal logs which were inadvertently run with a matrix setting of "LIME." The section of interest is quite certainly a sandstone.

    a. Based on your understanding of the response of epithermal neutron porosity devices, construct a correction chart for use in converting the log porosity values to "true" sandstone porosity. This correction should be of the form: $\phi_{true} = \phi_{log} + \Delta\phi$. Plot $\Delta\phi$ vs. ϕ_{log} for values of ϕ_{log} between 0 and 30 PU in 5-PU steps.

    b. If the log readings are running between 15 and 30 PU, can you simply shift the values to obtain a good porosity value?

2. The thermal porosity device response can be characterized rather well by the use of the migration length, L_m. As you recall, $L_m^2 = L_s^2 + L_d^2$, where

    $L_d^2 = \dfrac{D}{\Sigma}$. (D is the thermal diffusion coefficient, and Σ is the macroscopic thermal neutron absorption cross section.)

    a. What apparent "limestone" porosity would you expect the tool to read in a fresh-water-saturated 40-PU sandstone?

    b. Neglecting, for the moment, the effects of hydrogen displacement, what porosity would you expect it to read if the fresh water were replaced by salt-saturated water (260 Kppm NaCl)?

    c. However, you cannot neglect the effect of hydrogen displacement in this case, since the density of the salt solution is 1.18 g/cm³. When you take this into account, in addition to the salinity, what is the apparent "limestone" porosity? (Hint: First calculate an equivalent porosity due to the lower hydrogen concentration in the saltwater.)

3. An epithermal neutron porosity device basically measures the slowing-down length, as described in the text. Suppose the counting rate ratio from which the slowing-down length is derived is 10% in error for one reason or another. How much error does this translate to, in porosity units, for a 5% porosity sandstone? A 30% porosity limestone?

4. Using the log of Fig. 12–22, confirm that the lithology of the zone at 12500′–12540′ is limestone. What is the average porosity?

5. What is the neutron-density separation to be expected in a 10% porous sandstone with a 50% gas saturation? The density of the gas can be taken as 0.25 g/cm^3. Compute ϕ_n in limestone units and ϕ_d in limestone units.

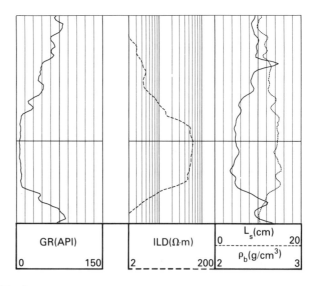

Figure 12–29. Log example showing the density and slowing-down length measured in a gas sand.

6. Fig. 12–29 shows a log of the density and slowing-down length of a gas zone, along with a deep resistivity device. The water resistivity is known to be 0.08 Ω·m. First estimate the porosity of the zone, using Fig 12–17. How does the saturation derived from R_t compare with that obtained from the density-L_s cross plot?

7. What is the epithermal porosity reading (sandstone units) expected to be in a 20% porous shaly sandstone, when the rock matrix is composed of 50% (by volume) of illite?

13
PULSED NEUTRON DEVICES

INTRODUCTION

The availability of pulsed neutron sources allowed the development of several valuable formation evaluation techniques: thermal die-away logging and spectroscopy of neutron-induced gamma rays. At the heart of these techniques is the neutron generator discussed in Chapter 11. In thermal die-away logging, the pulsing capability of this generator is fundamental to the determination of the formation thermal neutron absorption properties. A limited chemical analysis of the formation can be obtained from a combination of the controlled injection of high energy neutrons into the formation and spectroscopy of the neutron-induced gamma rays. By exploiting the high energy neutron reactions, we can determine the ratio of carbon to oxygen in the formation. Spectroscopy of gamma rays produced by subsequent thermal neutron capture reactions allows the detection of a dozen or so important elements present in the formation.

Pulsed neutron devices found their first logging application in the determination of water saturation in cased producing wells. The traditional method for obtaining S_w, of course, relies on electrical measurements. For the most part, however, producing wells contain metallic casings which render electrical measurements useless. Neutrons, however, penetrate the casing without much difficulty and consequently are applied to the problem of probing the formation in such situations.

Thermal neutron die-away devices respond to the macroscopic thermal capture cross section, which depends on the chemical constituents of the matrix and pore fluids. Chlorine, which is nearly always present in formation waters, has a large absorption cross section. Thus a measurement of the macroscopic absorption cross section (Σ) can provide the means (although indirect) for identifying saltwater and for measuring the water saturation.

To motivate the measurement of Σ, we will first look at the relationship between the concentration of thermal absorbers and the value of the macroscopic cross section. This will give an indication of the dynamic range of Σ expected in logging applications. The technique for determining the value of Σ from borehole measurements is examined, illustrating some its limitations. An introduction to interpretation of the Σ measurement and some applications are given.

In a second section, an overview of spectroscopy of neutron-induced gamma rays for formation chemical analysis is presented. The initial aim of this technique was to provide an alternate method of determining saturation behind casing. Rather than the more indirect approach of interpreting a measurement of Σ, it is possible to directly measure the ratio of carbon to oxygen atoms in the formation. If the lithology and porosity are known, this ratio can yield saturation. In a separate phase of the preceding measurement, spectroscopy of gamma rays, induced during the capture of neutrons, yields substantial information concerning the formation chemical composition. Some of the applications of this latter technique are explored in Chapter 19.

THERMAL NEUTRON DIE-AWAY LOGGING

Thermal Neutron Capture

As discussed in Chapter 11, neutron capture is one of many reactions which can take place during neutron interaction with matter. The cross section for capture varies as $1/v$ at low energies, where v is the neutron velocity. Thus, it is the predominant interaction mechanism at thermal energies and the only way neutrons are removed from the system. In the capture of neutrons, the target nucleus with atomic mass A transmutes into another isotope of the element with mass A+1. This "compound nucleus" is formed in an excited state that decays nearly immediately, with the emission of one or more gamma rays. The gamma ray energy can range up to a maximum of about 8 MeV.

To appreciate the magnitude of the thermal absorption cross section of a few common (and not so common) elements, refer to Table 13–1, which orders the elements by mass-normalized absorption cross section. The definition of the mass-normalized cross section follows from the macroscopic cross section. The third column in Table 13–1 gives the absorption cross section per nucleus, σ_a, in barns (10^{-24} cm^2). The macroscopic absorption cross section, Σ_a, is the product of σ_a and the number of nuclei per cubic centimeter:

$$\Sigma_a = \sigma \frac{N_o}{A} \rho.$$

The mass-normalized cross section, σ_m, is just Σ_a/ρ. This is a useful unit since a thousand times the mass-normalized cross section is numerically

ELEMENT		A Ave. Atomic Weight (Atomic Mass Units)	σ Ave. Atomic Absorption Cross Section (barns)	σ_m Mass-Normalized Absorption Cross Section (cm^2/gm)*	A_m Mass-Normalized Chlorine Absorption Equivalents
Gd	Gadolinium	157	49,000	188	333
B	Boron	10.8	759	42.3	75.0
Sm	Samarium	150	5,800	23.3	41.2
Eu	Europium	152	4,600	18.2	32.3
Cd	Cadmium	112	2,450	13.1	23.3
Li	Lithium	6.94	70.7	6.14	10.9
Dy	Dysprosium	163	930	3.45	6.11
Ir	Iridium	192	426	1.34	2.37
Hg	Mercury	201	375	1.13	2.00
In	Indium	115	193	1.01	1.80
Rh	Rhodium	103	150	0.878	1.56
Er	Erbium	167	162	0.583	1.034
Cl	Chlorine	35.45	33.2	0.564	1.00
Co	Cobalt	58.9	37.2	0.380	0.674
Tm	Thulium	169	103	0.367	0.651
Ag	Silver	108	63.6	0.355	0.630
Sc	Scandium	45.0	26.5	0.355	0.629
Hf	Hafnium	178	102	0.344	0.610
Re	Rhenium	186	88	0.285	0.505
Lu	Lutetium	175	77	0.265	0.470
Nd	Neodymium	144	50.5	0.211	0.374
H	Hydrogen	1.008	0.332	0.198	0.352
Kr	Krypton	83.8	25.0	0.180	0.319
Mn	Manganese	54.9	13.3	0.146	0.259
Cs	Cesium	133	29.0	0.131	0.233
Yt	Ytterbium	173	36.6	0.127	0.226
Se	Selenium	79.0	11.7	0.0892	0.158
N	Nitrogen	14.01	1.85	0.0795	0.141
Ti	Titanium	47.9	6.1	0.0767	0.136
Ta	Tantalum	181	21.0	0.0699	0.124
W	Tungsten	184	18.5	0.0606	0.107
V	Vanadium	50.9	5.04	0.0596	0.106
Ni	Nickel	58.7	4.43	0.0454	0.0806
K	Potassium	39.1	2.10	0.0323	0.0573
Fe	Iron	55.9	2.55	0.0275	0.0488
Na	Sodium	23.0	0.530	0.0139	0.0246
S	Sulfur	32.1	0.520	0.00977	0.0173
Ca	Calcium	40.1	0.43	0.00646	0.0115
Al	Aluminum	27.0	0.230	0.00513	0.0091
P	Phosphorus	31.0	0.180	0.0035	0.0062
Si	Silicon	28.1	0.16	0.0034	0.0061
Mg	Magnesium	24.3	0.063	0.00156	0.00459
Be	Beryllium	9.01	0.0092	0.00061	0.00109
C	Carbon	12.0	0.0034	0.00017	0.00030
O	Oxygen	16.0	0.00027	0.0000102	0.000018

*1,000 σ_{mi} is equal to the number of c.u. contributed per gram of Element i per cc of bulk material.

Table 13–1. Thermal absorption cross section of elements ordered by the mass-normalized cross section. Courtesy of Schlumberger.

equivalent to the number of capture units (cu) contributed per gram of element per cm³ of bulk material.

It is seen that, in terms of the mass normalized absorption cross section, chlorine is quite prominent on a list of otherwise relatively scarce elements, at least for petrophysical applications. Exceptions are boron, which is often associated with clays, and cadmium, which is used in the construction of epithermal neutron detectors. The next most important and frequently encountered element is hydrogen, which has an atomic cross section two orders of magnitude less than chlorine. However, because of its concentration in water, it plays a greater role than most other elements.

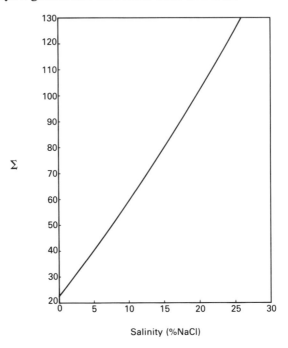

Figure 13–1. The macroscopic thermal neutron absorption cross section of water (in capture units) as a function of salinity. The salinity is given in weight percent of dissolved NaCl.

Figure 13–1, shows how the macroscopic cross section (in capture units) of saltwater varies as a function of the NaCl concentration. The relatively large capture cross section for fresh water (22 cu) is due primarily to the hydrogen, but the addition of NaCl, which has, in its crystalline form, a capture cross section of about 750 cu, increases the capture cross section of the liquid dramatically. Due to the lack of effective absorbers in most rock matrices, their capture cross sections are generally less than 10 cu. Thus the capture cross section of a formation will depend primarily on the salinity of the interstitial water, porosity, and water saturation. Water saturation is an important ingredient in the Σ of a formation, since the capture cross section of hydrocarbons is about the same as that of fresh water. Neutron die-away logs are generally presented on a scale of 0 to 60 cu, which reflects the anticipated range of variation of this formation parameter. A 30% porous formation containing only saturated saltwater (26% NaCl by weight), will have a Σ in excess of 40 cu, depending on the absorbers associated with the rock matrix. The presence of a significant oil saturation will consequently reduce the observed value of Σ.

Measurement Technique

In order to determine the macroscopic thermal cross section of a formation, the quantity that is measured is the lifetime of thermal neutrons in an absorptive medium. The practical realization of a device for this measurement is contingent on the availability of a pulsed source of high energy neutrons. This allows the periodic production of a population of neutrons whose absorption is then monitored.

The basic mode of operation consists of pulsing the source of 14 MeV neutrons for a brief period. This forms a cloud of high energy neutrons in the borehole and formation, which becomes thermalized through repeated collisions. The neutrons then disappear at a rate which depends upon the thermal absorption properties of the formation and borehole. As each neutron is captured, gamma rays are emitted. The local rate of capture of neutrons is proportional to the density of neutrons, which decreases with time. Thus the rate of production of capture gamma rays decreases in time. Measurement of the decay of the gamma ray counting rate reflects the decay of the neutron population. For this reason the logging tool generally consists of a detector of gamma rays in addition to the source of 14 MeV neutrons.

The operation of such a device is summarized in Fig. 13–2. Neutrons of high energy are shown, shortly after emission, forming a cloud of high energy neutrons in the first snapshot. As time passes, the neutrons make collisions, lose energy, and spread further from the point of emission, which is shown in the series of snapshots. The characteristic dimension of the cloud of neutrons, just prior to thermalization, is the slowing-down length. When the neutrons reach thermal energies and the cloud has attained a size which is related to the migration length, they begin to be absorbed through the thermal capture process. As time passes, the density of neutrons in the ever-expanding cloud decreases, depending upon the rate of absorption, as indicated in the sketch of the neutron density as a function of time in Fig. 13–2.

In a manner analogous to radioactive decay, the time-dependent behavior of the capture of thermal neutrons can be predicted. The reaction rate for thermal neutron absorption is given by the product of the macroscopic absorption cross section Σ_a and the velocity of the neutron v. So, for a system of N_N neutrons, the number absorbed in a time dt is:

$$dN_N = -N_N \Sigma_a v \, dt \, . \tag{1}$$

When integrated, this yields:

$$N_N = N_i \, e^{-\Sigma_a vt} \, , \tag{2}$$

which relates the number present at time t to the initial number N_i at time zero. The decay time constant is equal to $1/v\Sigma_a$.

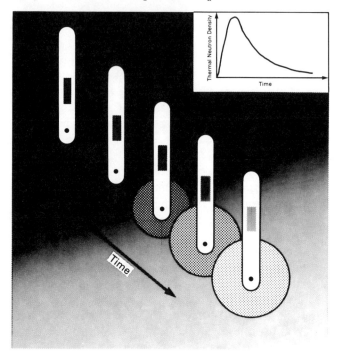

Figure 13–2. Evolution of the thermal neutron population produced by a pulsed neutron device.

The value of the capture cross section Σ_a is listed in Table 13–2 for a number of materials of interest. Included in the table is the decay time constant associated with each value of Σ_a. It has been computed from the relationship:

$$\tau_d = \frac{1}{v\Sigma_a} = \frac{4550}{\Sigma_a} \, [\mu s] \, . \tag{3}$$

The derivation of the decay-time relationship was based upon a simple model which starts with a cloud of thermal neutrons at time zero. It is of interest to know how much time elapses, after the burst of 14 MeV neutrons, before the cloud of thermalized neutrons is present. Estimates based on the average number of collisions required to reduce the energy of the neutrons to the thermal region and the variation of the mean free path between collisions indicate that the thermalizing time scale is on the order of 1 μsec, much shorter than typical decay times shown in Table 13-2.

Capture Cross Sections and Decay Times.		
MATERIAL	$\Sigma(cu)$	$\tau_d(\mu s)$
Quartz	4.26	1086
Dolomite	4.7	968
Lime	7.07	643
20 PU Lime	10.06	452
Water	22	206
Salt Water (26%% NaCl)	125	36

Table 13-2. Capture cross section and decay time constants for various materials.

The simple derivation of the capture gamma ray time-dependence misses an important aspect encountered in actual measurement: the thermal neutron diffusion effect. The diffusion of neutrons in a homogeneous medium arises from spatial variation in the neutron density (or flux). Intuitively, we feel that the thermalized neutron cloud will have such a spatial variation; initially, the flux of thermal neutrons will be highest near the source and will decrease with increasing distance from it. Physically, the diffusion of neutrons is to be expected since in regions of high flux the collision rate will be high, with the result that neutrons will be scattered more frequently toward regions of lower collision density. This results in a net current of neutrons from regions of higher flux to those of lower flux. The rate at which the diffusion occurs depends on the diffusion coefficient and the gradient of the neutron flux (see Chapter 12). The implicit assumption in the derivation of Eq. (2) was that somehow the global behavior of the neutron density could be monitored at all times. In this case, diffusion is of no consequence. However, in the actual logging application, decay of the neutron population is monitored only in the vicinity of the detector, and thus it is a local measurement.

At any observation point, the local thermal neutron density decreases because the neutrons are diffusing and being captured. To quantify the effect of the diffusion component on the local decay time constant, it is necessary to use the time-dependent diffusion equation (see appendix to this chapter). The result is that the apparent decay time of the local neutron population contains two components:

$$\frac{1}{\tau_a} = \frac{1}{\tau_{int}} + \frac{1}{\tau_{diff}}, \qquad (4)$$

where τ_{int} is the intrinsic decay time of the formation (i.e., that expected from global monitoring of absorption alone), and τ_{diff} is the diffusion time, which corresponds to the density reduction of neutrons as a function of time due to the diffusion of neutrons away from the center of the cloud. The value of τ_{diff} depends on the distance from the source emission point and the diffusion coefficient. The practical result of the diffusion effect is that, without correction, the measured Σ of a formation will appear greater than the intrinsic value due to the diffusion rate of the thermal neutron population in the vicinity of the detector. The effect will also be larger at low porosity, since the diffusion coefficient D decreases with increasing porosity (see Fig. 11–12).

Instrumentation and Interpretation

In logging devices, numerous schemes are used for controlling the period during which the 14 MeV neutrons are produced and the period during which the gamma rays are measured. Some devices use dual-detector systems in an attempt to correct for the disturbance that can be introduced by the borehole, as well as to provide an estimate of the porosity.

One of the challenges in making the thermal decay measurement is the separation of the borehole and formation signals. Although one would like to monitor the neutron population of the formation only, capture gamma rays from the borehole are in fact detected also. This component is of little interest for the determination of the formation properties. Figure 13–3 shows the approach used in one device.[1] A short time after the neutron burst, a series of timed gates accumulates the gamma ray counting rate as it decays.

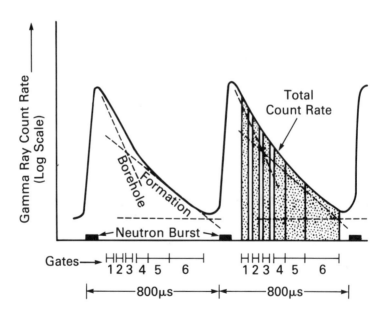

Figure 13–3. The gamma ray time distribution illustrates the decay of the borehole portion of the signal and the portion due to the formation. Adapted from Schultz et al.[1]

This cycle is repeated every 800 μsec. After 1250 of these cycles, a background signal is determined, which is then appropriately subtracted from the preceding sequence of pulsed measurements.

An example of the two-component behavior is shown in the data of Fig. 13-3. The decay rate of gamma rays is shown for two detectors at two spacings from the source. In this experimental setup the borehole absorption (Σ_{bh}) exceeds the formation absorption (Σ_{for}), giving rise to the rapid decay of the borehole component. Two exponentials have been fitted to the data. The curve with the longer decay time constant is related to Σ_{for} and could also have been determined from the composite decay curve by waiting a sufficient period of time (400 μsec in this case) before determining the rate of decay. In another situation in which the borehole fluid is fresh and the formation fluid saline, so that $\Sigma_{for} > \Sigma_{bh}$, the fastest-decaying component will be associated with the formation. In this case, waiting a sufficiently long time before determining the decay constant will result in obtaining only the Σ_{bh} value. In such a case, the decomposition of the decay curve into its components is seen to be an advantage.

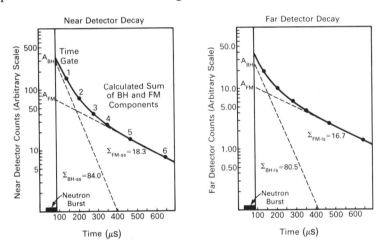

Figure 13-4. Laboratory data from a dual-detector device illustrating the two-component time decay. Comparison of the observed formation Σ with the intrinsic value of 15.7 cu shows that the diffusion effect is smaller for the far detector. Adapted from Schultz et al.[1]

The diffusion effect can also be seen in the data of Fig. 13-4. The 42-PU sandstone formation in which the measurements were made had a matrix Σ independently determined to be 11.1 cu. The borehole is filled with saltwater, and the actual Σ of the formation filled with fresh water is 15.7 cu. On the left, the nearer of the two detectors sees the formation component decay with an apparent Σ of 18.3 cu while, on the right, the detector farthest from the source yields an apparent Σ of 16.7. As expected, this latter value is quite close to the intrinsic value, but the effect of diffusion is still seen.

Table 13-3 compares some values of intrinsic and observed value of Σ with a particular measurement system. Note, in particular, the large discrepancies in the matrix materials; part are due to absorptive impurities in the samples and part to the diffusion effect. Other measurement systems with different detector spacings and timing schemes may be expected to have different diffusion components.

Intrinsic and Measured Σ		
Material	Σ_{int}(cu)	Σ_{meas}(cu)
Sandstone	4.32	8 - 13
Limestone	7.1	8 - 10
Dolomite	4.7	8 - 12
Anhydrite	4.7	8 - 12
Gypsum	18	18
Water (fresh)	22	22
Oil	22	16 - 22
Shale		20 -60

Table 13-3. Comparison of intrinsic and observed values of Σ. Adapted from Clavier et al.[10]

Regardless of the engineering details and variations in the pulsing modes, the devices described above are used for the determination of water saturation, particularly in cased wells. The most common important thermal neutron absorber is chlorine, which is present in most formation waters. Hence a measurement of the parameter Σ resembles the usual open-hole resistivity measurements. It can distinguish between oil and saltwater contained in the pores. If the porosity is known, gas/oil interfaces can be localized. When salinity, porosity, and lithology are known, the water saturation S_w can be computed.

Despite the complexity of the physics of the measurement and its engineering implementation, Σ has a particularly simple mixing law. In the simplest case of a single mineral, the measured value Σ consists of two components, one from the matrix and the other from the formation fluid:

$$\Sigma = (1-\phi)\Sigma_{ma} + \phi\Sigma_f .$$

This simple equation can be written because we are dealing with macroscopic cross sections; by definition, they combine volumetrically. To determine water saturation, the fluid component is broken further into water and hydrocarbon components:

$$\Sigma = (1-\phi)\Sigma_{ma} + \phi S_w \Sigma_w + \phi(1-S_w)\Sigma_h .$$

The graphical solution of this particularly simple equation for S_w is shown in Fig. 13-5. In order to use this approach, the values of Σ_w, Σ_h, and Σ_{ma} must be known or determined from logs. The presence of a water zone in the logged interval simplifies this task.

The presence of shale, which may contain thermal absorbers such as boron, seriously disturbs this simple interpretation scheme, but a number of references indicate methods for dealing with the problem.[2] For saturation determination, the interpretation of the measurement is questionable if the water salinity is less than 100,000 ppm and if the porosity is below 15%. This is particularly true in shaly zones, due to the limited range of Σ-variation with saturation.

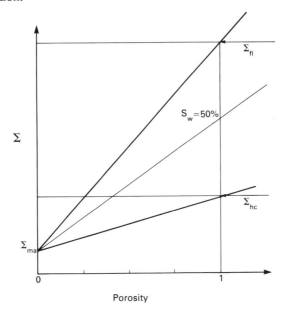

Figure 13–5. A graphical solution of water saturation from the measurement of Σ, when porosity is known.

Perhaps the most successful application of this type of measurement is in the time-lapse technique. In this procedure, the change in saturation between two runs in a producing reservoir can be determined directly from the difference between the two measured Σ values, the difference in Σ_w and Σ_h, and the porosity value, i.e.:

$$S_{w1} - S_{w2} = \frac{\Sigma_1 - \Sigma_2}{\phi(\Sigma_w - \Sigma_h)} .$$

Uncertainties in the quantities such as Σ_{ma}, Σ_{cl}, and clay volume disappear in this differential measurement technique.

To illustrate the behavior of a pulsed neutron log, refer to Fig. 13–6, which shows carbonate and shaly sand sections of the simulated reservoir. Notice that in this presentation format the Σ and GR anticorrelate. Thus, high capture cross sections correlate with high GR and thus, presumably, clay. Also the presentation of the Σ curve is in the same manner as other porosity tools (increasing to the left, or to the bottom, depending on how you hold the log). Also it indicates higher water saturation in a manner similar to the

292 Well Logging for Earth Scientists

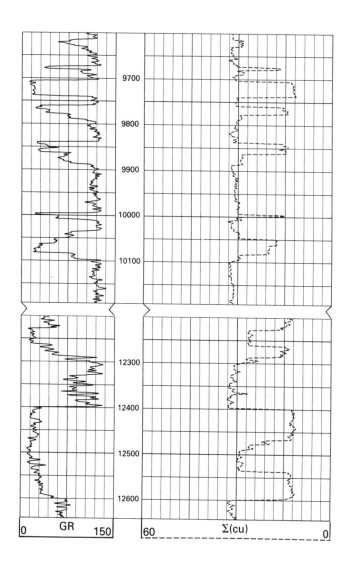

Figure 13–6. A typical presentation of a Σ log in the simulated reservoir.

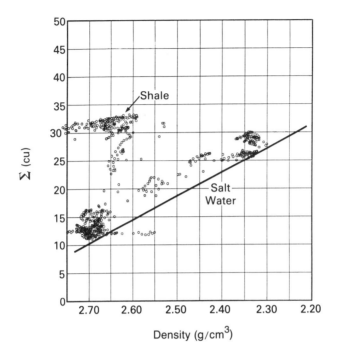

Figure 13–7. A cross plot of Σ and ρ_b from a zone of the simulated reservoir. Since the interval used contains no hydrocarbon, the determination of Σ_w is possible.

resistivity logs. The zone below 12200', in Fig.13–6, has been plotted as a function of the corresponding density values and shown in Fig. 13–7. A trend line of probable 100% water saturation is shown. The very low porosity portions are easy to identify, as is the cluster of "shale" points.

PULSED NEUTRON SPECTROSCOPY

Induced gamma ray spectroscopy tools are a slightly more complex family of pulsed neutron tools. They exploit the identification of gamma rays resulting from neutron interactions with the formation nuclei rather than simply the time behavior of the detected gamma ray flux. These reactions are of the three general types discussed previously: capture, inelastic, and particle reactions. In all three cases, the resultant nucleus is left in an excited state which decays, yielding gamma rays with energies which are characteristic of the particular nucleus which has undergone de-excitation. In one type of tool, inelastic interactions are separated from captures by timing.[3,4] The inelastic interactions, which occur only above an energy threshold that varies from nucleus to nucleus, are produced by relatively high energy neutrons. Thus gamma rays detected during the neutron burst and immediately thereafter, before the neutron energy has fallen too far, are most likely to be inelastic. At sufficient time after the neutron generator is turned off, capture gamma rays will begin to appear as the neutrons thermalize.

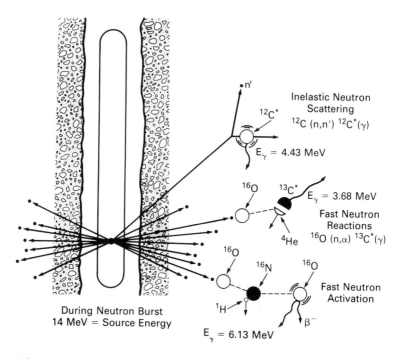

Figure 13–8. Inelastic reactions of interest in borehole logging applications. Adapted from Hertzog.[4]

Figure 13–8 schematically indicates the important reactions which are observed during the inelastic phase of operation of a pulsed neutron spectroscopy device. In one case, carbon is excited by the fast neutron and emits a gamma ray of 4.43 MeV. Two other kinds of reactions with oxygen are shown. In one, the oxygen is transmuted to nitrogen. It subsequently decays and emits a gamma ray of 6.13 Mev. The third case is more complicated, in that the oxygen is transmuted to ^{13}C, which then emits a gamma ray of 3.68 MeV.

An actual gamma ray spectrum in the inelastic mode, taken in an oil-filled sandstone, is shown in Fig. 13–9. With the help of the notations in the figure, it is possible to identify the peaks of oxygen and carbon; the distortion results from the use of an NaI detector. Because of this distortion, standard spectra are carefully determined in the laboratory for elements with significant gamma ray emission spectra. The set of inelastic standards used in the analysis of a measured spectrum is shown in Fig. 13–10. In order to quantify the relative amounts of carbon and oxygen (and other elements), the spectrum

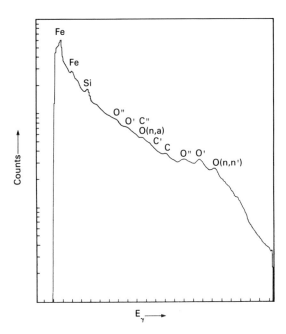

Figure 13–9. An experimental determined spectrum produced by inelastic reactions with C and O in an oil-filled sandstone. Adapted from Hertzog.[4]

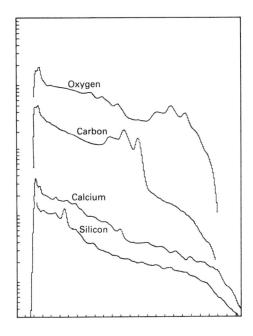

Figure 13–10. Standard spectra for determining the relative signals due to carbon and oxygen in the spectrum of Fig. 13–9. Adapted from Hertzog.[4]

measured in the borehole is compared to a linear sum of weighted standard spectra. The weights applied to each of the standards is varied until the sum is, in a least-squares sense, the best fit to the observed spectrum. The weights then represent the relative concentration of the elements in the standard set. Another measurement system takes a more direct approach and uses the counting rate in windows at appropriate locations in the spectrum to determine the C/O ratio.[5]

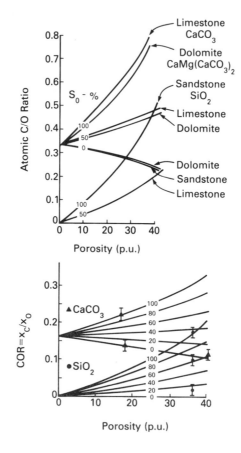

Figure 13–11. The ratio of elemental carbon to oxygen as a function of porosity and oil saturation compared to the tool response COR. Adapted from Hertzog.[4]

The interest in making the inelastic measurements (which is the only way to produce gamma rays from carbon and oxygen) can be understood from a glance at Table 13–4. There is a great contrast between the atomic carbon and oxygen densities in water and in oil. The upper portion of Fig. 13–11 shows the effect of oil saturation on the elemental C/O ratio as a function of porosity for limestone, dolomite, and sandstone. The lower portion illustrates

	Bulk Density	Atomic Densities [6.023×10^{23}]	
Formation	Density (g/cc)	Oxygen (/cc)	Carbon (/cc)
Limestone	2.71	0.081	0.027
Dolomite	2.87	0.094	0.031
Quartz	2.65	0.088	--
Anhydrite	2.96	0.087	--
Oil	0.85	--	0.061
Water	1.0	0.056	--

Table 13–4. Comparison of atomic densities of carbon and oxygen (in units of Avogadro's number) in formation and fluids of interest. From Westaway et al.[3]

the actual tool response; the relative spectral weights of carbon and oxygen are shown as a function of porosity and oil saturation. From this figure, it is clear that for clean formations of a given lithology the interpretation is relatively straightforward. However, it can be complicated, for example, in a calcite-cemented sandstone which could be confused with the presence of hydrocarbon. Also, an inherent difficulty in the measurement is immediately obvious. At low porosities the dynamic range of the C/O ratio, as a function of water saturation, shrinks to zero. Examples of interpretation of this ratio in more complex situations can be found in References 3 and 5.

Another mode of operation requires waiting sufficiently long after the burst to avoid the detection of the inelastic events. In this time regime, the gamma rays result from the capture of thermal neutrons. The spectra measured in this mode are much like those shown in Fig. 13–9. The identification and quantification of elements from the capture reactions is carried out by the use of standards in the same manner as described for the inelastic spectra.

In one type of logging equipment,[4] the timing procedure for making inelastic and capture measurements is shown in Fig. 13–12. Its features include inelastic data acquisition during the neutron burst, a background measurement period immediately following, and then, after the neutrons are thermalized, a phase that records the capture gamma rays.

In addition to providing a direct saturation measurement, the induced gamma spectroscopy tools are also very important for lithology determination in complex situations. From induced capture gamma rays, a large number of elements can be identified: hydrogen, silicon, calcium, iron, sulfur, chlorine, and others. Lithology identification can be made by comparing the yields of particular elements. For example, anhydrite is easily identified by the strong gamma ray yield from the sulfur and calcium comprising this mineral. Limestone can be distinguished from sandstone by comparing the silicon and calcium yields. Numerous references show examples of this type of procedure, as well as interpretation schemes proposed by some service companies.[6,7] Chapter 19 further discusses the application of this measurement to the determination of formation mineralogy.

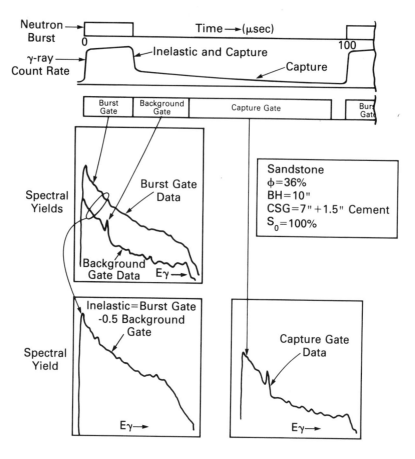

Figure 13-12. Illustration of the timing requirements for determining gamma ray spectra due to inelastic and thermal capture events. Included are two typical spectra obtained from the two phases measured in an oil-saturated sandstone. Adapted from Hertzog.[4]

Although the foregoing description of the capabilities of inelastic and capture spectroscopy sounds very promising, there are a few problems inherent to making this type of measurement in the borehole. One of the most important is that introduced by statistics. In the inelastic mode, the production of gamma rays characteristic of carbon and oxygen is governed by the cross sections, which are extremely small. Thus for realizable neutron generator outputs, the detected gamma ray flux is much smaller than desired for reliable continuous logging. Another problem is associated with the distortion in the gamma ray spectrum produced by the NaI detector. This type of detector does not have adequate resolution to do more than a crude analysis of the wealth of gamma rays emanating from a typical formation.

High resolution spectroscopic devices, on the other hand, may provide enormous advances in mineralogical identification. Such devices use solid state gamma ray detectors having energy-resolving power, orders of magnitude better than NaI. To illustrate the energy resolution afforded by the Ge detector, see Fig 13–13, which shows the spectrum obtained with a Ge detector in an ore sample containing TiO_2 at a concentration of 0.2% by weight. In addition to the gamma rays associated with Ti, a large number of other isotopes have been identified. It can be seen that the identification and quantification of the multitude of gamma ray lines is clearly one of the problems of this future well logging technique.

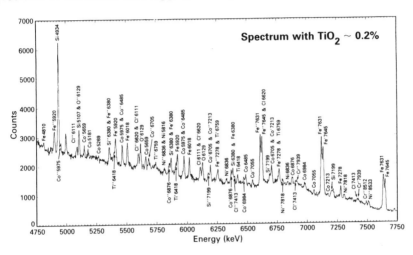

Figure 13–13. Spectrum measured from thermal neutron absorption in an ore sample containing 0.2% by weight of TiO_2. From Schweitzer and Manente.[9]

APPENDIX

A Solution of the Time-dependent Diffusion Equation

The time-dependent diffusion equation for the density of thermal neutrons, n, is given by:

$$\frac{\partial n}{\partial t} = S - nv\Sigma + \frac{D}{v}\nabla^2 n,$$

where v is the thermal speed. [Compare it to Eq. (1) of Chapter 12 which was for the steady-state case; the time derivative was zero.] After the burst, the source term becomes zero, and the governing equation is:

$$\frac{\partial n}{\partial t} = -nv\Sigma + \frac{D}{v}\nabla^2 n,$$

or

$$\frac{1}{n}\frac{\partial n}{\partial t} = -v\Sigma + \frac{D}{v}\frac{\nabla^2 n}{n}. \quad (A-1)$$

It is the term $\frac{D}{v}\nabla^2 n$ which is absent from the global analysis performed earlier. By comparing Eq. (A-1) with the result from the global model, an apparent decay time of the neutron population, τ_a, can be anticipated. The global behavior of the neutron population was given by:

$$N = N_o e^{-\frac{t}{\tau}} = N_o e^{-\Sigma vt},$$

from which the decay time, τ, can be identified as:

$$\frac{dN}{dt}\frac{1}{N} = -\Sigma v = \frac{1}{\tau}.$$

In this manner, the apparent decay time constant of the neutron population, τ_a, can be seen from Eq. (A-1) to be the sum of two terms: an intrinsic time constant and a diffusion time constant. The intrinsic time constant is given by:

$$\frac{1}{\tau_{int}} = v\Sigma,$$

and the diffusion time is defined as:

$$\frac{1}{\tau_{diff}} = -\frac{D}{v}\frac{\nabla^2 n}{n},$$

so that the local apparent time relationship is given by:

$$\frac{1}{\tau_a} = \frac{1}{\tau_{int}} + \frac{1}{\tau_{diff}},$$

which was given in Eq. (4). Thus the local decay rate contains two components: one related to the absorptive properties of the medium, and the second the result of neutron diffusion. In order to evaluate the diffusion component of apparent decay and to demonstrate that its value is position-dependent, a solution of Eq. (A-1) is required. In the case of a point source in an infinite homogeneous medium, it should bear some resemblance to the solution we saw in Chapter 12, but with an exponential time dependence with the characteristic time τ_a in it. The exact relation is found to be:

$$n(r,t) = n_o e^{-\frac{t}{\tau_{int}}} \frac{e^{-\frac{r^2}{4(L_s^2 + \frac{D}{v}t)}}}{(L_s^2 + \frac{D}{v}t)^{3/2}},$$

where the intrinsic decay time τ_{int} is given by $\frac{1}{v}\Sigma$.

Using this solution, the time constant τ_{diff} associated with the diffusion can be computed from:

$$\frac{1}{\tau_{diff}} = -\frac{D}{v}\frac{\nabla^2 n}{n}.$$

For the case of radial dependence only, the operator ∇^2 can be written as:

$$\nabla^2 = \frac{1}{r^2}\frac{\partial}{\partial r}\left(r^2\frac{\partial}{\partial r}\right).$$

Applying this to the solution given above, we find:

$$\frac{1}{\tau_{diff}} = \frac{3\frac{D}{v}}{2(L_s^2 + \frac{D}{v}t)} - \frac{\frac{D}{v}r^2}{4(L_s^2 + \frac{D}{v}t)^2}.$$

This demonstrates that the diffusion component of the apparent decay time will be dependent on the position of observation as expected. Also, since the neutron flux gradient will change with time, the rate of diffusion will decrease with time, which explains why τ_{diff} is not a time "constant."

The object of the die-away measurement is to determine the formation or intrinsic time decay constant. Since the expression for the apparent time decay of the neutron population contains the position, r, at which the observation is made, there seems to be some hope for locating the observation point at a position at which the apparent τ will equal the intrinsic value. This condition will be true at a distance r, given by:

$$r^2 = 6(L_s^2 + D/v\ t).$$

However, this expression contains L_s, which will vary with porosity and, of course, time. Thus no observation position will be entirely suitable. This may explain the controversy surrounding the magnitude of the diffusion correction;[8] it is tool-design dependent.

The appropriate spacing in 20-PU limestone for an observation time at 100 μsec can be estimated using a slowing-down length of 12 cm, and a diffusion coefficient, D of 0.75 cm, and the thermal neutron velocity of 0.22 cm/μsec. This yields a distance of about 55 cm. In the logging application, consisting of a borehole and formation, it is generally observed that the farther the spacing, the closer will the measured value of τ approach the intrinsic value (see Fig. 13–4).

References

1. Schultz, W. E., Smith, H. D., Jr., Verbout, J. L., Bridges, J. R., and Garcia, G. H., "Experimental Basis for a New Borehole Corrected Pulsed Neutron Capture Logging System," Paper CC, CWLS–SPLA Symposium, 1983.

2. Hoyer, W. A., ed., *Pulsed Neutron Logging*, SPWLA, Houston, 1979.

3. Westaway, P., Hertzog, R., and Plasek, R. E., "The Gamma Spectrometer Tool, Inelastic and Capture Gamma-Ray Spectroscopy for Reservoir Analysis," Paper SPE 9461, SPE Fifty-fifth Annual Technical Conference and Exhibition, 1980.

4. Hertzog, R. C., "Laboratory and Field Evaluation of an Inelastic-Neutron Scattering and Capture-Gamma Ray Spectroscopy Tool," Paper SPE 7430, Fifty-third Annual Technical Conference, SPE of AIME, 1978.

5. Oliver, D. W., Frost, E., and Fertl, W. H., "Continuous Carbon/Oxygen Logging: Instrumentation, Interpretive Concepts and Field Applications," Paper TT, SPWLA Twenty-second Annual Logging Symposium, 1981.

6. Gilchrist, W. A., Jr., Quirein, J. A., Boutemy, Y. L., and Tabanou, J. R., "Application of Gamma Ray Spectroscopy to Formation Evaluation," Presented at SPWLA Twenty-third Annual Logging Symposium, 1982.

7. Flaum, C., and Pirie, G., "Determination of Lithology from Induced Gamma Ray Spectroscopy," Paper H, SPWLA Twenty-second Annual Logging Symposium, 1981.

8. Hearst, J. R., and Nelson, P., *Well Logging for Physical Properties*, McGraw-Hill, New York, 1985.

9. Schweitzer, J. S., and Manente, R. A., "In-Situ Neutron-Induced Spectroscopy of Geological Formations with Germanium Detectors," *American Institute of Physics Conf. Proc.*, Vol. 125, 1985.

10. Clavier, C. L., Hoyle, W. R., and Meunier, D., "Quantitative Interpretation of Thermal Neutron Decay Time Logs," *JPT*, Vol. 23, 1971.

Problems

1. Figure 13–1 shows Σ as a function of salinity. Since Σ has a linear volumetric mixing law, how do you explain the nonlinearity shown in the figure?

2. Show that the constant of proportionality between τ and $\frac{1}{\Sigma}$ is 4330 μsec, by considering the speed of thermal neutrons and the dimensions of capture units.

3. Using the data of Table 13–1, compute Σ for NaCl whose density is 2.17 g/cm^3. Express the answer in capture units.

4. You are using a pulsed neutron device to log a well of a limestone section in a secondary recovery project. Fresh water injection wells are located in the vicinity. The original formation water was known to contain 120 kppm NaCl.

   a. Construct an interpretation chart to convert Σ to water saturation as a function of density in the limestone section.

   b. In the lowest section of the limestone, the porosity is 20 PU and the temperature is 100°C. The LLD reading is 2.75 $\Omega\cdot$M, and Σ is 12 cu. Calculate S_w from the laterolog and Σ. What is a reasonable explanation for the discrepancy?

   c. Quantify the cause of the discrepancy.

5. Figure 13–14 shows a cross plot of Σ and density in a sandstone reservoir containing a water zone, a hydrocarbon zone, and a gas zone. The points associated with the gas zone have been identified. Construct on the figure an interpretation chart similar to Fig. 13–5.

   a. What is the estimated value of Σ_{ma}?
   b. What do you estimate Σ_w to be?

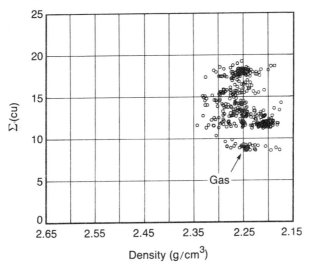

Figure 13–14. A cross plot of Σ and ρ_b for Problem 5.

6. In the well of Problem 5, the water resistivity is known to be 0.18 $\Omega \cdot$m in a zone at 115°F. Assuming that the water contains NaCl only, what value of Σ_w does this imply?

   a. What is the value of S_w in the water zone, using this new input?

   b. The induction log clearly sees this zone as 100% water; however, the drilling fluid is oil-based mud. What does this say about the relative depth of investigation of the Σ measurement?

14
NUCLEAR MAGNETIC LOGGING

INTRODUCTION

The subject of nuclear magnetic logging has been placed in this sequence since it concerns a fundamental nuclear property of matter: the nuclear magnetic moment. However, since the resonance technique for exploiting the existence of the nuclear magnetic moment involves electromagnetic techniques alone, it perhaps should be included as a final section on induction logging since its discoverers baptized it "nuclear induction."[1,2]

To introduce the subject of nuclear magnetic logging, rather than get involved with the powerful analytical technique of nuclear magnetic resonance used by physicists and chemists, we will be concerned first with the description of the proton magnetometer. This commonly used geophysical exploration tool will be seen to be directly related to one type of nuclear magnetic logging tool. A description of the magnetometer's physical principles will suffice to put the logging measurements into perspective.

Nuclear magnetic logging is used to determine the free fluid index (FFI) of a formation. This quantity, similar to porosity, is related to the density of protons which are free to reorient in response to an applied magnetic field. Thus hydrogen associated with shale or that in a layer of fluid close to the rock surface will be excluded from the measurement. It is a measure of the "free" fluid and thus is useful in predicting movable hydrocarbon, for example. In addition to FFI, details of the time dependence of the buildup of nuclear magnetization of the formation can yield parameters which can be used as good estimators of formation permeability.

For an understanding of the details of the logging measurement, a review of gyroscopic behavior and the interaction of magnetic moments with magnetic fields is included. A discussion of the earliest nuclear induction experiments introduces a description of the current logging tools.

NUCLEAR RESONANCE MAGNETOMETERS

Most nuclei have a magnetic moment. From the classical point of view each nucleus is equivalent to a tiny magnetic dipole. It is to be expected, then, that in the presence of an externally imposed magnetic field, the dipoles would tend to line up in the direction of the field lines. As we continue further with this classical description we must be note that each nucleus has, in addition to a magnetic moment, an angular momentum. The angular momentum can be described by a vector which is oriented along the axis of rotation. The magnetic moment and the angular momentum are coaxial.

Two important implications exploited in nuclear magnetic resonance measurements follow from these two properties of nuclei. The first is that the existence of the magnetic moment allows electromagnetic energy to be absorbed by the magnetic dipole, by changing the orientation of the magnetic dipole moment with respect to the external magnetic field. The second is that the existence of the angular momentum along the same axis as the dipole moment will tend to resist any change in the orientation of the angular momentum vector.

The interaction between the magnetic moment and the applied field produces a torque, which in turn produces a precession of the angular momentum vector about the axis of the applied field. This precession is analogous to the precession of a gyroscope whose vector of angular momentum is off-axis from the force of gravity. In the case of the nucleus, the precession frequency is governed by the intrinsic magnetic moment and the applied external magnetic field. It is known as the Larmor frequency.

Hydrogen, since it corresponds to a single proton, is the simplest example of a nucleus of a chemical element which possesses both spin and a magnetic moment. Oxygen, on the other hand, has no magnetic moment. Water is a substance which readily exhibits nuclear magnetic polarization when a field is applied to it, and it can be used as the sensitive element of a magnetometer.

The operating principle of the magnetometer consists of applying a magnetic field, roughly 100 times the magnitude of the earth's field, to the water sample, which may be simply a bottle of water. After a period of a few seconds, some of the magnetic moments of the protons are aligned with the external field, which is oriented nearly perpendicular to the earth's field. This alignment of proton spins and magnetic moments in the direction of the applied magnetic field produces a net magnetic moment in the water bottle.

When the applied magnetic field is removed, the induced magnetic moment will begin to precess about the remaining field, i.e., the earth's magnetic field. The frequency of precession is proportional to the local magnetic field strength. The precession of the induced bulk magnetic moment of the sample will induce a sinusoidal voltage in the same coil, as was used previously to establish the magnetic field. This effect is referred to as nuclear free induction. The measurement of the local geomagnetic field consists, then, of determining the frequency of the voltage induced in the coil.

At this stage some questions may come to mind. For example, what has this to do with a logging tool? Perhaps you are more curious about the operation of the magnetometer. Why does the polarizing field have to be applied for several seconds, rather than, say, an hour? Once the field is removed, does the Larmor precession continue indefinitely? How can this effect be exploited in logging, and what has it to do with obtaining an evaluation of a formation for the production of hydrocarbons?

WHY NUCLEAR MAGNETIC LOGGING?

As we have seen from the description of the proton magnetometer, there exists a very powerful technique for identifying the free precession of protons. Of what interest is this for well logging? The answer is more apparent if one knows that hydrogen is the only nuclear species encountered in earth formations that can be easily detected by the nuclear induction technique. The details of detectability will be discussed in a later section, but the first requirement is that a nucleus have a nuclear angular momentum and magnetic moment. Many of the common elements do not have sufficient numbers of isotopes which possess these attributes, notably carbon, oxygen, magnesium, sulfur, and calcium. However, iron and potassium have weak magnetic moments, as do sodium and aluminum. Nonetheless, these will be seen to be much less detectable than hydrogen.

For this reason, proton free-precession measurements in earth formations will reflect exclusively hydrogen. Because of the technique used to measure this free precession, the only hydrogen detectable will be that associated with liquids, either water or hydrocarbons. The measurement will not be sensitive to the hydrogen associated with shale.

It is not surprising, then, that one of the possible deductions from a nuclear magnetic measurement in a wellbore is related to the porosity of the formation. In itself this is not of great interest, since other porosity-sensitive measurement devices are available. However, it will be seen that the nuclear magnetic porosity estimate, or FFI, corresponds only to the portion of the fluid which is movable.

A special technique has also been evolved to measure, under the proper experimental circumstances, the residual oil saturation. This technique is extremely valuable in predicting oil left in place after water flooding, for estimates of secondary production efficiency and the feasibility of further recovery techniques.

We shall also see that some attempts have been made to correlate nuclear magnetic logging measurements with surface to volume measurements of porous samples. The motivation for seeking this correspondence is to be able to predict the permeability of the formation. More recent work has shown that the details of the buildup of nuclear magnetization of the formation can yield parameters which provide good estimators of formation permeability.

A LOOK AT MAGNETIC GYROSCOPES

The free precession of hydrogen nuclei in a magnetic field, exploited in nuclear magnetic logging, results from the fact that the nuclei possess both a magnetic moment and angular momentum. Before considering the case of the nuclei, let us start with the connection between angular momentum J, and the magnetic moment μ, for the case of a circulating electron at some distance r from the center of its orbit, shown in Fig. 14–1. In this circular path, its instantaneous velocity is v, and the angular momentum of this system is perpendicular to the plane of the orbit. It is given by:

$$J = m_e vr,$$

where m_e is the mass of the electron.

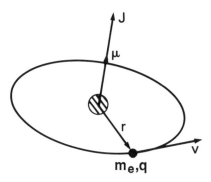

Figure 14–1. The angular momentum and magnetic moment of a charged particle in circular orbit. From Feynman.[13]

To compute the magnetic moment of this simple system, we use the expression for a circular current-carrying loop: It is equal to the current times the area of the loop. The current which is represented by the circulating charge is just the amount of charge per unit time passing any given point:

$$I = q \frac{v}{2\pi r}.$$

Since the area of the loop is πr^2, the magnetic moment is:

$$\mu = \frac{qvr}{2},$$

and is directed in the same direction as the angular momentum. Thus we can write:

$$\mu = -\frac{q_e}{2m_e} J ,$$

where q_e is the charge of the electron (taken to be negative). When the quantum mechanical description of the electron is used, the relation becomes:

$$\mu = -g \left[\frac{q_e}{2m_e}\right] J ,$$

where g is a factor characteristic of the atom.

Although the above calculation was for an orbital electron, it also holds for the case of a spinning charge distribution. For the pure spin of an electron, the g factor is 2. In the case of the proton, we could expect the magnetic moment to be given by:

$$\mu = g \left[\frac{q_e}{2m_p}\right] J .$$

However, for the proton the g factor is not 2 but rather 2.79 times greater. Frequently the constant $g\left[\frac{q_e}{2m_p}\right]$ is replaced by the symbol γ and is referred to as the gyromagnetic ratio.

The Precession of Atomic Magnets

The consequence of having a magnetic moment coaxial with angular momentum is that such an atomic particle will precess in an applied magnetic field. A magnetic field of strength B will exert a torque on the magnetic moment μ with a magnitude:

$$\tau = \mu \times B .$$

However, the torque will be resisted by the angular momentum, and instead a precession of the angular momentum vector will take place around the direction of the magnetic field vector, B, such that the rate of change of the angular momentum is equal to the torque applied. The position of the angular momentum vector is shown at two instants in time in Fig. 14–2 to illustrate the variables necessary to define the precessional angular velocity ω_L.

In a time Δt, the angle of precession is shown to be $\omega_L \Delta t$. From the geometry, the change of angular momentum J is given by:

$$\Delta J = (J \sin\Theta)(\omega_L \Delta t) .$$

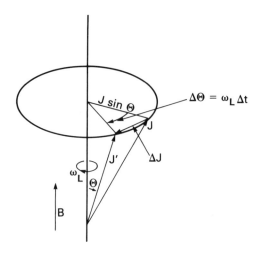

Figure 14–2. Precession of the angular momentum of an object with a magnetic moment subjected to a torque from an external magnetic field of strength B. From Feynman.[13]

The rate of change of the angular momentum is:

$$\frac{dJ}{dt} = \omega_L J \sin\Theta \;.$$

This must equal the torque ($\mu B \sin\Theta$) so that:

$$\omega_L = \frac{\mu}{J} B \;.$$

Since the ratio of μ to J is the gyromagnetic ratio, γ, the angular precession (or Larmor) rate is:

$$\omega_L = \gamma B \;.$$

When this is evaluated for a nucleus, the Larmor frequency is given by:

$$f_L = \frac{\omega_L}{2\pi} = .76 \, g \, B \; \left[\frac{\text{kHz}}{\text{gauss}}\right] .$$

For the case of the proton, g is about five, and the strength of the earth's magnetic field is about 0.5 Gauss, which means that the Larmor frequency is somewhere around 2000 Hz.

Paramagnetism of Bulk Materials

Consider a sample of some substance which has particles with magnetic moments, such as the hydrogen atoms in water. What happens, on a gross scale, when a magnetic field is applied? Before it is applied, presumably the orientation of the magnetic moments is at random. After the magnetic field is applied, there will be more of the moments aligned in the direction of the field than away from it, and the object as a whole will be magnetized to some degree.

The magnetization M is defined as the net magnetic moment per unit volume. So if each particle has an average magnetic moment $<\mu>$, the magnetization will be:

$$M = N<\mu>,$$

if there are N particles per unit volume. The net magnetization is proportional to the applied field. The constant of proportionality is known as the magnetic susceptibility. For an evaluation of the detectability of the induced magnetic moment, the derivation of the factors involved in the magnetic susceptibility is given below.

The quantum mechanical picture of spin assigns spin values to nuclear particles in multiples of one half. The second observation is that orientation of particles with spin and magnetic moment, in an external magnetic field, is limited to a number of discrete values equal to $2(I + ½)$, where the spin of the particle is I. Thus for a particle of spin 1, there are three possible orientations, as illustrated in Fig. 14–3. These multiple orientations will have an impact on the total magnetization, which can be induced by an external magnetic field.

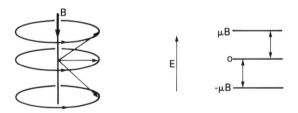

Figure 14–3. Possible orientations of a particle of spin 1 in an external magnetic field.

If we consider the quantum mechanical picture of the case of particles (like hydrogen) with spin ½ and a magnetic moment μ_o, then there are only two possible energy states: those with spins aligned, and those with spin and magnetic moment opposed to the imposed magnetic field. The energy associated with the alignment, E, is either $+ \mu_o B$ or $- \mu_o B$. According to the Boltzmann equation, the probability that an atom is in one state or the other is proportional to $e^{-(E)/kT}$. At normal temperatures, the probability of the occurrence of either state can be estimated. This is done by using only the first term of the expansion of the preceding exponential expression, coupled with knowledge of the fact that the number of atoms in either state, before application of the field, is about one half. Thus the probability of the occurrence of the upper energy state is $½(1 - \frac{\mu B}{kT})$, and the probability of the occurrence of the lower energy state is $\frac{1}{2}(1 + \frac{\mu B}{kT})$.

This can be done more exactly: In the applied magnetic field, the number of atoms with spin up is:

$$N_{Up} = ae^{+\mu_o B/kT},$$

and the number with spin down is:

$$N_{Down} = ae^{-\mu_o B/kT}.$$

The constant a is determined from the condition:

$$N_{Up} + N_{Down} = N,$$

where N is the total number of atoms per unit volume. From this we obtain:

$$a = \frac{N}{e^{+\mu_o B/kT} + e^{-\mu_o B/kT}}.$$

The average magnetic moment $<\mu>$ is given by the difference between the atoms aligned up and down:

$$<\mu> = \mu_o \frac{N_{Up} - N_{Down}}{N}.$$

This can be evaluated and expressed as M by:

$$M = N\mu_o \frac{e^{+\mu_o B/kT} - e^{-\mu_o B/kT}}{e^{+\mu_o B/kkT} + e^{-\mu_o B/kT}}.$$

Fig. 14–4 illustrates this behavior, which is seen to be linear for values of the interaction energy small compared to kT. Under this condition, the above expression can be reduced to:

$$M = \frac{N\mu_o^2 B}{kT}.$$

The constant $N\mu_o^2/kT$ is known as the magnetic susceptibility of the sample. Now we have obtained the formalism for estimating the induced magnetism for a substance placed in a magnetic field. In order to get an appreciation of the delicate nature of the nuclear induction technique, let us first look at the question of how many nuclei are actually participating in the production of the induced magnetic moment.

Consider the case of protons in a sample of water. Because of the presence of the applied field, a certain number of protons will be in the up state and a certain number in the down state. The ratio of these two numbers can be found from the Boltzmann distribution to be:

$$\frac{N_{Up}}{N_{Down}} = \frac{e^{\mu_o B/kT}}{e^{-\mu_o B/kT}}.$$

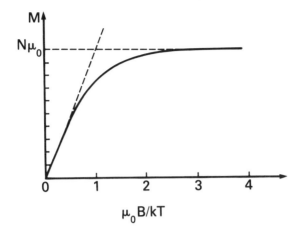

Figure 14-4. The variation of induced magnetic moment as a function of the magnetic field strength. From Feynman.[13].

In the case of a small magnetic interaction energy, compared to the thermal energy kT (which is always the case except at very low temperatures), this reduces to:

$$\frac{N_{Up}}{N_{Down}} \approx 1 + 2\frac{\mu_o B}{kT}.$$

If this is evaluated for a relatively large magnetic field (10,000 Gauss) at room temperature, we find that about 10 out of 10^6 nuclei are contributing to the internal polarization of the sample. A rough calculation for water indicates that the induced field is about 10^{-6} Gauss. Thus it can be seen that the direct detection of the magnetic moment will be somewhat difficult. It is for this reason that resonance electrical methods are used.

The next section examines the sensitivity of detection in the case of nuclear induction. It will give a better idea of why the only nuclei of practical detectability in borehole experiments are the protons of hydrogen.

SOME DETAILS OF NUCLEAR INDUCTION

We saw earlier that the application of a magnetic field produces an induced magnetic moment in the sample which is proportional to the applied field strength B. The mechanism behind this induced magnetic moment is the redistribution of the protons between higher and lower energy states. But how quickly is this redistribution of states achieved?

To examine this process qualitatively, let us define n as the difference between the number of states per unit volume in the upper state and those in the lower energy state:

$$n = N_{Up} - N_{Down}.$$

We now assume some probabilities, P_+ and P_-, which give the probability per unit time that a nucleus will make an upward or downward transition. These two probabilities, yet undefined, can be shown to be unequal. This can be seen by considering the condition of equilibrium which will be attained at some later time; the transitions between upper and lower levels must be equal. This can be expressed as:

$$P_-N_{Up} = P_+N_{Down},$$

or by taking the ratio:

$$\frac{P_+}{P_-} = \frac{N_{Up}}{N_{Down}}$$

$$\approx 1 + 2\frac{\mu_o B}{kT}.$$

Consider now the approach to equilibrium: An upward transition increases n by 2, and a downward transition decreases it by 2. Thus on a unit time basis we can write:

$$\frac{dn}{dt} = 2N_{Down}P_+ - 2N_{Up}P_-.$$

If we now replace the two probabilities, which already differ from one another by so little, with a suitable average P, we have:

$$\frac{dn}{dt} = -2P(N_{Up} - N_{Down}) = -2Pn.$$

The solution to this is simply:

$$n = n_o e^{-t/2P} = n_o e^{-t/T_1},$$

where the time constant T_1 is referred to as the spin-lattice relaxation time. Fig. 14–5 shows this behavior, along with a cartoon of the classical view of the polarization. The bulk nuclear magnetization of the sample is represented by a spinning gyroscope, oriented at right angles to the applied magnetic field. As time passes, the gyroscope is seen to align in the direction of the applied field. The curve at the bottom represents a measurement of the induced magnetization in the direction of the applied field. At the onset of the experiment there is no net magnetization in the direction of the applied field, because of the random orientation of the magnetic moments. The parameter T_1 quantitatively describes the rate of change of magnetization in a sample, which can only occur as the nuclei give up or absorb quanta of energy from the surroundings, such as the kinetic energy of other molecules. The surroundings which can absorb or transmit this energy are collectively called the lattice, since the earliest workers in this field were solid state physicists.

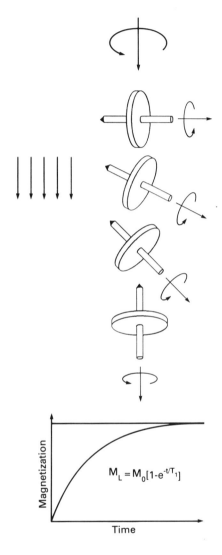

Figure 14–5. Illustration of the T_1 behavior which characterizes the rate of change of magnetization of a substance with time in the presence of a constant external field. The magnetic gyroscope is shown to come into alignment with the external field with a characteristic time constant known as the spin-lattice relaxation time.

The value of the spin-lattice relaxation time varies considerably with the type of nucleus and environment. It can yield information concerning the environment. For our primary interest, the proton, the nuclear spin can only be coupled to the environment through a local fluctuating magnetic field. Such a fluctuating field may be caused by rotational or translational motion of the molecules, which alters distances to other magnetic moments in the material. For liquids, the value of the spin-lattice relaxation time can vary between 10^{-2} and 10^2 sec, although the presence of paramagnetic ions in the solution can reduce it considerably. For the case of pure water, it is on the order of 2 to 3 seconds.

In addition to the spin-lattice relaxation time, T_1, another time constant of interest, can be defined. It is related to the dephasing of the nuclear spins that are the result of local field inhomogeneities. To understand this phenomenon, we look first at a particular experimental procedure of the nuclear induction measurement. As indicated in Fig. 14–6, a spherical sample is placed in a large uniform steady-state magnetic field. After a time longer than T_1, the magnetic moment M of the sample will line up with this constant field. We have already seen that the magnitude of this induced magnetic field is rather small and that to observe it another technique will be needed.

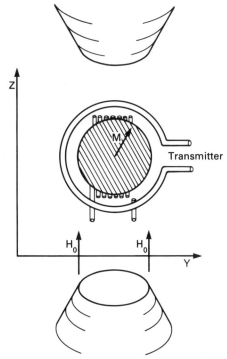

Figure 14–6. Experimental setup for observing the decay of nuclear induction. Adapted from Bloch.[1]

One of the common laboratory techniques for observing the induced magnetic moment is to produce a much weaker alternating magnetic field at right angles to the principal field. The frequency of this alternating field (produced by a coil, encircling the sample, as indicated in the figure) is chosen to be exactly the Larmor frequency of the magnetic moments in the principal field H_0.

To see the effect of this second, alternating field, it is convenient to make use of a rotating coordinate system which has an angular frequency equal to the Larmor frequency. In this system, initially, the sample magnetization is aligned with the z-axis as shown, on the left, in Fig. 14–7. The alternating magnetic field produced by the additional coil will appear stationary in the new frame of reference, illustrated on the right.

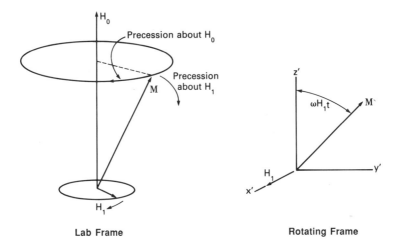

Figure 14–7. Two views of the motion of the magnetization vector M produced by the stationary polarizing field H_0 and perturbed by the rotating field H_1. In the rotating frame, M is seen to rotate away from the z'-axis linearly with time.

Viewed in this frame, the magnetic vector M will begin to rotate about the new applied field and to deviate from its initial direction. Of course, the deviation from its initial z-axis direction will induce Larmor precession, but in the rotating frame of reference it will appear stationary and will just begin to tilt. The tilting, in fact, is due to the magnetic vector trying to precess (although at a much lower frequency due to the weak oscillating field strength) around the new field. The rotation of M away from the z-axis can be controlled by the strength and length of time during which the oscillating field is left on.

In order to maximize the signal from the precessing bulk magnetic moment, M, the rotation angle is made 90°, as shown in the upper portion of Fig. 14–8. The magnetization vector, seen in the laboratory frame, below, begins to rotate in the x-y plane as indicated. If this magnetization vector is contained within a coil, it will induce, because of the changing flux linkage, an alternating signal. Now the induced signal in the coil should ideally pick up a sinusoidal signal as M rotates in the x-y plane, which gradually decays with a time constant T_1 as M begins to align, because of thermal relaxation, with the z-axis. However, it is observed to decay with another time constant T_2^*, the result in part, of the dephasing of the spins. One of the mechanisms for this behavior is the so-called spin-spin interaction. The other, an often more important reason in rock samples is the result of slight inhomogeneities of the primary field H.

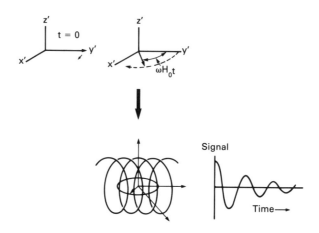

Figure 14-8. The condition for obtaining maximum signal from the nuclear induction experiment. The magnetization vector has been rotated by 90° away from the z' axis. The coil which contains the rotating system will have a decaying sinusoidal voltage induced as sketched. Adapted from Fukishima and Roeder.[14]

The source of the decaying field can be seen by considering the state of the magnetized vector M in the rotating system after the application of the 90° pulse. It is shown, viewed from above, in three panels of Fig. 14-9. The first shows the x-y projection of the magnetization vector M before the pulse. As it is oriented along the z-axis, it has no projection in this plane. The second panel shows the vector at the instant it has been rotated by 90°. As it is viewed in the rotating frame, it appears stationary. However, if there are slight field inhomogeneities at various parts of the specimen, then each of these may have a slightly different precessional frequency. Some will rotate faster than others in this plane, and in time, the spins from these various regions will tend to separate, as shown in the third panel.

After sufficient time, these spins will be completely dephased and the sinusoidal signal in the pickup coil will have completely disappeared. The waveform in Fig. 14-8 shows a schematic representation of the free precession decay. The envelope will often appear as an exponential decay with a characteristic time constant T_2^*. This decay time will be composed of the spin-spin interactions (T_2) as well as a component due to local field inhomogeneities. At the same time, since the primary magnetic field has remained on, the z-component will increase from its initial value of zero to a final value of M, with a characteristic time constant T_1.

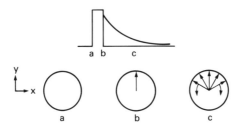

Figure 14-9. Behavior of the components of the magnetization vector M during the three phases of free induction decay.

It is of interest to calculate the detection sensitivity of such a measurement. For simplicity, we take the case of a coil enclosing a sample in a uniform magnetic field of strength H_o, where a magnetic moment M has been produced in the sample at right angles to the field H_o. The flux F, which will be intercepted by the coil as the magnetic moment M rotates at the Larmor frequency, will depend on the coil details: its area, A, and number of turns, N_t. The flux may be written as:

$$F \propto N_t \, A \, M \, \sin(\omega t) .$$

From the law of induction, the voltage, V, induced in the coil, will be proportional to the time rate of change of this flux:

$$V \propto \frac{dF}{dt} \propto N_t \, A \, M \, \omega \, \cos(\omega t) .$$

The magnitude of M in the pulsed experiment described above does not depend on the strength of the polarizing pulse which rotated the magnetization vector but instead is just proportional (where the constant of proportionality is the magnetic susceptibility, χ) to the initial polarizing field H_o. Thus the signal induced can be written as:

$$V \propto N_t \, A \, \chi \, H_o \, \omega \, \cos(\omega t).$$

The quantum mechanical version of susceptibility depends on the spin I and the gyromagnetic factor γ (accounting for the multiple orientations of the magnetic moment and consequent decrease of magnetization). It is given by:

$$\chi = N \frac{I(I+1)}{3kT} (\gamma \hbar)^2 .$$

Substituting the frequency relation for H_e (i.e., $\omega = \gamma H_o$) gives the final result for the voltage:

$$V \propto N_t \, A \, N \frac{I(I+1)}{3kT} \hbar^2 \gamma \omega^2 \cos(\omega t) \, . \tag{1}$$

An evaluation of Eq. (1), along with a few more details, indicates that a sample of 2 cm^3, completely enclosed by the detection coil, will generate a signal of a few millivolts for a strong field of 1800 Gauss.

It is the relationship in Eq. (1), the product of $\gamma \omega^2$, which is used to establish sensitivity charts such as that of Table 14–1. The only free ion of any consequence (in solution) is that of Na associated with saltwater. From the table it can be seen to be about 30% more detectable than hydrogen at a constant frequency (and consequently requiring a polarizing field strength about four times greater), on the basis of equal numbers of nuclei. Since the concentration of Na, even in highly salt-saturated solutions, is rather small compared to hydrogen, it would be of marginal detectability in the type of experimental setup described above. The sensitivity to Na of a logging tool, described in the next section, is even poorer because of its mode of operation.

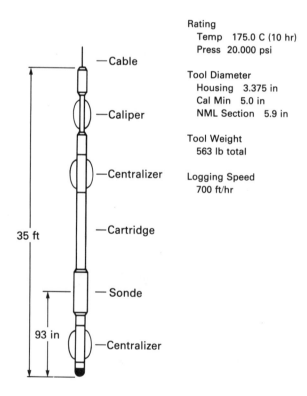

Figure 14–10. Tool configuration for a logging sonde which utilizes free induction decay. Courtesy of Schlumberger.

OPERATION OF A CONVENTIONAL NUCLEAR MAGNETIC LOGGING TOOL

Fig. 14–10, shows the basic elements of a conventional logging tool configuration.[3] Its active element consists of a coil through which a large amount of current is passed. Due to the nature of its winding, it produces a magnetic field roughly perpendicular to the earth's magnetic field. This serves to align a certain fraction of the protons in water, oil, and gas, within the depth of investigation, if the polarizing field is left on for a long enough time.

After a sufficient period (depending on the mode of tool operation), the current is turned off in the polarizing coil. The same coil is then used to receive the induced signal from the previously aligned protons as they precess at the Larmor frequency around the earth's magnetic field. The signal received has a frequency of about 2 kHz and is found to be damped with a time constant of roughly 50 msec, as opposed to the 2- or 3-sec damping which is observed in large volumes of pure water. Fig. 14–11 schematically illustrates, at several moments of elapsed time, the behavior of the magnetic moment induced by the polarization field H_p. It is seen to start out with the individual moments in phase. As time elapses, the phase begins to get out of step, and the overall vector begins to reorient itself along the earth's field, H_e.

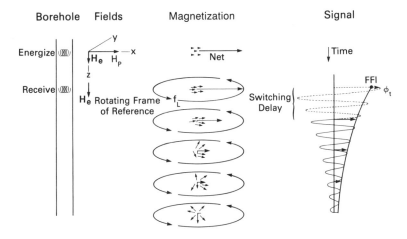

Figure 14–11. Schematic representation of the generation of the observed nuclear magnetic logging tool signal and its extrapolation to the start of the free induction decay. The portion of the signal during the switching decay is not actually observed.

The signal that is observed is indicated schematically in Fig. 14–12. Note that there is a delay of some 20 msec between the end of the polarizing pulse and the beginning of the observation. This is both an annoyance and a benefit. During this period, signals with very short T_2 components decay and do not affect the later measurements. These may be the result of the signal from the mud, which must be doped with magnetite (to keep the borehole hydrogen from overwhelming the measurement), or from the hydrogen in the shale or silt, or from very high viscosity fluid.

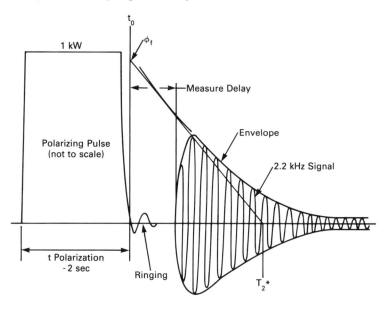

Figure 14–12. Extraction of the free fluid index, ϕ_f, from the free induction decay curve.

The annoyance comes from the uncertainty in obtaining the primary objective of the measurement, which is to extrapolate the envelope which is decaying with a time constant T_2^*, back to the end of the polarizing pulse. This is also shown in Fig. 14–12. The value of the envelope at this point is called the free fluid index, or FFI. After tool calibration and environmental corrections, it corresponds to the volume fraction of movable fluids in the formation.

Following from the previous analysis, the major factors concerning the signal strength from the logging tool can be described. The basic relation for the induced voltage is the same:

$$V \propto N_t \, A \, M \, \omega ,$$

where $\omega = \gamma H_e$. H_e is the strength of the earth's magnetic field. However, M is obtained by the polarizing pulse and will thus depend on the field H_p produced by the logging tool. Thus the final dependence for the signal strength is:

$$V \propto \frac{1}{kT} \gamma^2 \omega H_p = \frac{1}{kT} \gamma^3 H_e H_p . \qquad (2)$$

The field strength H_e is fixed by the earth's field, and H_p is limited by downhole power considerations. This indicates that the sensitivity to other isotopes will depend on γ^3. The data of Table 14–1 show that the sensitivity of this measurement to Na is about 1/10 that of hydrogen, under the condition of equal numbers of nuclei. The concentration differences actually encountered in borehole applications make it of little practical concern.

Isotope	v_o for 10,000-gauss Field (MHz)	Natural Abundance (%)	Relative Sensitivity for Equal Numbers of Nuclei		μ (in units of μ_N)	I (in units of $\hbar$)
			At Constant Field	At Constant Frequency		
n_1	29.167	—	322	.685	-1.91315	1/2
H^1	42.576	99.9844	1.000	1.000	2.79268	1/2
H^2	6.357	1.56×10^{-2}	9.64×10^{-2}	.409	.85738	1
B^{11}	13.660	81.17	.165	1.60	2.6880	3/2
C^{13}	10.705	1.108	1.58×10^{-2}	.251	.70220	1/2
N^{14}	3.076	99.635	1.01×10^{-3}	.193	.40358	1
N^{15}	4.315	.365	1.04×10^{-3}	.101	-.28304	1/2
O^{17}	5.772	3.7×10^{-2}	2.91×10^{-2}	1.58	-1.8930	5/2
F^{19}	40.055	100	.834	.941	2.6273	1/2
Na^{23}	11.262	100	9.27×10^{-2}	1.32	2.2161	3/2
Al^{27}	11.094	100	.207	3.04	3.6385	5/2
P^{31}	17.236	100	6.64×10^{-2}	.405	1.1305	1/2
Cl^{35}	4.172	75.4	4.71×10^{-3}	.490	.82091	3/2
Mn^{55}	11.00	—	.361	5.41	5.050	7/2
Co^{59}	10.103	100	.281	4.83	4.6388	7/2
Sn^{119}	15.87	8.68	5.18×10^{-2}	.373	-1.0409	1/2
Tl^{205}	24.57	70.48	.192	.577	1.6115	1/2
Pb^{207}	8.899	21.11	9.13×10^{-3}	.209	.5837	1/2
Free Electron	27.994	—	2.85×10^{-8}	658	-1836	1/2

Table 14–1. NMR properties of selected isotopes. Indication of relative detectability at constant magnetic field and at constant Larmor frequency are given. From Davis.[17]

Eq. (2) indicates that corrections will have to be made to the signal amplitude before the FFI can be extracted. The first correction deals with the magnetic inclination and is necessary when the tool's magnetic field and the earth's field are not at right angles. The signal amplitude is a function of the cosine of the magnetic declination. Another correction is for temperature: It has an impact on the magnetic susceptibility of the formation material. The borehole size will also influence the signal strength, since a relatively constant volume of material is "seen" by the receiver coil. The formation signal will be a maximum when the borehole size and tool diameter are of equal size, and it decreases (since no signal is allowed to come from the mud) as the borehole size increases. These corrections are applied to the FFI signal as part of the logging service.

Although the extraction of the FFI is the primary task of the nuclear magnetic logging tool, it will be seen later that the time constants T_1 and T_2 are also of some interest for other applications. For this reason there are three different modes of operation for the tool: All three provide an estimate of FFI.

The first to be considered is the T_2-continuous mode. In this case the tool is moved continuously at about 600'/hr, and a long polarizing pulse is applied to the coil. The pulse duration is about four or five times the expected value of T_1 and thus assures complete magnetic polarization of the formation. This is followed by a period in which the decaying Larmor precession signal is recorded continuously. The object is to extract, from these exponential decaying curves, the initial value of magnetization at time zero. This is the type of operation indicated in Fig. 14–12. An idealized log obtained in this mode is shown in Fig. 14–13. In addition to the extrapolated value of FFI, there are two additional traces: the observed value of T_2^* and the Larmor frequency.

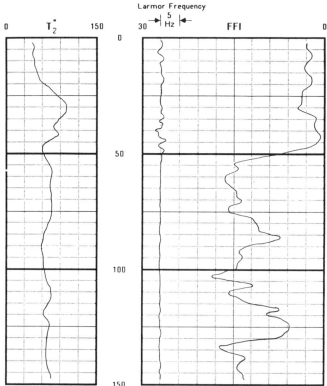

Figure 14–13. Example of an NML log presentation from the free induction decay mode of operation. The extrapolated free fluid index and the Larmor frequency are shown as continuous measurements, as is the apparent signal decay time constant T_2^*.

There are two additional modes which utilize T_1 rather than T_2^* to estimate FFI. The operating principle is shown in Fig. 14–14. The top figure shows a series of polarizing pulses applied to the formation, each of greater duration than the preceding. Between the application of the variable-width polarizing pulses, the free induction decay envelopes are recorded, as shown in the middle diagram. Extrapolating the series of decay envelopes, characterized by a time constant T_2^*, back to time zero sketches out the buildup envelope (FFE) of the magnetization, which has an associated time constant, T_1. In the T_1-continuous mode, three successive polarizing pulses are applied to the coil, each double the length of the previous. From the three decaying envelopes (extrapolated to the end of the polarization pulses), the growth curve of magnetization can be established, and instead the three estimates of FFI are obtained. A typical log output of this mode is shown in Fig. 14–15.

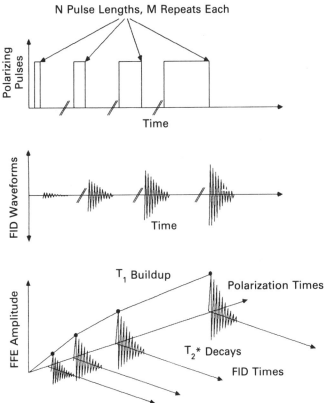

Figure 14–14. Method for determining the T_1 behavior of a formation by measuring the buildup of magnetization by extrapolating the T_2^* decay between polarization pulses of different duration. Adapted from Kenyon et al.[8]

For more refined measurements, a stationary T_1 mode can be used. In this case, the tool is positioned in the formation and is stationary during a succession of measurements. A sequence of multiple pulses, as well as repeats of the sequences, can be used to perform signal averaging. In the best of circumstances, the components of T_1 as well as T_2^* can be discerned.

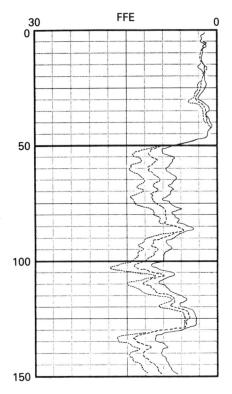

Figure 14-15. A log presentation derived from the T_1 mode of operation showing the approach to FFI from three portions of the buildup curve.

INTERPRETATION

Factors Affecting T_1 and T_2

The behavior of the thermal relaxation time, T_1, depends on proton mobility and the sources of local magnetic fields. For this relaxation to proceed there must be interactions between the spin and the lattice molecular motion. The relaxation time is long both for very low molecular mobility (i.e., a very viscous fluid) or when there is very high mobility. This latter case is the result of the interaction time, which is too short to affect the energy transfer. This behavior is illustrated in Fig. 14-16 which schematically depicts the behavior of T_1 (and T_2) as a function of the correlation time τ_c. The correlation time is related to the time scale of the fluctuating local magnetic fields within the sample. In this plot, water would be found to the extreme left and solids to the right.

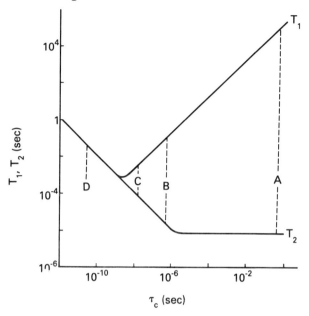

Figure 14–16. The behavior of T_1 and T_2 as a function of the correlation time. From Pople.[4]

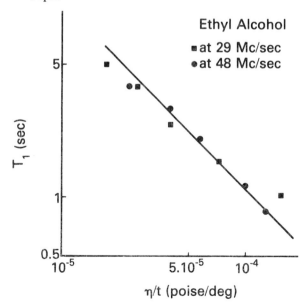

Figure 14–17. The variation of T_1 as a function of viscosity and temperature for ethyl alcohol. From Pople.[4]

It has also been found that $\frac{1}{T_1}$ should vary roughly as $\frac{\eta}{T}$ (viscosity/temperature) for a simple model based on Brownian motion or from the concept of diffusion time.[4] The behavior is illustrated in Fig. 14–17 for the case of alcohol. A more interesting case is shown in Fig. 14–18, which indicates the variation of T_1 for various crude oil samples of different viscosities.[5]

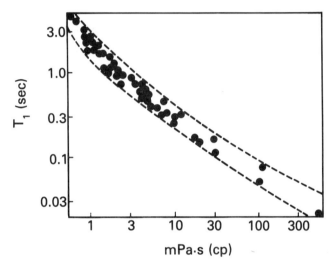

Figure 14–18. The measured relaxation times for crude oil samples as a function of their viscosity. From Brown.[5]

Up to this point in the discussion, the fluctuating magnetic fields that the protons experience are caused by variations in the intermolecular distances. The fields are provided by the other protons. But paramagnetic ions can be a more important source. In solution, paramagnetic ions can considerably shorten T_1 and T_2. Fig. 14–19 shows the variation in T_1 as a function of the concentration of three types of paramagnetic ions in a water solution.

Local magnetic fields at the surfaces of the rocks are a third major factor in the determination of the effective T_1 and T_2 for fluids in porous formation. If a proton is near a surface, there is a large probability that it will interact with paramagnetic sites on the surface and thus have its precessing phase disturbed. For this to happen, the proton need not be bound to the surface. The quantity governing the interaction is the fluid diffusion length ($\sqrt{DT_1}$) and its relation to the pore size.

Surface Interactions

It has been observed that the thermal relaxation time of water in porous media is decreased from the value expected in a bulk sample of water. In the case of uniform quartz grains, the measurement of T_1, shown in Fig. 14–20, indicates a reduction of about a factor of two from what was observed for the bulk sample of water. This is thought to be the result of an increase in the correlation time for the random motion of the water molecules as well as the presence of paramagnetic sites on the pore walls.

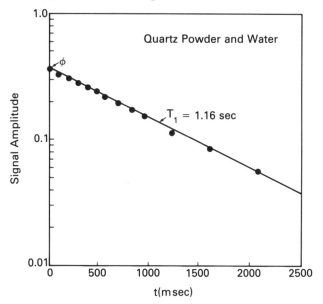

Figure 14–19. The change in relaxation time of solutions as a function of the concentration of paramagnetic ions. From Pople.[4]

A model (KST) developed to explain this type of observation considers the sample to consist of volumes similar to that indicated in Fig. 14–21.[6] In the central region of the pore space, the T_1 is as expected from the bulk sample. However, around the surface of the pore, to some thickness h, the relaxation time is a new value, r_s. In this model, the difference in relaxation rates between the observed T_1 and the value for bulk water (T_B) is attributed to the fractional volume contained in this surface layer:

$$\frac{1}{T_1} - \frac{1}{T_B} = \frac{V_s}{V_B} r_s . \qquad (3)$$

In this approach, a fairly solid relationship is established between the observed T_1 and the surface-to-volume ratio of laboratory samples. This is of interest in making an estimate of the permeability of the formation.

In contrast with Fig. 14–20, real rocks show a slightly more complicated behavior. Fig. 14–22 shows the results for a determination of T_1 for a sandstone. The quite apparent nonexponential decay has been decomposed into the sum of three different decays. The distribution of the values of the three components is related to the granular nature of the sample, perhaps to the pore size distribution.[7] A more recent study has shown that the relaxation curves can be much better described by a stretched exponential representation, one which allows a continuous variation of T_1 rather than a two or three component decay.[8]

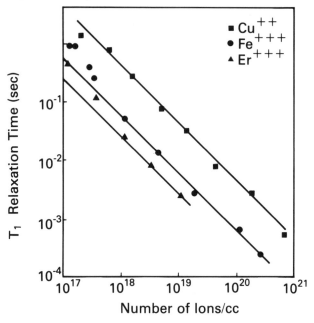

Figure 14–20. Measurement of the relaxation time of a water-saturated quartz powder. From Seevers.[10]

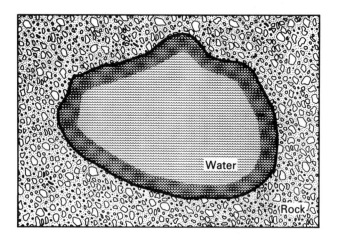

Figure 14–21. The KST model for the NMR behavior of liquid contained in a pore space. V_s is the volume fraction contained in the layer next to the pore surface, and V_B is the total fraction fluid volume.

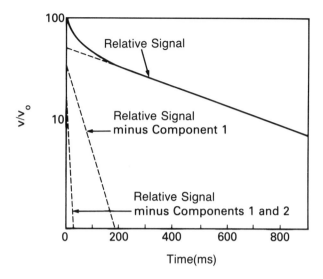

Figure 14-22. The measured relaxation time for a sandstone core sample. From Timur.[15]

APPLICATIONS

Free Fluid Index

The primary measurement for the nuclear magnetic logging tool is the volume fraction of free fluid (FFI). Since oil and water have about the same proton content, both species contribute to this measurement under usual logging conditions. The difference between porosity deduced from other logging methods (neutron-density, for example) and the FFI will indicate the volume fraction of irreducible water, ϕS_{wirr}. This is because the FFI is not sensitive (due to the dead time of 20 msec) to fluids bound to shales or matrix. In fact, one of the checks on the proper tool operation is that it gives an FFI of zero in shales.

Since the FFI is not sensitive to bound hydrogen, it can give fairly good porosity information in shales and hydrated matrices such as gypsum. The value of S_{wirr} finds use in empirical relations relating permeability to S_{wirr}.[9]

Permeability from Surface/Volume Information

A more direct method of using the nuclear magnetic logging tool for permeability is given by the work of Seever[10] and Timur.[11] The basis is the Kozeny expression for the permeability:

$$k = \frac{\phi^3}{T(1-\phi)^2 S^2},$$

where ϕ is porosity, S is the specific surface area (pore surface/volume of rock), and T is a textural or tortuosity factor.

The connection with nuclear magnetic logging comes from the KST model described earlier. It relates the observed spin-lattice relaxation time T_1 to the volume of water in intimate contact with the pore surfaces. Eq. (3) can be recast in terms of the Kozeny equation by noting that V_s, the fractional volume of water close (within a distance h) to the pore surface, is given by:

$$V_s = S(1 - \phi)h.$$

Since the interface fractional volume is assumed to be much smaller than unity, we can also let $V_B = \phi$. Thus the rate equation may be rewritten as:

$$\frac{1}{T_1} - \frac{1}{T_B} = \frac{S(1-\phi)hr_s}{\phi},$$

or

$$\frac{\phi}{(1-\phi)S} = hr_s \frac{T_1 T_B}{T_B - T_1}.$$

When this expression is squared and substituted into the Kozeny equation, it results in the following:

$$k = \frac{(hr_s)^2}{T} \phi \frac{(T_1 T_B)^2}{(T_B - T_1)^2},$$

so that permeability can be found as a function of ϕT_1^2. Recent work by Kenyon et al.[8] has shown that an improved estimator of permeability can be obtained from the quantity $\phi^4 T_1^2$.

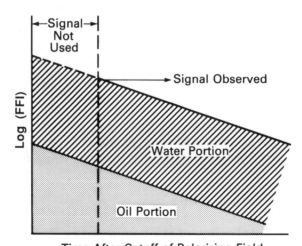

Figure 14–23. Visualization of the contributions to the nuclear magnetic logging signal from a formation containing a mixture of oil and water in the pore space. From Neuman and Brown.[16]

Measurement of Residual Oil

Nuclear magnetic logging can be used to measure residual oil in place if the signal from the water contained in the pores is "killed" in the same manner as the borehole signal. Normally the decaying signal is composed of two components in an oil/water mixture, as illustrated in Fig. 14–23. By using paramagnetic ions, such as manganese, which dissolve in water, the water decay time can be made short enough to have disappeared in the 20 msec dead time of the tool, as illustrated in Fig. 14–24.

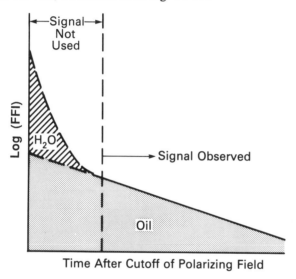

Figure 14–24. The signal resulting from the same oil-water mixture as Fig. 14–23, after the addition of paramagnetic ions to the mud system and the allowing of sufficient time for penetration into the formation. The water signal has been effectively "killed." From Neuman and Brown.[16]

The success of such an operation involves the use of a water-soluble compound which surrounds the manganese. Such a substance is known as EDTA. Once the EDTA is mixed into the mud system, the mud filtrate must penetrate some 4–5" into the permeable formation in order to kill the signal from the formation water. This involves a so-called wiper run to remove mudcake and allow invasion of the newly doped mud filtrate. Fig. 14–25 shows the log obtained before and after the addition of the manganese EDTA, compared to the total formation porosity. In the first logging run, the FFI is seen to give a porosity roughly 5 PU less than the total formation porosity. This difference is attributed to the bound fluid. In the second run, after the EDTA has presumably killed the formation water signal, the FFI indicates that portion of porosity which corresponds only to the residual oil. It is roughly constant, at 5 PU throughout the zone.

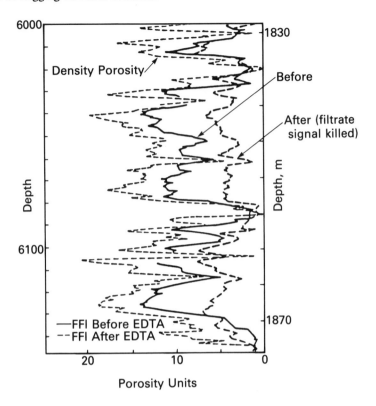

Figure 14-25. Successive FFI logs of an oil zone before and after the addition of paramagnetic ions to eliminate the water portion of the formation signal. From Neuman and Brown.[16]

As a final application of nuclear magnetic logging, we consider the case where resistivity cannot distinguish between low salinity water and oil. The identification of steam injection points can be obtained in this case from the FFI if the viscosity of the oil is high, because there is a general relation between T_1 (and T_2) and the viscosity (see Fig. 14-18). For heavy oils, the signal seen by the measurement will sense only the water. Thus the FFI can be used to pick zones which will produce excessive amounts of water.

A NEW APPROACH

As noted earlier, for the current logging tool to be able to see a signal from the formation, steps must be taken to "kill" the borehole signal. This is done by adding paramagnetic ions to the mud system and circulating it to produce a uniform mixture. Detractors of this method point out the expense and time involved in such a procedure.

A new tool concept which avoids these problems, in principle, has been proposed by Jackson.[12] His technique is called the inside-out NMR. Instead of using the earth's magnetic field for producing the precession of the protons, two opposed permanent magnets are located in the tool for this purpose. The opposition of these two dipole magnets produces a radial magnetic field in the plane halfway between the two magnets. The variation of the radial magnetic field at a distance r from the magnets is shown in Fig. 14–26 as a function of the magnet radius a and separation h. It is normalized by the axial field strength H_o of a single magnet.

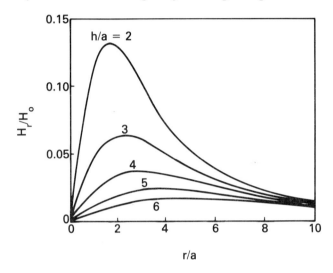

Figure 14–26. The radial variation of the magnetic field of two infinitely long opposed magnetic dipoles. The dimensionless radial variable is the distance from the center of the magnets divided by radius of the magnets. From Jackson.[12]

In order to get a reasonable signal, the field strength must be large and constant. It is clear from Fig. 14–26 that there will be a region of roughly toroidal form, around the sonde, in which the field is relatively constant. With this type of configuration, the more classical pulsed NMR experiments may be performed, and the signal should not be influenced by the borehole.

The idealized operation is shown in Fig. 14–27. An oscillating current in the coil, with a frequency of the Larmor frequency of the toroidal region, is used for the appropriate length of time to flip the net magnetization vector M_o by 90°. Once the flipping pulse is turned off (which can be done in 0.5 msec, as opposed to 20 msec in the conventional tool, since large polarizing signals are not needed), the coil functions as the receiver to record the precessing signal about H_r, as indicated in Fig. 14–27.

The primary advantage of such a system is the avoidance of mud doping. An additional aspect is the availability of the signal without delay. This enables a more precise determination of the FFI, since the uncertainties of extrapolating to the end of the polarizing pulse are avoided. The behavior of the T_1 curve can be examined at early times and deconvolved to give a pore size distribution.

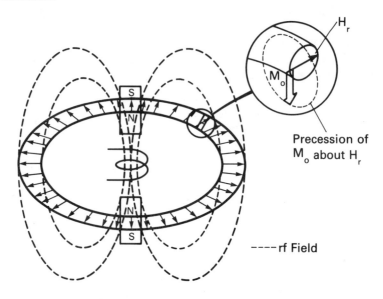

Figure 14–27. Illustration of the behavior of the induced magnetization vector after it has been rotated by 90°. From Jackson.[12]

The defects of the prototype tool center on the small signal strength, which requires lengthy data accumulation times. This is done to take advantage of signal averaging. Also the depth of investigation is rather small, and is linked to the size and strength of the permanent magnets which are already too large in diameter for practical applications. However, because of the promise that it holds, the technique is worthy of further instrumentation refinements.

References

1. Bloch, F., "Nuclear Induction," *The Physical Review*, Vol. 70, 1946.

2. Bloch, F., Hansen, W. W., and Packard, M., "The Nuclear Induction Experiment," *The Physical Review*, Vol. 70, 1946.

3. Herrick, R. C., Couturie, S. H., and Best, D. L., "An Improved Nuclear Magnetism Logging System and Its Application to Formation Evaluation," Paper SPE 8361, SPE Annual Technical Conference, 1979.

4. Pople, J. A., Schneider, W. G., and Bernstein, H. J., *High-resolution Nuclear Magnetic Resonance*, McGraw-Hill, New York, 1959.

5. Brown, R. J. S., "Proton Relaxation in Crude Oils," *Nature*, Vol. 189, 1961.

6. Korringa, J., Seevers, D. O., and Torrey, H. C., "Theory of Spin Pumping and Relaxation in Systems with a Low Concentration of Electron Spin Resonance Centers," *The Physical Review*, Vol. 127, 1962.

7. Loren, J. D., and Robinson, J. D., "Relations between Pore Size Fluid and Matrix Properties, and NML Measurements," Paper SPE 2529, SPE Annual Technical Conference, 1969.

8. Kenyon, W. E., Day, P. I., Straley, C., and Willemsen, J. F., "Compact and Consistent Representation of Rock NMR Data for Permeability Estimation," Paper SPE 15643, SPE Annual Technical Conference, 1986.

9. Timur, A., "An Investigation of Permeability, Porosity and Residual Water Saturation Relationships for Sandstone Reservoirs," *The Log Analyst*, July–Aug., 1968.

10. Seevers, D. O., "A Nuclear Magnetic Method for Determining the Permeability of Sandstone," Paper L, SPWLA Transactions, 1967.

11. Timur, A. T., "Nuclear Magnetism Studies of Carbonate Rocks," Paper N, SPWLA, 1972.

12. Jackson, J. A., "Nuclear Magnetic Well Logging," *The Log Analyst*, Sept.–Oct., 1984.

13. Feynman, R. P., Leighton, R. B., and Sands, M. L., *Feynman Lectures on Physics*, Vol. 2, Addison-Wesley, Reading, Mass., 1965.

14. Fukishima, E., and Roeder, S., *Experimental Pulse NMR: A Nuts and Bolts Approach*, Addison-Wesley, Reading, Mass., 1981.

15. Timur, A., "Pulsed Nuclear Magnetic Resonance Studies of Porosity, Movable Fluid, and Permeability of Sandstones," *Journal of Petroleum Technology*, Vol. 21, 1969.

16. Neuman, C. H., and Brown, R. J. S., "Applications of Nuclear Magnetism Logging to Formation Evaluation," *Journal of Petroleum Technology*, Vol. 34, 1982.

17. Davis, J. C., Jr., *Advanced Physical Chemistry, Molecules, Structure and Spectra*, Ronald Press, New York, 1965.

15
INTRODUCTION TO ACOUSTIC LOGGING

INTRODUCTION

Although the acoustic properties of rocks are of interest for their own sake, academic interest was not responsible for the development of acoustic borehole logging. The requirements of hydrocarbon exploration and evaluation were the stimuli for the introduction of this third category of physical measurements into well logging. Unlike resistivity measurements, which could be used directly for hydrocarbon detection, and nuclear measurements, which were initially directed at the determination of porosity, acoustic logging started out as a companion to seismic exploration. The first part of this chapter relates the history of how acoustic measurements finally came to be an integral part of wireline logging, after a shaky start as a seismic adjunct.

As a basis for understanding some of the applications of acoustic measurements, both realized and promised, a review of elastic properties of materials is presented. The relationship between the various types of elastic parameters used to describe materials is summarized and related to the velocity of propagation of shear and compressional waves in elastic media. A rudimentary logging device for the simplest and most common logging measurement, the interval transit time of an acoustic compressional wave, is presented, along with the basic empirical data relating transit time to porosity, which changed the entire application of acoustic logging. These data indicate the need for acoustic models of rocks to explain observations in detail.

A SHORT HISTORY OF ACOUSTIC LOGGING IN BOREHOLES

The use of acoustic energy to produce an image of the subsurface has had a long history. Reflection seismic prospecting consists of using a low frequency acoustic source at the surface to create down-going pulses of energy which are partially reflected by layers more or less directly below the source. A surface array of detectors, usually close to the source, detects the reflected waves. This technique was the outgrowth of a number of years of refraction seismic work, in which the source and receivers were separated by large distances compared to the depth of reflectors. Some of the earliest refraction work was done in Germany during World War I. It was used for the location of enemy artillery positions by a type of triangulation. This approach was extended to the search for petroleum and had its first application in the detection of a salt dome around 1920.

The reflection seismic technique was immediately put to use in the petroleum exploration business and had its first successful demonstration in mapping a previously discovered salt dome. Its first solo success came in 1925 with the mapping of a petroleum deposit in Oklahoma. From then on it became a much-used tool, and in some cases a standard exploration technique.

The biggest interpretation problem for the early seismic pioneers was the correlation between time and depth. Even with knowledge of the speed of sound in a variety of rock types, it was impossible to be certain of the depth of a given reflector. To deal with this problem, velocity surveys began to be conducted in 1927. These consisted of the setting off of explosions at the surface and the recording of the arrival times at known depths in a well. The business of velocity surveys increased rapidly.

Seeing the opportunity for additional revenue, Schlumberger Well Surveying Corporation entered the business of velocity surveys in the mid-1930s, by making available, for rent, the trucks and cables for these operations. At about the same time, Conrad Schlumberger patented a device for measuring the velocity of sound over a short interval of rock traversed by the borehole.[1] The cover page of the patent is shown in Fig. 15–1. What is not apparent from the drawing is that the source of down-hole acoustic energy was the horn of a Model A Ford. The initial version of this device was tested but failed to work. It was an idea ahead of its time. The achievement of a workable device required the improved technology that was to become available at the end of World War II.

The desire to know local formation velocities prompted the development, nearly simultaneously, of borehole logging devices by three oil companies: Humble, Magnolia, and Shell. Two of the companies produced a two-receiver device, while the third used only a single receiver. At that time, Schlumberger had one enterprising individual who invented and implemented a sort of inverted velocity survey.[2,3] Instead of drilling many shot holes at the surface, he proposed putting the explosive in the well and recording

Introduction to Acoustic Logging

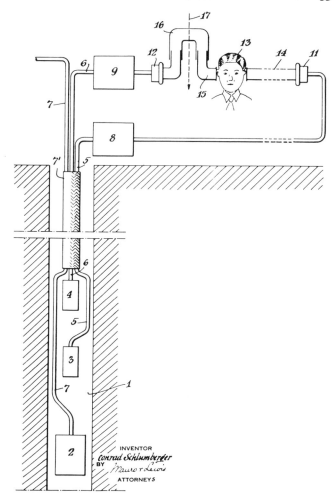

Figure 15-1. Front cover of a 1935 patent for obtaining the interval transit time.

arrivals at the surface. The first source was the side-wall sample taker (an explosive device which drove a steel cup into the formation to extract a sample of the rock), but it did not create enough acoustic energy to be detectable very easily at the surface. He later tried the conventional perforating gun, using dummy Bakelite bullets, which worked for depths shallower than about 3500′

The inverted velocity survey then evolved into the precursor of the sonic log when the perforating gun was suspended from 400′ of special low velocity cable. At the top of this special cable hung a geophone, and 400′ higher a second geophone. This device allowed the determination of the velocity continuously, up and down the well, with a check on each 400′ section. However, the idea was abandoned due to the unwieldly nature of the apparatus.

In 1955, Schlumberger entered the velocity logging business in earnest by acquiring the Humble velocity logging patent. The design was modified to produce a practical logging tool without any notion of sonic log interpretation. Fortunately, during the same years researchers at Gulf were measuring the relation between velocity and porosity on real and synthetic rock samples. The publication of this data led to the well-known Wyllie time-average relation, which provided the needed breakthrough to make velocity logging an important tool in well logging and formation evaluation.

APPLICATION OF BOREHOLE ACOUSTIC LOGGING

The conventional approach to acoustic logging uses the transmission method. It consists of a transmitter of acoustic energy and a receiver at some distance from it. The acoustic energy is in the frequency range of about 20 kHz and is transmitted in short bursts rather than continuously. The detected acoustic signal has been transmitted through a portion of the rock formation surrounding the borehole. This method commonly measures the velocity of compressional and (more recently) shear acoustic waves. The most common use of the velocity measurements is to relate them to formation porosity and lithology. However, as discussed in Chapter 17, it is also possible to detect and estimate abnormally high pore pressures and to estimate mechanical rock properties from the acoustic measurements.

An additional parameter that can be measured in the transmission mode, besides the velocities, is the attenuation. Although not as common as the simple interval transit time measurement, it is exploited with variable success in the detection of fractures. On the other hand, the attenuation method has been used routinely to evaluate the quality of the cement bond to the steel casing.

The second method of borehole acoustic measurement is the reflection mode. In this case the source of acoustic energy is at much higher frequencies, on the order of 500 kHz. Usually the transmitter and receiver are a single element. Depending upon the application, the transit time between emission and reception, or attenuation, of the signal is measured. One of the applications of the reflection technique is to obtain an acoustic image of the borehole wall. This method has achieved considerable success in the detection and evaluation of fractures. Another application, described in Chapter 17, is also for the evaluation of cement bonding.

REVIEW OF ELASTIC PROPERTIES

Two important types of energy transport mechanisms are supported by elastic media: compressional waves and shear waves. Fig. 15–2 illustrates the notion of the compressional wave for a system of masses suspended or linked to one another by springs. This might represent the atoms in a crystalline structure with the electrostatic repulsion replacing the springs. If the far end of this mass-spring assembly is moved rapidly to the right, causing a compression, and then is suddenly stopped, the compression will propagate to the right with a velocity v_p. The traveling of this compressional disturbance will cause local stresses (forces) and local displacements, as noted in the bottom portion of the figure. For this type of wave, the particle displacement and disturbance propagation are in the same direction.

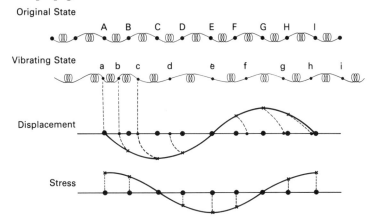

Figure 15-2. Representation of an elastic medium by coupled springs and masses. A vibration induced in the material is seen to induce variable displacements and stresses on individual sites.

To illustrate the notion of a shear wave in which the particle displacement is at right angles to the motion of the disturbance, refer to Fig. 15-3. In this case the one-dimensional mass-spring assembly of Fig. 15-2 is replaced by a two-dimensional structure with spring crossmembers. The element of material composed of four linked masses is shown, on the right, to undergo a shearing action. The result of this motion is that one of the diagonal springs is elongated and the other is compressed. Once the shearing force is removed, the small system will realign itself to relieve the compression of one spring and the tension of the other.

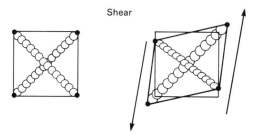

Figure 15-3. Representation of shear forces in an elastic material which involves consideration of the cross-coupling of mass elements.

In Fig. 15–4 a series of these elements is shown linked together. A shearing distortion is seen to be applied at the left end. If the left edge is held fixed while the shearing force is released, then the shear distortion will propagate to the right with a velocity v_s.

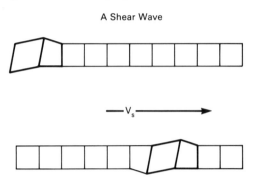

Figure 15–4. The propagation of a wave of induced shear forces.

In order to quantify these two elementary types of wave propagation, it is necessary to review the parameters used to characterize elastic media. The first is Young's modulus, Y (sometimes E). In Fig. 15–5, the effects of stretching a bar under uniform tension are illustrated. From Hooke's law the elongation Δl will be proportional to the force F applied to the stretching.

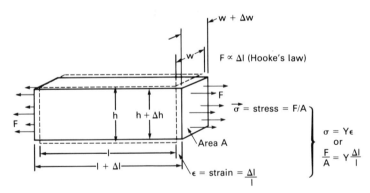

Figure 15–5. Deformation of a bar under uniform horizontal tension. Adapted from Feynman et al. [10]

The stress σ, under which is bar is placed, is defined as the force F, divided by the cross-sectional area A, and Young's modulus is the constant which links this stress to strain, or fractional change in length in the direction of the stress:

$$\frac{F}{A} = Y \frac{\Delta l}{l} .$$

Introduction to Acoustic Logging   345

The stress on the illustrated bar will also produce contractions in the width $\frac{\Delta W}{W}$, and contractions in height $\frac{\Delta h}{h}$. Both of these contractions will be proportional to the elongation and will be reductions, if the elongation is taken to be positive. This can be written as:

$$\frac{\Delta W}{W} = \frac{\Delta h}{h} = -\sigma \frac{\Delta l}{l},$$

where σ (sometimes known as ν) is Poisson's ratio. Thus the elongation in one axis multiplied by Poisson's ratio will yield the contraction in the other two axes when a uniform stress is applied. These two constants, Young's modulus and the Poisson ratio, are the only two parameters necessary to completely specify the elastic properties of a homogeneous, isotropic material. However, there are a variety of other commonly used elastic constants that also describe the same medium. They arose from either the experimental methods used to study materials or from theoretical considerations.

One of the other common constants used is the bulk modulus K (sometimes B). It is the measure of the compressibility of a material put under a uniform confining pressure. The defining relation for the bulk modulus is:

$$p = -K \frac{\Delta V}{V},$$

where the bulk modulus K is seen to relate the pressure p, necessary to change the volume by the fractional amount $\frac{\Delta V}{V}$. Fig. 15–6 indicates how this parameter, an obvious one for measurement in the laboratory, can be related to Young's modulus and Poisson's ratio.

The forces on the bar, which is under uniform pressure, are considered in only one direction at a time. The analysis first examines the change in the length of the bar from the three components. The confining force F_1 will produce a change in length given by $\frac{\Delta l}{l} = -\frac{p}{Y}$. It is negative because the length will decrease. The result of the decrease in length will, as the second sketch in the sequence shows, produce an increase in the height of the bar.

In a second step, consider the change in length of the bar due to the pressure P_2 applied to the top and bottom of the bar. The fractional change in height of the bar will produce a lengthening of the bar:

$$\frac{\Delta l}{l} = \sigma \frac{\Delta h}{h}.$$

The fractional change in height due to the force F_2 is just as in the last equation, $\frac{p}{Y}$, thus:

$$\frac{\Delta l}{l} = \sigma \frac{\Delta h}{h} = \frac{\sigma p}{Y}.$$

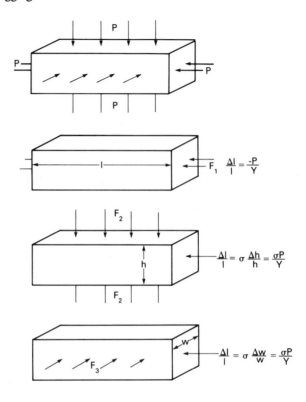

Figure 15-6. Deformation of a bar under uniform pressure. Adapted from Feynman et al. [10]

In the third sequence, the effects of the pressure on the sides, which produces a change in width and a subsequent change in fractional length, can be seen to be, by analogy:

$$\frac{\delta l}{l} = \sigma \frac{\Delta w}{w} = \frac{\sigma p}{Y} \ .$$

Thus the total fractional change in length due to these three components of the pressure is:

$$\frac{\Delta l}{l} = -\frac{p}{Y}(1 - 2\sigma) \ .$$

From this, the total volume strain $\frac{\Delta V}{V}$ can be found; it is three times the length change just established:

$$\frac{\Delta V}{V} = 3\frac{\Delta l}{l} = -3\frac{p}{Y}(1 - 2\sigma) \ .$$

From this expression, it is seen that a constraint on Poisson's ratio is that it must be less than ½.

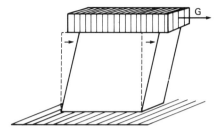

Figure 15–7. Deformation of a cube under uniform shear. Adapted from Feynman et al. [10]

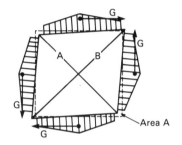

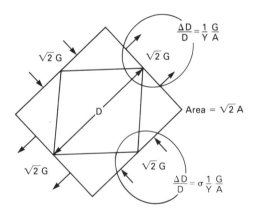

Figure 15–8. Resolution of forces on cube into equivalent stretching and compressive forces. Adapted from Feynman et al. [10]

The relation of the shear modulus to the two initial constants is considered next. In Fig. 15–7, an example of pure shear is shown. A tangential force G is applied to a surface area A, to produce a uniform shear. Fig. 15–8 shows that two pairs of shear forces can be considered equivalent to a pair of compressional and stretching forces, shown in the lower portion of the figure. The shear modulus μ relates the shear strain Θ to the shear stress $\dfrac{G}{A}$, which is the tangential force per unit area. Fig. 15–9 shows the definition of the shearing angle Θ, which is the ratio of the maximum displacement δ, divided by the length of the cube under shear. With reference to these definitions, the relations are:

$$\Theta = \frac{\delta}{l} = \frac{1}{\mu}\frac{G}{A}.$$

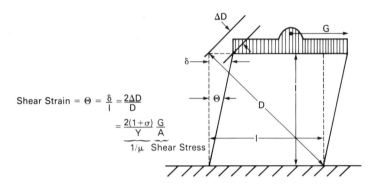

Figure 15–9. Definition of shear strain. Adapted from Feynman et al. [10]

To relate the shear modulus to the previous elastic constants, consider Fig. 15–8, which shows the equivalent diagram for shear. The change in the diagonal lengths needs to be evaluated. Along the diagonal indicated, the elongation that results from the stretching force is given by the Hooke relation:

$$\frac{\Delta D}{D} = \frac{1}{Y}\frac{G}{A}.$$

The compressional force will produce a similar elongation. It is given by:

$$\frac{\Delta D}{D} = \sigma\frac{1}{Y}\frac{G}{A}.$$

Thus the total change in the one diagonal length is given by:

$$\frac{\Delta D}{D} = \frac{1}{Y}(1+\sigma)\frac{G}{A}.$$

The total shear strain will be twice this quantity, since the same relations will be found for the other diagonal.

Thus the shear strain relation can be written as:

$$\Theta = \frac{\delta}{l} = \frac{2\Delta D}{D} = \frac{2(1+\sigma)}{Y}\frac{G}{A}.$$

From the relation given above, the identification of the shear modulus μ can be made:

$$\mu = \frac{Y}{2(1+\sigma)}.$$

The third constant used to describe elastic media is the so-called Lame constant, λ. In conjunction with the shear modulus μ, it is used to relate the stress and strain tensors in a compact representation.[4,5] An additional advantage is that the compressional and shear velocities have rather simple mathematical expressions when this formulation is used.[6]

WAVE PROPAGATION

Fig. 15–10 shows an idealized representation of a compressional wave. The periodic pressure variation of period T (sec) has a frequency f of $\frac{1}{T}$. It is shown by increased particle density, which is seen to be separated by the wavelength λ. The separation λ between pressure disturbances is related to the frequency of the disturbance by the velocity of propagation v_c of compressional waves, by the well-known relation:

$$\lambda f = v_c .$$

Assuming that the material represented in the figure can be described by an appropriate Young's modulus Y and density ρ, we will derive, for a special case of one-dimensional propagation, the velocity of the compressional wave, v_c.

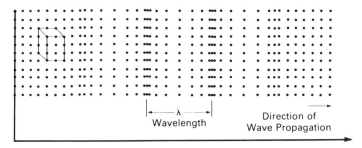

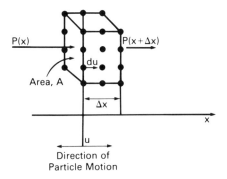

Figure 15–10. One-dimensional wave propagation in a very stiff material.

The approach is to consider a small element of volume with thickness Δx and area A, and write the equation of motion for it. First note the frame of reference which is indicated in Fig. 15-10. The x-axis position will denote the location of the element of volume, and a superimposed axis u will denote the movement of the particles about their rest position x. The equation of motion (F = M a) can be expressed as:

$$F = \rho \, A \, \Delta x \, \frac{d^2u}{dt^2},$$

where the force F is the difference in force experienced by the two faces of the volume:

$$F(x) - F(x + \Delta x).$$

This in turn can be related to the pressure difference by dividing both sides of the equation by the cross-sectional area A:

$$P(x) - P(x + \Delta x) = \rho \Delta x \frac{d^2u}{dt^2}.$$

To put this equation in terms of Young's modulus, we now relate the pressure on the two faces to the change in the length of the unit volume:

$$P(x) = Y \frac{\Delta l}{l} = Y \frac{du}{\Delta x} \Big]_x,$$

$$P(x + \Delta x) = Y \frac{du}{\Delta x} \Big]_{x + \Delta x}.$$

From the preceding two equations, and the definition of the derivative, the pressure gradient is given by:

$$\frac{dp}{dx} = Y \frac{d^2u}{dx^2}.$$

Using this relationship, the equation of motion can now be written as:

$$Y \frac{d^2u}{dx^2} \Delta x = \rho \Delta x \frac{d^2u}{dt^2},$$

or

$$\frac{d^2u}{dt^2} = \frac{Y}{\rho} \frac{d^2u}{dx^2},$$

which is recognized as a wave equation. The velocity of wave propagation is given by the square root of the ratio of elastic constants:

$$v_c = \sqrt{\frac{Y}{\rho}}.$$

This velocity of propagation, however, is a special case, because for the general case it is:

$$v_c^2 = \frac{Y}{\rho} \frac{1-\nu}{(1+\nu)(1-2\nu)} ,$$

which involves Poisson's ratio (ν here). Thus the special case of wave propagation which we considered is appropriate for a very stiff material ($\nu = 0$), so that no bulging occurs perpendicular to an applied compression.

The velocity of a shear wave is given by a much simpler expression: $\sqrt{\mu/\rho}$. The derivation of this fact is much like the preceding but a bit more complicated. However, if we use the approach with the Lame constant, it can be found almost by inspection.[7]

Table 15-1 gives a number of useful relations between the elastic constants and the two types of wave velocities. From this table some constraints on the elastic constants can be deduced, as well as relations between shear and compressional velocity.

	Y, ν	B, μ	λ, μ	ρ, v_c, v_s
Lame, λ	$\dfrac{Y\nu}{(1+\nu)(1-2\nu)}$	$\dfrac{(3B-2\mu)}{3}$	λ	$\rho(v_c^2 - 2v_s^2)$
Shear, μ	$\dfrac{Y}{2(1+\nu)}$	μ	μ	ρv_s^2
Young's, Y	Y	$\dfrac{9B\mu}{(\mu+3B)}$	$\dfrac{\mu(3\lambda+2\mu)}{(\lambda+\mu)}$	$\dfrac{\rho v_s^2(3v_c^2-4v_s^2)}{v_c^2-v_s^2}$
Bulk, B	$\dfrac{Y}{3(1-2\nu)}$	B	$\lambda + \dfrac{2}{3}\mu$	$\rho(v_c^2 - \dfrac{4}{3}v_s^2)$
Poisson, ν	ν	$\dfrac{3B-2\mu}{2(3B+\mu)}$	$\dfrac{\lambda}{2(\lambda+\mu)}$	$\dfrac{v_c^2 - 2v_s^2}{2(v_c^2 - v_s^2)}$
v_c^2	$\dfrac{Y(1-\nu)}{\rho(1+\nu)(1-2\nu)}$	$\dfrac{(B+\dfrac{4}{3}\mu)}{\rho}$	$\dfrac{(\lambda+2\mu)}{\rho}$	v_c^2
v_s^2	$\dfrac{Y}{\rho\, 2(1+\nu)}$	$\dfrac{\mu}{\rho}$	$\dfrac{\mu}{\rho}$	v_s^2

Table 15-1. Relationships among elastic constants and wave velocities. Adapted from White.[7]

From the definition of the bulk modulus:

$$B = -\frac{p}{\frac{\Delta V}{V}} ,$$

we can immediately say that $B > 0$. With reference to the table, one implication is that:

$$v_c^2 - \frac{4}{3} v_s^2 > 0.$$

Thus we should expect v_c to exceed v_s by at least 15%. However, another estimate of this contrast of velocities can be obtained by examination of the constraints on the Poisson ratio, v.

The upper limit on v can be obtained from an inspection of the bulk modulus as a function of Young's modulus and Poisson's ratio. This expression:

$$B = \frac{E}{3(1-2v)}$$

shows that $v < \frac{1}{2}$. The lower limit can be obtained from the expression for Poisson's ratio as a function of the bulk modulus and the shear modulus:

$$v = \frac{3B - 2\mu}{2(3B + \mu)}.$$

If B is set to zero, the minimum value of v is -1. However, this implies an expansion of the perpendicular dimensions of a sample material for a stretching in the other. A real material, like a rock, will not exhibit this bizarre behavior. A reasonable limiting behavior would be no change in dimension in the direction perpendicular to the stretching. Thus a practical limit for v is zero. The expression for velocities in terms of v shows that for the minimum value of Poisson's ratio, the compressional velocity will exceed the shear velocity by 40%.

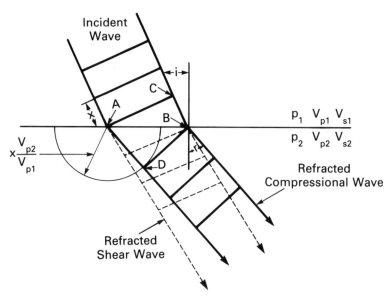

Figure 15-11. The ray representation of transmitted acoustic energy. Refraction is shown at the boundary between two materials of different acoustic properties.

The reflection and refraction of acoustic waves can be visualized by use of Huygen's principle, familiar to students of optics. Fig. 15-11 shows the interface of two media which are characterized by two different densities, ρ_1 and ρ_2, and compressional and shear velocities denoted by v_p and v_s, which are different in the two regions. The parallel rungs of the ladder correspond to maxima in the incident periodic pressure disturbance. At point A, the pressure disturbance will create an outgoing wave in medium 2. This will be characterized by a speed v_{p2}, which in this case is taken to be greater than that of the speed in region 1. (For this example, we consider only the P wave [compressional] propagation, but the same applies for the shear [or S] wave, which is generated at this interface.) As the wave front at point C, in the first medium, travels the distance x to the interface marked at point B, the compressional disturbance in medium 2 will have expanded to a radius larger than x by the factor $\frac{v_{p2}}{v_{p1}}$. Thus the point marked D will be in phase with the new wave commencing at B. The net result is that the incident wave, which made an angle i, with respect to normal incidence, will be seen to leave the interface with a new angle r. The relation between the two is controlled by the velocity contrast in the two media:

$$\frac{\sin i}{\sin r} = \frac{v_{p1}}{v_{p2}},$$

which is known as Snell's law. One interesting aspect of this relationship is that if the angle of incidence i becomes large enough, then the refracted wave will travel parallel to the interface surface. This critical incidence angle, i_{crit}, is given by:

$$\sin i_{crit} = \frac{v_{p1}}{v_{p2}}.$$

Waves which are critically refracted and travel along the boundary are referred to as head waves. As they travel along the interface, they radiate energy back into the initial medium. It is this phenomenon which allows the detection, by an acoustic device centered in the borehole, of acoustic energy which has propagated primarily in the formation.

RUDIMENTARY ACOUSTIC LOGGING

Fig. 15–12 illustrates the necessary elements of a device for the measurement of the compressional velocity of a formation. It consists of a transmitter of acoustic energy and a receiver. This device is centered in a borehole filled with fluid which has a compressional velocity of approximately 6000'/sec. The ray diagram, sketched at the side, indicates two paths for the acoustic energy, one in the mud and the other refracted in the formation at the critical angle. The compressional velocity in the formation is somewhere between 10,000 and 20,000'/sec. The measurement consists of recording the time from the transmitter firing until the first detectable signal arrives.

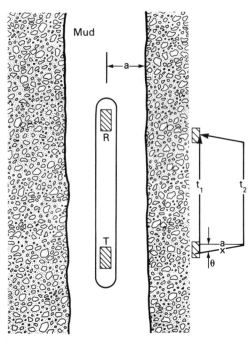

Figure 15–12. A rudimentary borehole device for measuring the interval transit time Δt. Shown, to the side, are the ray paths of the acoustic energy refracted from the borehole into the formation and back to the receiver.

Even for such a rudimentary device, some design precautions must be taken. The spacing between transmitter and receiver must be large enough so that the acoustic energy traveling in the borehole mud does not arrive before the signal from the formation. The ray diagram of Fig. 15–12 shows the parameters of interest for obtaining the minimum spacing d between transmitter and receiver to obtain this condition. The mud travel time must be greater than the travel time in the formation plus the two-way travel time in the mud. The mud travel time t_1 is the separation distance d divided by the mud velocity v_1. The travel time in the formation t_2 is approximately $\frac{d}{v_2}$. The two-way travel time in the mud to the formation can be found from the length x. Using the definition of the critical angle, the distance x is given by:

$$x = \frac{a}{\sqrt{\frac{v_2^2 - v_1^2}{v_2^2}}}.$$

The final result for the separation distance d is that:

$$d > \frac{2a\sqrt{v_2^2 - v_1^2}}{v_2 - v_1}.$$

RUDIMENTARY INTERPRETATION

The work of Wyllie and others was the breakthrough necessary to put acoustic logging into a central role in log interpretation.[8,9] It occurred at a time when porosity logs were almost unknown; only a rudimentary neutron device was available. Their measurements of acoustic velocities of materials, including core samples, gave log interpreters a way to convert the measured travel time into something other than integrated travel time for seismic corrections. It gave them a way to relate it to porosity of the rock.

The data which launched this revolution are shown in Fig. 15–13. The measurements consist of velocity (presented as the reciprocal and referred to as the interval transit time, or Δt) of sandstone core samples of different porosities. These were carried out under several different conditions. The upper two trends show that the dry, unconfined samples had the slowest velocity, which increased when saturated with water. However, it is the bottom trend which is of interest. This corresponds to a few water-saturated samples which were confined by pressure during the measurement. The reciprocal velocities for these points follow the expression:

$$\frac{1}{v} = \frac{1}{v_{solid}}(1-\phi) + \frac{1}{v_{fluid}}\phi, \quad (1)$$

or

$$\Delta t = \Delta t_{solid}(1-\phi) + \Delta t_{fluid}\phi,$$

which has come to be known as the Wyllie time-average equation. It is a linear relationship between the interval transit time and the porosity.

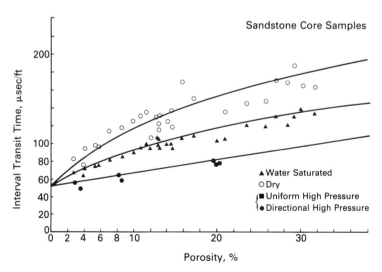

Figure 15–13. Laboratory data showing the relationship between interval transit time and porosity of core samples. Adapted from Wyllie et al.[8]

What is remarkable about Eq. (1) is that it was devised by Wyllie for artificial samples of alternating layers of aluminum and Lucite, as shown in Fig. 15–14. It is reasonable for such a laminated model that the total time delay is given by a sum of the delays in the various lengths of formation and fluid which scale as the porosity. Why this should be so for pressure-confined sandstone core samples is not so clear.

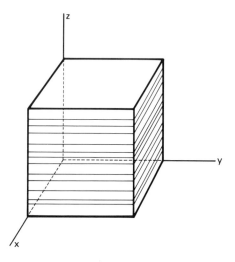

Figure 15–14. A geometric representation of the formation model implied by the time average expression for the evaluation of porosity.

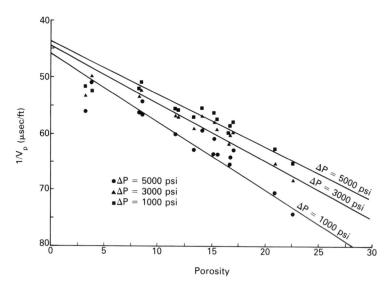

Figure 15–15. Laboratory data showing the dependence of interval transit time on the differential pressure applied to rock samples. From Pickett.[11]

In the course of their investigations, Wyllie and others, discovered that travel time depended on other parameters. Some of these were the matrix material, cementation, type of fluid in the pores, and pressure. As an example, Fig. 15-15 shows the interval travel time in dolomites as a function of porosity for various effective (overburden minus pore pressure) pressures. It is clear that the slope is dependent not only on the fluid velocity but also the difference between pore and overburden pressure. To cope with these realities, we will have to look next at some models of rocks which have been developed to explain these velocity variations.

References

1. Schlumberger, C., *Procede et appareillage pour la reconnaisance de terrains traverses par un sondage*, Republique Francaise Brevet d'Invention, numero 786,863, June 17, 1935.
2. Kokesh, F. P., "The Development of a New Method of Seismic Velocity Determination," *Geophysics*, Vol. 17, 1952.
3. Kokesh, F. P., "The Long Interval Method of Measuring Seismic Velocity," Presented at the Society of Exploration Geophysicists, New York, 1955.
4. Turcotte, D. L., and Schubert, G., *Geodynamics*, J. Wiley & Sons, New York, 1982.
5. Hearst, J. R., and Nelson, P. H., *Well Logging for Physical Properties*, McGraw-Hill, New York, 1985, pp. 285-290.
6. Tittman, J., *Geophysical Well Logging*, Academic Press, Orlando, 1986.
7. White, J. E., *Underground Sound: Application of Seismic Waves*, Elsevier, Amsterdam, 1983, pp. 12-17.
8. Wyllie, M. R. J., Gregory, A. R., and Gardner, L. W., "Elastic Wave Velocities in Heterogeneous and Porous Media," *Geophysics*, Vol. 21, 1956.
9. Wyllie, M. R. J., Gregory, A. R., Gardner, L. W., and Gardner, G. H. F., "An Experimental Investigation of Factors Affecting Elastic Wave Velocities in Porous Media," *Geophysics*, Vol. 23, 1958, p. 459.
10. Feynman, R. P., Leighton, R. B., and Sands, M. L., *Feynman Lectures on Physics*, Vol. 2, Addison-Wesley, Reading, Mass., 1965.
11. Pickett, G. R., *Formation Evaluation*, unpublished lecture notes, Colorado School of Mines, 1974.

Problems

1. Formation compressional velocities generally range between 10,000'/sec and 20,000'/sec, while the compressional velocity of the mud is about 6000'/sec.

    a. What is the variation in the critical angle, in degrees, for these extremes?

    b. For a transmitter-receiver pair with a nominal 3' separation, centered in a 10"-diameter borehole, what is the actual distance of formation traversed by the first arrival compressional wave in the case of the largest critical angle determined in part a?

2. To understand the timing constraints on the Δt measurement and the difficulties of performing this measurement successfully, consider the connection between the frequency of acoustic emission and the space between transmitter and receiver for obtaining the compressional transit time to a given degree of accuracy.

    a. Using the original data of Wyllie (see Fig 15–13) determine a representative figure for the precision of the Δt measurement required for determining porosity to within 1 porosity unit.

    b. For a source-to-receiver distance of 3', what is the lower limit of the frequency of emission that can be established to meet the 1-PU resolution, assuming that the detection system is capable of determining the first arrival somewhere in the first quarter of a cycle of the sinusoidal emission wavetrain?

    c. What is the frequency required if the first arrival is detected somewhere between 0.1 and 0.5 times the maximum amplitude of the detected wavetrain?

3. With reference to Fig. 15–12, determine the minimum spacing required for the first arrival to be a compressional formation signal rather than the direct mud arrival. The bore hole is 16" in diameter, with a 189-μsec/ft mud and a 120μsec/ft formation. In some shales the transit time may be as great as 150 μsec/ft. What is the minimum spacing required?

4. What is the conversion factor between the conventional units of transit time Δt, in μsec/ft and velocity in km/sec?

    Note that it is numerically equal to the constant c, which relates the expected Δt (in μsec/ft) of a fluid to $\sqrt{\frac{\rho}{K}}$ where K is the bulk modulus in GPa (gigaPascals = 10 kB = 10 kilo-Bar) and the density ρ is in g/cm^3, i.e.:

    $$\Delta t = c \sqrt{\frac{\rho}{K}}.$$

16
ACOUSTIC WAVES IN ROCKS

INTRODUCTION

The previous chapter notes the use of the measured interval transit time for the determination of porosity. It also suggests that, in addition to porosity, other rock properties such as lithology and environmental effects such as pressure might be important factors determining the acoustic velocity. What are the important factors which affect the acoustic velocity of rocks? How can these factors be related to elastic constants which are representative of the rock? These are some of the questions to be addressed in this chapter.

As background, we begin with a description of the experimental procedure used in investigating acoustic rock properties. This is followed by a review of some of the laboratory data accumulated by a wide variety of researchers who have studied the acoustic properties of rocks. This work shows that the velocity of acoustic propagation is a rather more complex phenomenon than would be implied by the simple Wyllie time-average equation. As a consequence of the laboratory observations, some models of rocks which attempt to describe actual rock acoustic properties have been developed. The approach used in several models is described. Returning to logging applications, the final portion of the chapter is an introduction to some aspects of acoustic waves in boreholes and to the propagation of acoustic energy at the interface between materials with different properties.

A REVIEW OF LABORATORY MEASUREMENTS*

An examination of the literature shows that compressional acoustic velocity in rocks depends primarily on six factors: porosity, composition or lithology, the state of stress, temperature, fluid composition for a saturated porous rock, and the rock texture.[1,2] Of all these items temperature is the least important and is not considered here. Before we examine some of the experimental data, it is worth examining how acoustic properties of rocks are measured.

Fig. 16–1 shows a typical laboratory apparatus for the measurement of acoustic velocities of rock samples. It consists of a p-wave or s-wave transducer (usually piezoelectric) which is energized by an electrical pulse. At the other end of the sample is a similar transducer which produces an output voltage that is the result of strains induced in the transducer. The time delay between the firing pulse and the received pulse is converted to the appropriate acoustic velocity.

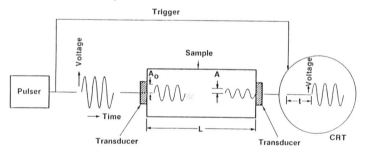

Figure 16-1. The laboratory setup for the measurement of acoustic properties of rock samples. From Timur.[1]

One of the important influences on acoustic velocity is the effective state of stress of the rock. Although this condition is rarely known with any precision for an actual formation, it is a laboratory expedient to approximate it with a differential pressure which is the difference between the confining pressure and the pore pressure. To study the effect of stress on velocities, the sample can be placed in a special cell similar to the one illustrated in Fig. 16–2. It can be seen that the sample is completely enclosed by a rubber jacket which is then surrounded by a fluid whose pressure can be varied to produce the equivalent of an overburden pressure. There is an additional connection, seen at the lower right of the figure, which allows communication with the fluid saturating the porous sample. Thus the pressure of the pore fluid may also be changed, and the velocity can be determined as a function of the difference of the two pressures.

The effect of porosity on the compressional velocity of a number of

---

* For much of this chapter, inspiration (and figures) were drawn from the draft copy of "Acoustic Logging," by A. Timur, from *SPE Petroleum Production Handbook*,[1] and *Wel Logging II: Resistivity and Acoustic Logging*, by J. R. Jordan and F. Campbell.[2]

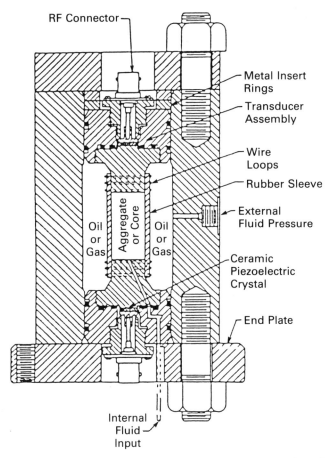

Figure 16-2. Details of a differential pressure cell for the simulation of overburden and pore pressure. Adapted from Wyllie.[3]

sandstone samples can be seen in Fig. 16-3. The porosity is plotted as a function of the measured travel time and the straight line is the time-average fit to the data. The matrix and fluid travel times used in the fit are given in the figure. These measurements were made with no confining pressure, and the scatter may represent other factors which are important, such as texture. The effect of matrix is clearly seen in Fig. 16-4, in which porosity is plotted versus travel time for quartz and calcite core samples. The matrix velocity of calcite is somewhat greater than that of quartz, as indicated in the figure. A summary of some of the fluid and matrix velocities for both shear and compressional waves is shown in Table 16-1.

To study the effect of total confining pressure on velocities, measurements are made with the pore fluid at atmospheric pressure, and the confining pressure (the pressure in the fluid surrounding the rubber jacket of Fig. 16-2) is varied. The result of this type of measurement in samples of

362   Well Logging for Earth Scientists

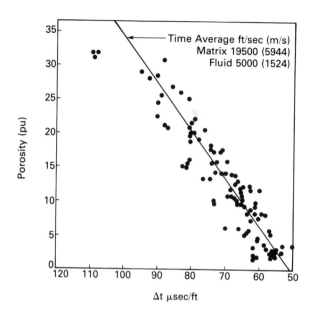

Figure 16-3.   Measurements of the effect of porosity on compressional transit time for sandstone samples. From Wyllie.[4]

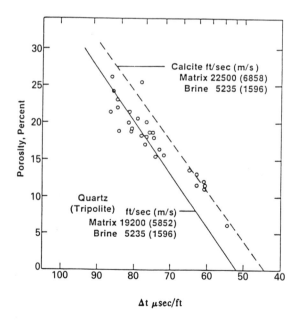

Figure 16-4.   Effect of rock composition on the relationship between porosity and transit time. From Timur.[1]

Acoustic Waves in Rocks   363

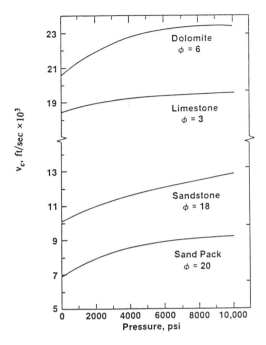

Figure 16-5. Compressional velocity for several rock samples as a function of confining pressure. From Timur.[1]

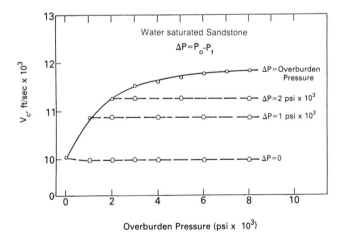

Figure 16-6. The effect of simulated differential pressure on compressional velocity. From Wyllie.[3]

the three major types of lithology is shown in Fig. 16–5. Substantial changes in velocity are noted: nearly 20% in the case of the sandstone samples, but less than 10% for the limestone.

The separate effects of confining pressure and pore pressure are investigated by making velocity measurements for differences of pore and overburden pressure. An example of this type of data is shown in Fig. 16–6. The upper trace shows the velocity variation as a function of overburden pressure with the pore fluid at atmospheric pressure, as was the case for the data in Fig. 16–5. The other three traces, however, are for cases where the pore fluid pressure tracked or was slightly less than the overburden pressure. The nearly horizontal line for $\Delta P = 0$ indicates that, to first order, if the confining pressure and pore pressure are equal, then the average elastic properties of the sample have not changed compared to the unstressed state. The observations can be generalized by noting that the compressional velocity increases for increasing overburden pressure and decreases for increasing pore pressure.

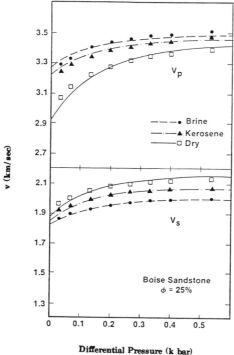

Figure 16-7. Effect of the saturation fluid on the compressional and shear velocity of 25% porous sandstone. From Timur.[1]

Although we have already seen that the pore fluid has an influence on the acoustic velocities, through the time-average representation of the data, it is no surprise to see the velocity differences in Fig. 16–7 for the three different types of pore fluids: water, kerosene, and air. However, the magnitude of the velocity change as a function of the differential pressure on the sample depends upon the fluid. What is even more striking, in Fig. 16-7, is the reversed behavior of the compressional and shear velocities.

The effect on the compressional and shear velocities of porous rocks under confining pressures has also been studied. Representative data is shown in Fig. 16-8. The first conclusion that can be drawn from this data summary is that the effect of water saturation will be largest for the compressional (or p) waves. Note that the shear wave behavior hardly changes under the two extreme conditions considered. This is because liquids do not support a shear wave. However, especially at low porosities, the compressional velocity is quite sensitive to water saturation. The difference in behavior for shear and compressional waves leads naturally to a technique for distinguishing gas-bearing formations. The ratio of the two velocities can be used as a gas indication.

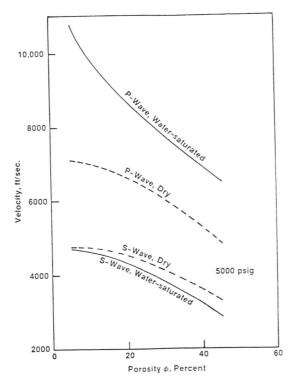

Figure 16-8. A summary graph of the effect of the extremes of water saturation on compressional and shear wave velocities as a function of porosity. The shear wave velocity increases slightly with the density decrease that results from replacing water with gas. The compressional velocity decreases when gas replaces water because of the change in the rock bulk modulus. From Timur.[1]

A qualitative explanation of this effect can be obtained from the basic parameters which govern velocity. The shear velocity is given by $\sqrt{\frac{\mu}{\rho}}$. At a fixed porosity, if water is exchanged for low density gas, the shear modulus will not change, but the density will decrease, thus producing the inversion seen in the figure. In the case of the compressional wave, one expression for velocity is: $\sqrt{\frac{B + (4/3)\mu}{\rho}}$. The same argument holds for μ, which does not change. However, the argument for density must be overcompensated by a change in the bulk modulus. At a fixed porosity, the replacement of liquid, which has a very large bulk modulus, by gas, which is very compressible, will certainly reduce the contribution of the pore fluid to the overall bulk modulus of the formation. According to the data, this reduction will dominate the velocity, regardless of the porosity. Just how the compressibility of the fluid influences the overall rock compressibility is the kind of question which is answered by rock models, to be discussed in the next section.

The data just considered treat the two extremes of either full water or gas saturation. What about those cases in between? What is the effect of partial saturation? That there is little sensitivity of the velocity ratio to saturation has been confirmed by both laboratory and model calculations, which are shown in Fig. 16–9. These curves indicate a large change in the compressional velocity as soon as the gas saturation attains 10%, and little afterwards As expected, there is no sensitivity indicated for the shear velocity.

The final data to be considered here, which deal with the effect of texture on velocity, are shown in Fig. 16–10. In this data, the p-wave and s-wave velocity in Troy granite are shown as a function of the differential pressure on the sample. For the s-wave, there is little or no change in velocity as a function of the differential pressure and no noticeable difference between the dry and water-saturated sample. However, for the p-wave the result is rather dramatic. For the dry rock, the velocity increases by about 50% for a modest differential pressure. This has been attributed to tiny microfissures in the Troy granite, which tend to close under pressure. From this example, it is clear that the texture of the rock must be included in any comprehensive rock model.

ACOUSTIC MODELS OF ROCKS

Classical wave theory predicts none of the effects discussed above. The velocities of the compressional and shear waves are given by simple combinations of the material moduli and bulk density. Realistic rock models will have to provide a means of extracting appropriate elastic constants for fluid-saturated rocks: a method of predicting the average elastic properties

Acoustic Waves in Rocks 367

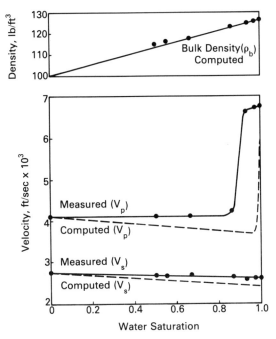

Figure 16-9. The effect of partial water saturation on the compressional and shear velocity. The change in bulk modulus occurs dramatically with a slight introduction of gas. From Timur.[1]

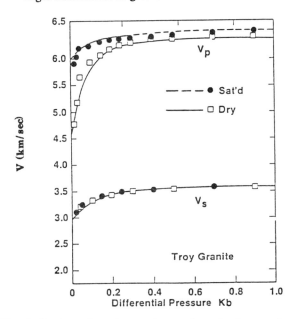

Figure 16-10. Indication that the texture of rock has a large impact on compressional velocity. From Timur.[1]

from a knowledge of the elastic properties of the constituents and some details of the rock fabric.

Our goal in using rock models is to be able, from some set of reasonable parameters, to predict the behavior of acoustic velocities under a variety of conditions, such as confining pressure, differential pressure, and pore fluid properties. One of the first attempts at such a model was made by Gassman.[5] His model for porous rocks consisted of a porous skeleton filled with fluid and was based on the assumption that the properties of the fluid and skeleton materials are known. The major simplification incorporated is that the relative motion between the fluid and skeleton during acoustic wave propagation is negligible.

The basic definitions used in the model are shown in Fig. 16–11. Following the convention of Gassman, the symbol for bulk modulus used here is k. We start with a knowledge of the material properties of the skeleton material, in particular its density ρ_s and bulk modulus k_s. The known properties of the evacuated skeleton are its porosity ϕ, its density $\overline{\rho}$, its bulk modulus $\overline{k}$, its shear modulus $\overline{\mu}$, and its plane wave modulus $\overline{M}$. The plane wave modulus is taken to be $\overline{M} = \overline{k} + \frac{4}{3}\overline{\mu}$ and is of interest since the compressional velocity can be obtained from it easily (see Problem 1). The fluid properties are denoted by its density ρ_f and bulk modulus k_f. The object of the model is to obtain the average properties of the saturated rock: its density ρ, bulk modulus k, shear modulus μ, and plane wave modulus M.

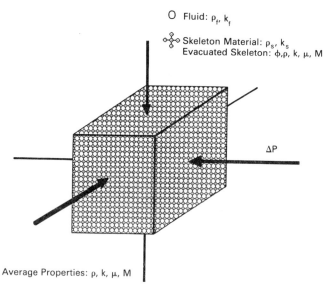

Figure 16-11. Model of a porous rock according to Gassman. It is considered to consist of a rigid frame and a saturating fluid. The elastic constants of the frame material and fluid are known. The model predicts the elastic constants of the ensemble.

The bulk density is given simply by:

$$\rho = \phi\rho_f + (1-\phi)\rho_s,$$

and the shear modulus will be the same as for the skeleton $\bar{\mu}$. However, the average bulk modulus is not obvious but can be determined from the following analysis.

In order to determine the bulk modulus, we need first to refer to the defining equation:

$$k = -\frac{\Delta P}{\frac{\Delta V}{V}}.$$

Consider what the change in volume of our saturated skeleton is when a pressure ΔP is applied, as shown in Fig. 16–11. First, the applied pressure can be separated into two components:

$$\Delta P = \Delta\bar{P} + \Delta P_f,$$

where $\Delta\bar{P}$ is the portion of the overburden pressure which is actually supported by the skeleton, and ΔP_f is the portion of the applied pressure which is supported by the fluid. This applied pressure will induce a change in the volume ΔV, which is given by the sum of two contributions:

$$\Delta V = \Delta V_f + \Delta V_s,$$

which are the fluid compression and the actual skeleton compression.

The first of these two is given quite simply by:

$$\Delta V_f = -\phi V \frac{\Delta P_f}{k_f},$$

which follows directly from the definition of the bulk modulus. The skeleton change has two components: one the result of the pressure portion actually borne by the skeleton and the other the result of the pressure of the fluid. This is given by:

$$\Delta V_s = -(1-\phi)V\frac{\Delta P_f}{k_s} - \frac{V\Delta\bar{P}}{k_s}.$$

The first term is the shrinkage of the skeleton constituents, and the second term is the shrinkage of the entire skeleton framework. Thus the total volume change is given by the sum of these two:

$$\frac{\Delta V}{V} = \left[-\frac{\phi}{k_f} - \frac{1-\phi}{k_s}\right]\Delta P_f - \frac{1}{k_s}\Delta\bar{P}.$$

The other relation that is available is:

$$\frac{\Delta V}{V} = -\frac{1}{k_s}\Delta P_f - \frac{1}{k}\Delta\bar{P},$$

which follows from the pressure breakdown specified initially and the definition of the bulk modulus. After algebraic manipulation, it can be shown that the plane wave modulus M is given by that expected for the dry skeleton plus an additional term:

$$M = \overline{M} + \frac{(1-r)^2}{\frac{\phi}{k_f} + \frac{1-\phi}{k} - \frac{\overline{k}}{k_s^2}} , \qquad (1)$$

where $r = \frac{\overline{k}}{k_s}$.

In order for this model to be of any use, some relationship between the bulk modulus of the dry frame $\overline{k}$, and the rock matrix k_s must be established. Initially, Gassman was able to do this for a very special case in which he approximated a loose sand by a set of cubic packed spheres.[6] By considering the grain contact area as a function of applied stress, he could relate the compressibility to the particle size and grain properties.

At first inspection, Eq. (1) is unsatisfying. One of the observational facts which it must predict is the dependence of the compressional velocity v_c on the differential pressure. This dependence, however, is hidden in the term $\frac{\overline{k}}{k_s}$. Reformulations of the Gassman theory for cubic packed spheres by White and Pickett show this dependence explicitly.[7,8] This formulation contains five parameters in addition to effective stress for the prediction of porosity from Δt.

A more complete model was developed by Biot.[9] The additional approach to reality was to allow relative motion between fluid and the rock skeleton. The fluid motion is assumed to obey Darcy's law (which relates fluid flow to differential pressure, viscosity, and permeability) so that additional terms of permeability and fluid viscosity appear in the analysis. The Biot theory predicts a frequency dependence for compressional wave velocity. However, in the low frequency limit, the theory reduces to that of Gassman. Further models which deal specifically with the rock texture have been developed by Kuster and Toksoz, for example.[10] They derive average elastic parameter for a matrix containing spheroidal cavities of differing aspect ratios. This model has been quite successful in fitting the data, similar to that in Fig. 16–10.

Despite the existence of detailed but unwieldy acoustic rock models, in practical logging interpretation the Wyllie time-average equation is widely used to determine porosity from measurements of Δt. The justification, if anything beyond simplicity is required, is due to Geertsma.[11] He showed that in the limit of low porosity, the Biot theory predicts a simple porosity dependence for the inverse compressional velocity:

$$\frac{1}{v_c} \propto a + b\phi .$$

ACOUSTIC WAVES IN BOREHOLES

The picture of borehole acoustic logging presented earlier (see Fig. 15–12) is very simplified. What is implied in this figure is that something approximating the transmitted pulse is also received. However, that this is not the case can be seen clearly in Fig. 16–12 which shows an actual acoustic waveform recorded with a logging device in a borehole. In this particular example, three distinct packets of energy are seen. The first is indicated to be the result of the compressional wave moving through the formation. The second, with a much larger amplitude, is the result of a slower shear wave moving through the formation. Finally, arriving some 1500 μsec later is a wave of large amplitude, known as the Stoneley wave, which is entirely due to the presence of the borehole, as we shall see later. Although the transmitter pulse is brief and fairly uncomplicated, the received signal is quite complex, because of the various reflections and interferences which can be produced as a result of waves traveling in the borehole.

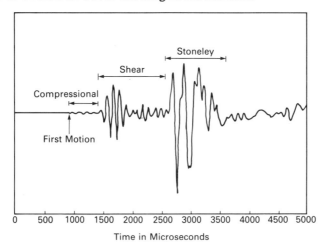

Figure 16-12. A typical acoustic waveform recorded in a borehole. Three distinct arrivals are indicated. Courtesy of Schlumberger.

In order to understand the two basic arrivals of Fig. 16–12 the wavefronts produced in a two-dimensional medium of mud and formation have been computed for different times after emission in Fig. 16–13. The case taken here is for a fast, or "hard," formation in which the compressional velocity in the solid is greater than the shear velocity in the solid, which is greater than the compressional velocity in the mud, i.e., $v_c > v_s > v_{mud}$. The first frame, at 40 μsec after emission, shows a spherical wavefront traveling out from the source toward the formation. At 70 μsec, the pressure wave has hit the borehole and three things have happened: There is a wave reflected back into the mud, a shear and a compressional wave have been created in the formation, and all of these have begun expanding.

372 Well Logging for Earth Scientists

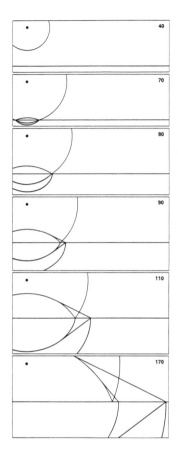

Figure 16-13. Simulation of the wavefronts in a two-dimensional medium consisting of mud and a fast formation. Courtesy of Schlumberger.

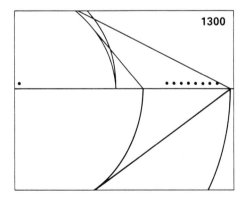

Figure 16-14. Much later in the simulation, when the wavefronts have passed the detector array. Courtesy of Schlumberger.

If we follow the compressional wave through the rest of the frames, we see that it goes on expanding in time. At somewhere around 90 μsec, the angle it makes with respect to the interface is 90°, and from that point on it travels down the interface with a speed of v_c. This is the so-called head wave. It can be thought to create a series of disturbances along the borehole wall which reradiate into the mud as a plane wave, resembling the wave of a boat. The angle of this "wake" will be related to the ratio of the formation compressional and mud velocities. Turning our attention to the shear wave, whose expansion is much slower because of its reduced velocity, we see that it does not make a right angle with respect to the interface until somewhere between 110 and 170 μsec, at which point it travels to the right with the shear velocity creating a wave in the borehole mud, just as in the compressional case.

Fig. 16–14 shows the situation after nearly 10 times as much time has elapsed (at 1300 μsec). The distance scale has been much enlarged and shows the location of 8 receivers. In this view, the compressional disturbance has already passed through all 8 receivers, and the next one to arrive will be the result of the shear wave. It will then be followed by the direct mud arrival, which will be followed shortly by the reflected mud pulse.

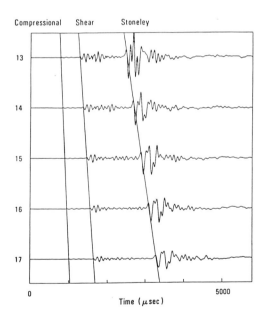

Figure 16-15. Recorded acoustic wavetrains recorded at 13' – 17' from the transmitter in a fast formation. The compression, shear, and Stoneley arrivals are clearly separated. The slopes of the lines indicating each of the three types of first arrivals is proportional to their velocity. Courtesy of Schlumberger.

374   Well Logging for Earth Scientists

Although in fast formations the shear signal can often be seen at large distances from the transmitter, this is not always the case. Returning to data recorded in a borehole, Fig. 16–15 shows waveforms at five different receivers at increasing distances from the transmitter. The shear arrivals are well-separated from the very low amplitude compressional arrivals, and the separation improves with increased transmitter-to-receiver distance. However, as shown in Fig. 16–16 for a soft formation, it is clear that there is no shear arrival present. In this example, the mud velocity is greater than the formation shear velocity.

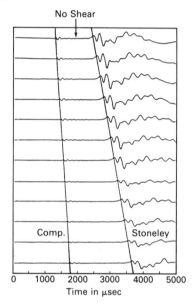

Figure 16-16. Acoustic wave trains recorded at 1" intervals in a slow formation where the shear arrival is seen to be absent. Courtesy of Schlumberger.

To understand the lack of shear development, refer to Fig. 16–17, which shows the wavefronts in the two-dimensional simulation of a soft formation at some time after emission. It is clear that there will be a compressional arrival. The compressional wavefront is perpendicular to the interface and travels along it at a speed of v_c, which is larger than v_{mud} in this case. However, the shear wave does not make a right angle since $v_s < v_{mud}$. The head wave does not appear, because the condition for constructive interference between the disturbance along the borehole wall and its transmission in the mud does not occur. Thus the "wake" will not develop.

The last aspect of the borehole sonic waveforms to be considered is the Stoneley wave. A good example of this is shown in Fig. 16–18. The full sonic waveform is shown from a tool in a borehole over a 200' section. The variations, due to lithology or porosity changes, are clearly seen in the shear and compressional arrivals. The Stoneley wave is also seen to have some variation in its arrival time, even though it is a pulse of energy which is traveling only in the borehole. What causes the variations in the Stoneley wave velocity?

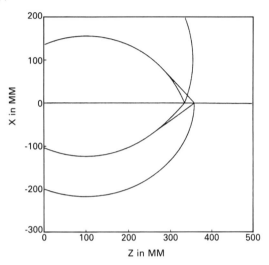

Figure 16-17. Results of the two-dimensional wavefront simulation for a slow formation. The shear velocity is such that a head wave never develops and thus is not observed in the borehole. Courtesy of Schlumberger.

Fig. 16–19 illustrates the basic notions of the Stoneley, or tube wave. In a liquid-filled tube which has very rigid walls, a low frequency pressure wave will travel as a nearly plane wave at the compressional velocity of the fluid. This type of phenomenon is responsible for the so-called water hammer. In this case, however, the sound is produced by the small distortion of the walls of the pipe carrying the water, which has been suddenly stopped. For a borehole, with semi-rigid confining walls, the speed of the pressure disturbance is related to the elastic constants of the wall as well as the fluid.

White has shown in the low frequency limit that the speed of this tube wave is given by:

$$v_{tube} = v_m \left[\frac{1}{1 + \frac{\rho_m v_m^2}{\rho v_s^2}} \right]^{1/2},$$

where v_m is the compressional velocity in the mud, v_s is the shear velocity in the formation, and ρ is the density of the formation.[7] Thus the shear velocity of the formation v_s can be obtained in the absence of a shear arrival from the speed of the tube wave if the density of the formation is known.

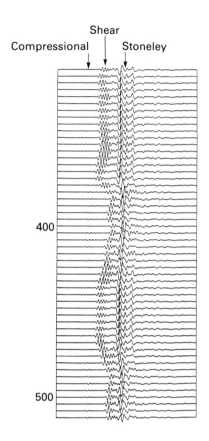

Figure 16-18. A sequence of acoustic waveforms recorded in the borehole with shear and Stoneley arrivals clearly shown. In addition to variations in the shear velocity, which reflect formation properties, the Stoneley wave is also seen to have velocity variations. Courtesy of Schlumberger.

Using the elastic constant relationships developed in Chapter 15, the tube wave velocity dependence on the mud bulk modulus B_m and the formation shear modulus μ can be found from the preceding expression to be:

$$v_{tube} = \left[\frac{1}{\rho_m \left[\frac{1}{B_m} + \frac{1}{\mu} \right]} \right]^{1/2}.$$

This illustrates the role of the borehole wall rigidity in modulating the mud velocity.

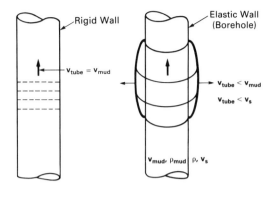

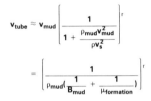

Figure 16-19. A simplified representation of the low frequency Stoneley, or tube wave.

References

1. Timur, A., "Acoustic Logging," in *Petroleum Production Handbook,* edited by H. Bradley, SPE, Dallas (in press).

2. Jordan, J. R., and Campbell, F., *Well Logging II: Resistivity and Acoustic Logging,* Monograph Series, SPE, Dallas (in press).

3. Wyllie, M. R. J., Gregory, A. R., Gardner, L. W., and Gardner, G. H. F., "An Experimental Investigation of Factors Affecting Elastic Wave Velocities in Porous Media," *Geophysics,* Vol. 23, 1958, p. 459.

4. Wyllie, M. R. J., Gardner, G. H. F., and Gregory, A. R., "Some Phenomena Pertinent to Velocity Logging," *JPT,* July 1961, pp. 629–636.

5. Gassman, F., "Elasticity of Porous Media," Vierteljahrschr. Naturforsch. Ges. Zurich, Vol 96, No. 1, 1951.

6. Gassman, F., "Elastic Waves Through a Packing of Spheres," *Geophysics,* Vol. 16, No. 18, 1951.

7. White, J. E., *Underground Sound,* Elsevier, Amsterdam, 1983.

8. Pickett, G. R., "The Use of Acoustic Logs in the Evaluation of Sandstone Reservoirs," *Geophysics*, Vol. 25, 1960.

9. Biot, M. A., "Theory of Propagation of Elastic Waves in Fluid-Saturated Porous Solids," *Journal of the Acoustic Society of America*, Vol. 28, 1956.

10. Kuster, G., and Toksoz, M. N., "Velocity and Attenuation of Seismic Waves in Two-Phase Media," *Geophysics* Vol. 39, 1974.

11. Geertsma, J. "Velocity-Log Interpretation: The Effect of Rock Bulk Compressibility," *SPEJ*, Dec. 1961.

Problems

1. Using Table 15–1, find an expression for compressional velocity v_c as a function of the bulk modulus and the shear modulus. The quantity $B + \frac{4}{3}\mu$ is known as the plane wave modulus.

   Show that the plane wave modulus can be expressed as $\lambda + 2\mu$, where λ is the Lame constant.

2. Consider two immiscible liquids of bulk modulus k_1 and k_2 which are mixed together with volume fractions V_1 and V_2. What is the mixing law for the bulk modulus of the mixture of two liquids? Write it in terms of the volume fractions of the two components.

3. It was seen that the Gassman equation predicts the bulk modulus k of a liquid-filled rock of porosity ϕ to be:

$$k = k_{df} + \frac{(1 - \frac{k_{df}}{k_{ma}})}{\frac{\phi}{k_f} + \frac{1-\phi}{k_{ma}} - \frac{k_{df}}{k_{ma}^2}},$$

   where the subscripts of the bulk moduli are as follows: df refers to the dry frame, ma to the matrix, and f to the fluid.

   a. Using the mixing law of Problem 2, write the above expression for bulk modulus in terms of the state of water saturation.

   b. Describe how you would obtain a value of k_{df} if your laboratory measurements limited you to a judicious choice of compressional velocity measurements. An explicit equation for the value of k_{df} may be more than you want to do, but please indicate the procedure.

4. Given the bulk modulus of a rock sample to be 141.7 GPa and a compressional velocity of 7.8 km/sec, what is an upper limit to be

expected for the shear velocity? What is the actual shear velocity if the density of the rock is 5.00 g/cm³?

5. Fig. 16–7 shows some data for a 25%-porosity sandstone that indicate the effect of the saturating fluid on compressional and shear velocities. The implied change in apparent formation Δt for the compressional arrivals is at a maximum for the case of no differential pressure.

   a. Using the Wyllie time-average approach, what value of Δt (in μsec/ft) would you ascribe to air? The velocities of typical materials in Table 16–1 may be of some use in this calculation.

   b. If, while logging a sandstone formation, you encountered a formation Δt corresponding to this minimum velocity and did not recognize the presence of gas, what porosity would you estimate it to be?

   c. The curves for shear waves are inverted on this figure compared to the compressional. Show, however, that they are in the anticipated order and that the magnitude of the separation is as expected for no differential pressure.

Non-Porous Solids		V_c	V_s
Anhydrite		20,000	11,400
Calcite		20,100*	
Cement (Cured)		12,000	
Dolomite		23,000	12,700
Granite		19,700	11,200
Gypsum		19,000	
Limestone		21,000	11,100
Quartz		18,900*	12,000
Salt		15,000*	8,000
Steel		20,000	9,500
Water-Saturated Porous Rocks In Situ	Porosity		
Dolomites	5-20%	20,000-15,000	11,000-7,500
Limestones	5-20%	18,500-13,000	9,500-7,000
Sandstones	5-20%	16,500-11,500	9,500-6,000
Sands (Unconsolidated)	20-25%	11,500-9,000	
Shales		7,000-17,000	
Liquids**			
Water (Pure)		4,800	
Water (100,00 mg/1) of NaCl		5,200	
Water (200,00 mg/1) of NaCl		5,500	
Drilling Mud		6,000	
Petroleum		4,200	
Gases**			
Air (Dry or moist)		1,100	
Hydrogen		4,250	
Methane		1,500	

* Arithmetic Average of Values Along Axes (Wyllie et al., 1956)[18]
**At Normal Temperature and Pressure

Table 16-1. Acoustic velocities for various materials. From Timur.[1]

17
ACOUSTIC LOGGING METHODS AND APPLICATIONS

INTRODUCTION

The measurement of the simple interval transit time of down-hole formations has been accomplished by a variety of logging devices. Despite the seeming simplicity of this measurement, which extracts only a part of the information contained in the complex acoustic waveform, a number of innovations were necessary to make valid measurements under a variety of borehole conditions. The techniques employed in conventional borehole acoustic logging tools are examined in this chapter. In particular, several methods for compensating for variable borehole size are discussed.

After reviewing the typical log presentation format, the performance of sonic logging tools is contrasted with other porosity-sensitive devices, most of which are more sensitive to borehole conditions. Despite their relative insensitivity to the borehole, sonic logging tools do have some limitations. Weak received signals can produce cycle-skipping, which is easily recognized on the log. Alteration of the formation near the borehole can effectively eliminate measurement of the virgin formation transit time.

Newer acoustic devices employ arrays of detectors, or detectors at extremely long spacing. Wavetrain recording and signal processing help them avoid some of the defects which mar conventional sonic logging under certain conditions. They also provide a measurement of the formation shear velocity and detection of the Stoneley waves, which allows applications beyond the conventional estimation of porosity. These applications include determination of lithology and fluid type, detection of overpressured zones and fractures, and estimation of the formation strength. Ultrasonic devices operating in the range of 1 MHz have opened the door to acoustic borehole imaging. These can be used in the detection of naturally occurring fractures and, in another application, the evaluation of the cementing of cased wells.

TRANSDUCERS

Acoustic measurements rely on the production of a pulse of pressure which is applied, through the borehole mud, to the formation. Two types of transducers have been in use both as acoustic source generators and receivers. One type is based on a magnetostrictive behavior of certain materials. For these, application of a magnetic field causes a volume reduction in the material. Consequently the sudden application of a magnetic field initiates a pressure pulse which is completed upon the removal of the magnetic field. This is accompanied by a subsequent volume relaxation.

The general form of the magnetostrictive transducer used in logging is a torus. The magnetic field is produced by supplying current to a coil which completely wraps the toroidal core material. Since the magnetostrictive material is also magnetized, it can operate as a receiver. Any impinging compressional acoustic energy will cause volume distortions in the core and thus vary the magnetic field which threads the coil windings. This changing magnetic field will produce a voltage at the terminals of the coil which is representative of the acoustic signal.

The second type of device in common use is based on ceramic materials, such as $BaTiO_2$, which have piezoelectric properties. This dielectric material responds to an applied electric field by changing its volume. Applying a voltage pulse between the inner and outer surfaces of the ceramic torus produces a subsequent fluctuation in its volume and thus the generation of a pressure disturbance. As a receiver, the incoming compressional wave distorts the ceramic, setting up a polarization charge, which appears as a voltage across the two sides of the torus.

The output power and operating frequency of both types of devices are limited by surface area and material properties. The dimensions dictated by logging sondes result in frequencies around 25 kHz. As a transmitter, the application of a voltage pulse results in a "ringing" at the central frequency which lasts for several periods.

CONVENTIONAL SONIC LOGGING

Conventional sonic logging, for this discussion, is taken to mean the determination of the transit time of compressional waves in the material surrounding the borehole by a device with two receivers. These are generally located at 3' and 5' from the transmitter. This type of device is illustrated in Fig. 17-1. The technique consists of measuring the difference in the arrival times of the acoustic energy at the two transducers. This difference divided by the span between the two detectors yields a transit time Δt (usually expressed in μsec/ft) for the formation. The depth of investigation of this measurement is somewhat difficult to define in the case of a uniform formation. Since only the transit time of the first detectable signal is being measured, the measurement will be sensitive only to the acoustic path which has the shortest time. This is generally the one parallel to the borehole wall and very close to its surface. The notion of depth of investigation will become meaningful only when we consider the problems of alteration, and damage (both imply a reduction in the formation velocity in the vicinity of the borehole) to the borehole wall.

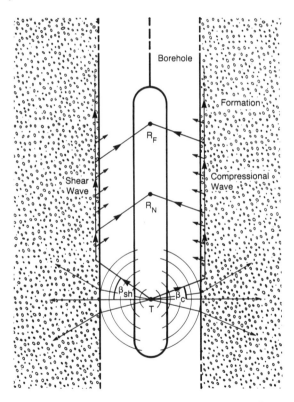

Figure 17-1. A standard sonic tool in the centered logging configuration. From Tittman.[1]

The first arrival transit time may characterize the undisturbed formation, depending on the source-to-detector spacings, the velocity contrast between the invaded (or altered) and undisturbed zone, and the thickness of this altered zone. Some authors have attempted to define a pseudogeometric factor for the first arrival time.² Their results, shown in Fig. 17–2, can be interpreted as giving the maximum thickness of an altered zone for which the measured Δt is still representative of the virgin formation. This depth of investigation increases with source-to-detector spacing and for increased velocity contrasts between the two zones. For the conventional sonic tool a typical value may be on the order of 6″.

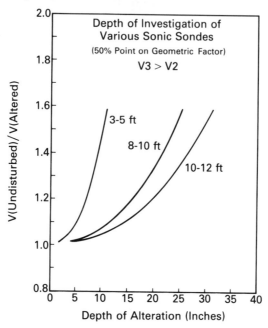

Figure 17-2. Estimates of the depth of investigation for three types of conventional sonic arrays. From Chemali et al.²

The typical presentation format for a standard sonic log is shown in Fig. 17–3. The formation transit time is presented in tracks 2 and 3. Increasing transit times (or slowness, as it is more chic to say in acoustic logging circles) are shown to the left, which is also the trend for increasing porosity. An additional trace consists of a series of little pips every so often. These are seen at the beginning of track 2 in the figure. Each pip represents 1 msec of integrated travel time and serves as a reminder of the origin of the sonic log: It was developed to correlate time with depth in seismic sections.

The conventional sonic log presentations for portions of the simulated reservoir are shown in Fig. 17–4. The bottom zone is from the carbonate section, and the upper from the shaly sand section. To emphasize the porosity sensitivity of the sonic log, comparison with the neutron and density logs over the same intervals is provided.

The use of two-detector devices, with normal spacing of 3′ and 5′ from transmitter to receiver, introduces some problems with the resolution of thin

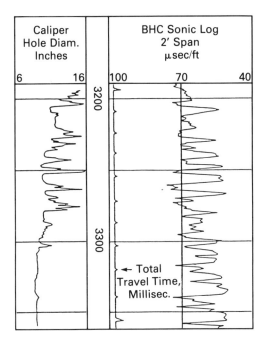

Figure 17-3. A standard acoustic log presentation format with the integrated travel time pips. From Timur.[3]

beds whose compressional velocity is different from the surrounding medium. This problem is considered in Fig. 17–5. which shows a fast limestone bed surrounded by slower shales. As indicated in part a, if the span between the two receivers exceeds the bed thickness, then the measured Δt will never attain the true value but some weighted average over a length which is equal to the difference between the span and the bed thickness. In part b of the figure, the span is shorter than the thickness of the bed, and for a short stretch, the value of Δt attains the true value of the fast formation.

One of the real advantages of sonic logging is its relative insensitivity to borehole size variations. Fig. 17–6 qualitatively compares borehole size effects for the density device, the neutron porosity device, and an acoustic tool. In the section of log shown, there is an enormous borehole irregularity over a 20' interval that is seen on the caliper trace. Although the $\Delta\rho$ curve for the density log is not shown, an experienced interpreter might question the value of ρ_b indicated in this region. In large irregularities like this, the skid of the measuring device cannot possibly follow the borehole profile and consequently rather large tool-formation stand-off can develop. The compensation can cope only with gaps which are generally less than 1". For cases which exceed this stand-off, the correction will not be sufficient, and the bulk density will indicate a value which may be much too low.

In the middle log, the neutron porosity can be seen to be nearly flat just below the shale section. However, at the point of largest caliper activity

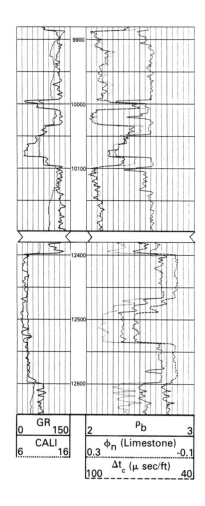

Figure 17-4.   A sample interval transit time log from the simulated reservoir model.

Acoustic Logging Methods and Applications 387

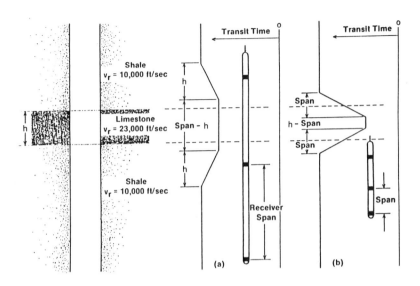

Figure 17-5. The effect of the spacing between two receivers on the measured acoustic travel time of a thin high velocity bed. From Timur.[3]

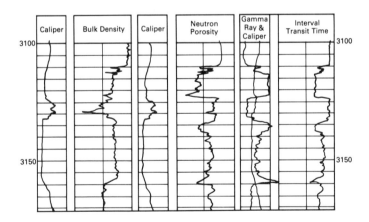

Figure 17-6. Illustration of the insensitivity of the acoustic measurement to extremely poor borehole conditions. Compared are the responses of a density and a neutron porosity tool; both show incorrect responses in the caved section of the borehole. From Timur.[3]

there is also a peak in the neutron porosity, which no doubt is the result of an inadequacy in its compensation scheme. The dramatic data, however, are in the third log, which shows the measured Δt in this zone. It is very steady despite the large caliper excursions. No doubt the porosity of this zone is quite constant, a fact that could not be deduced from either of the other two porosity devices. The question remains: How is the sonic borehole compensation achieved?

As Fig. 17-7 indicates, in a region of changing borehole size, a single source-detector sonic device will measure abnormally long transit times when the hole becomes enlarged. This is a result of the increased transit time from the transmitter across the mud to the formation and back to the receiver. A partial solution to this problem is obtained by use of one transmitter and two receivers. By determining the travel time to the two detectors and using the difference to determine the travel time, as indicated in the figure, the effect of the borehole diameter is eliminated except at the boundaries, where "horns" can appear on the log response.

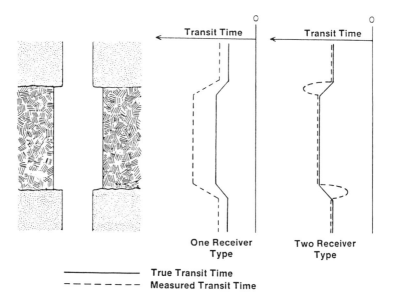

Figure 17-7. The effects on the integrated travel time at the boundaries of hole diameter changes. From Timur.[3]

The more general logging situation is shown in Fig. 17-8. Not only can there be changes in borehole diameter, but the tool is not necessarily centered in the borehole, because of deviation of the borehole and differential sticking of the tool string. As indicated, this more general case can be solved by the use of two transmitters and two pairs of closely spaced receivers. Two sets of differential travel time measurements are made: an up-going one and a down-going one. In the case sketched in Fig. 17-8, the up-going transit time will exceed the down-going case. By averaging the two results, the effect of existing unequal mud travel paths is eliminated, and the measurement reflects the travel time of the formation. Tools of this type are said to be borehole compensated (BHC).

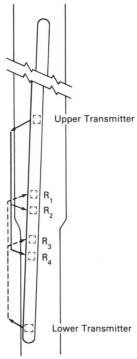

Figure 17-8. The use of four detectors to compensate for borehole size and tool tilt.

Another variation of this technique, used for a long spacing device to be discussed later, is shown in Fig. 17-9. In this case, two transmitters and two receivers are used to produce the same result as the six-transducer tool of Fig. 17-8. For the long spacing device, there are two receivers at the top of the tool and two transmitters at the bottom (a saving of two transducers). The measurement is made in two phases. At one position in the well, the bottom transmitter fires and the transit time between the two top receivers is measured. Shortly after, when the tool has moved so that the two transmitters are nearly in the position previously occupied by the two receivers, the two transmitters are fired in succession, and the two transit times (from the different transmitters) are measured to the lower detector. As the figure indicates, this is equivalent to the use of two transmitters and four receivers, and the technique is referred to as depth-derived borehole compensation.

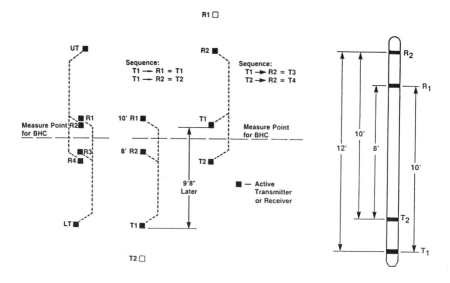

Figure 17-9. The principle of the depth-derived borehole compensation. Courtesy of Schlumberger.

Some Typical Problems

Despite the good performance of the compensated sonic tool, as noted above, there are a few situations which can cause problems. One of them results from the possibility that, in slow formations, with very large borehole sizes, the direct mud arrival will precede the formation arrival. In the conventional sonic tool, an amplitude rise in the detected pulse is sensed to determine the first arrival. However, it is not necessarily the result of a signal from the formation. Because of the generally large contrast between formation compressional velocities and mud velocities (generally the formation velocity can exceed the mud velocity by a factor of 2), the formation arrival and mud arrival separation can be increased by simply increasing the distance between transmitter and receiver. However, for a given spacing it is possible for the two signals to overlap, if the mud transit time to and from the formation is large (because of a very large borehole size). This notion is quantified in Fig. 17–10. The area of reliable Δt measurements is indicated for receivers at three different distances: The slower the formation, the smaller the borehole size must be in order to see the formation arrival before the direct mud arrival. The situation improves dramatically for increased spacing.

One serious environmental effect for the sonic device is that of damage or alteration of the material near the borehole wall. Generally this occurs in some clays, commonly known as swelling clays, which take on water, expand, and suffer changes in density as well as velocity. Another source of alteration can be induced stress-relief fracturing around the well bore, which

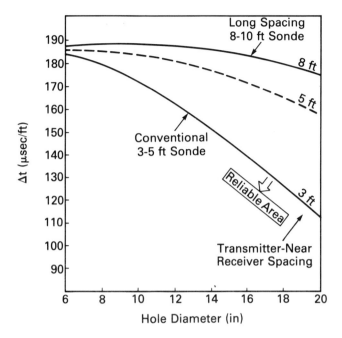

Figure 17-10. Areas of confidence for conventional two-receiver tools as a function of borehole diameter. From Goetz et al.[4]

can largely alter the acoustic properties of the material. A striking example of the former type of shale alteration is shown in Fig. 17–11, which shows the transit time measured in the same well two months apart. In general the transit times have increased by about 20 μsec/ft due to the shale alteration. In a case such as this, the fastest travel time is through the slower altered medium. This is due to the thickness or depth of alteration; the two-way travel time through it to reach the faster undamaged formation exceeds the time difference between them, and thus the first signal to arrive travels only through the altered zone.

An annoying feature which sometimes appears on acoustic logs is cycle-skipping, which is shown in Fig. 17–12. This condition is immediately recognized by the spiky nature of the Δt trace; apparent travel time changes on the order of 40 μsec/ft are visible. Fig. 17–13 indicates the origin of these problems: Either the timing circuitry is triggered by random noise, or the anticipated signal strength falls below that expected and the arrival is not detected until a full cycle (≈40 μsec) later.

Newer Devices

The preceding discussion concentrates on devices which extract the simple interval transit time, or compressional wave velocity. However, a few

392 Well Logging for Earth Scientists

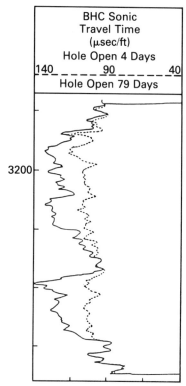

Figure 17-11. Log example of the effect of formation alteration observed between two logging runs with 75 days of elapsed time. From Timur.[3]

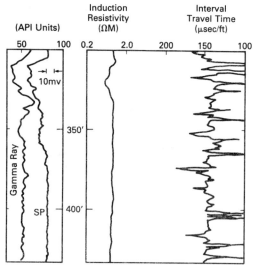

Figure 17-12. An example of cycle-skipping on the interval transit time log. From Timur.[3]

Acoustic Logging Methods and Applications   **393**

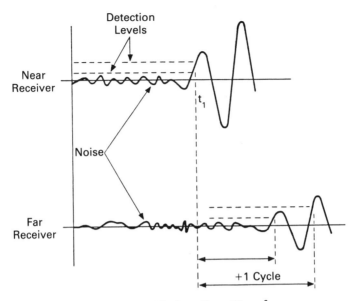

Figure 17-13. The origin of cycle-skipping. From Timur.[3]

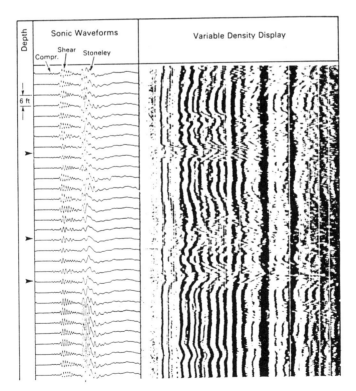

Figure 17-14. One method of identification of shear arrivals: the variable density plot. Courtesy of Schlumberger.

devices extract other information from the complex acoustic waveforms, such as the shear wave velocity. These newer devices, of varying sophistication, basically provide access to a large portion of the received wavetrain for signal processing rather than providing a crude analysis such as the first arrival time.

The measurement of compressional velocity was natural. Since this wave has a larger velocity, it always arrives at the detector first. The shear arrival can be masked if it arrives in the midst of the ringing portion of the transmitter signal, or it can be lost in some of the other modes produced by acoustic waves in boreholes. In the best cases, the problem of detecting the shear arrival is one of just looking for it after the compressional arrival. Such a trivial case is illustrated in Fig. 17–14. In the waveform shown, the shear arrival is clearly distinguishable. The accompanying VDL (variable density log) presentation, shows quite clearly the variations in arrival time for the compressional and shear waves: The positive amplitudes of the waveform are replaced by dark bands as one waveform after another is reduced in this manner.

From such a demonstration, it seems possible to extract the shear travel time by judicious time-windowing of the received waveform from the conventional sonic tools. However, the recorded waveform must be available for analysis. Until recently this has been a somewhat tedious specialty, but it has been done, on occasion, nevertheless. With the development of longer spacing sonic devices, shear velocity determination has become less difficult.

The stacked waveforms from six receivers placed from 3' to 16' from the transmitter, as shown in Fig. 17–15, indicate that the separation of shear from

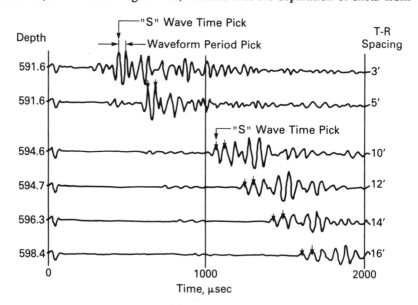

Figure 17-15. The effect of spacing on the separation of shear and compressional arrivals. From Timur.[3]

compressional arrivals is greatly aided by spacing. This is one of the motivations for the development of long spacing sonic sondes, devices which are currently coming into vogue. A second reason for their development is to combat the problem of the altered zone. As might have been guessed from the results of Fig. 17-10 and the discussion on depth of investigation, the longer the spacing, the more reliably one can measure the transit time of the faster, undamaged formation. A common detector configuration is at 8' and 10' from the transmitter. The results of such a device compared to the conventional 3'-5' spacing device are compared in Fig. 17-16. In this zone of nearly constant velocity, the conventional device is seen to read Δt values which are much too high.

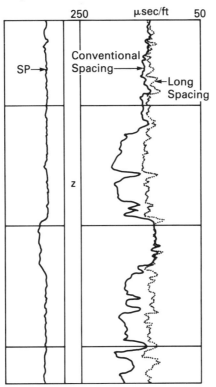

Figure 17-16. Comparison of the response of a conventional and long spacing sonic tool in an altered formation. From Timur.[3]

The quantification of this improvement can be found in Fig. 17-17. It indicates the reliable zone of measurement (to the left of the indicated curves) for the conventional and long spacing sonde. The change in velocity of the damaged zone which can be tolerated as a function of its thickness is indicated as a function of the formation travel time. In a formation characterized by a Δt of 100 μsec/ft, a 20 μsec/ft alteration can be tolerated up to 5" thick for the conventional sonde, but it can be up to 14" thick with the long spacing device. This tolerance to alteration also eliminates problems with direct mud arrival in very large borehole sizes.

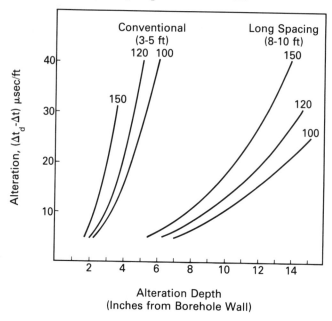

Figure 17-17. The effect of alteration as measured between the observed interval transit time and the formation transit time as a function of alteration depth into the formation. As expected, the long spaced tool is able to tolerate deeper alteration before noticeable effects appear. From Goetz et al.[4]

Acoustic array tools containing a battery of receivers, variable detector spans, and waveform digitization are making progress in extracting reliable shear measurements. An example of one of these devices with eight receivers is shown in Fig. 17-18. In an array tool such as this, waveform processing and signal extraction is aided by the possibility of stacking signals recorded at the same depth from different receivers for noise elimination as well as discrimination against other acoustic signals produced within the borehole. Typical waveforms from the eight receivers are shown in the upper portion of Fig. 17-19 with a clear indication of the compressional, shear, and Stoneley arrivals. To deal with cases less clear than this example, an elaborate signal processing scheme known as slowness-time coherence has been successful in extracting the various arrivals.[6] Basically, it measures the similarity of the eight wave forms by comparing a portion of wavetrain 1 to shifted portions of the other seven waveforms. Using this processing, a plot such as that shown in the bottom portion of Fig. 17-19 can be developed. The ordinate is the time (in msec) along the wavetrain of receiver 1. The abscissa is Δt or slowness, determined from the conversion of the delays applied to the other

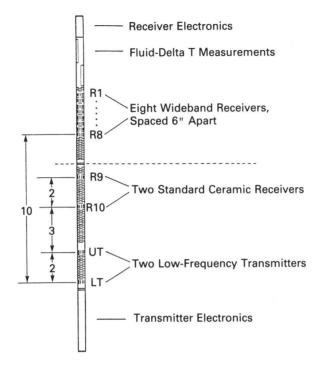

Figure 17-18. An eight-receiver sonic array tool. From Morris et al.[5]

receiver waveforms. The contours which clearly delineate the three arrivals indicate regions of largest similarity between the shifted waveforms and the original signal. A sample log section from this type of processing is shown in Fig. 17-20. In addition to the interval transit time of the compressional, shear, and Stoneley waves, a log of their amplitude can be produced which has application in the detection of fractures.

ACOUSTIC LOGGING APPLICATIONS

One of the first extensive uses of borehole sonic logs was for correlation. Wyllie and others observed that there was a strong correlation between sonic travel time and the porosity of consolidated formations. This resulted in the so-called Wyllie time-average equation discussed earlier. The laboratory data seemed to indicate that a volumetric mixing law held for the case of transit time. Knowing the matrix transit time and the fluid transit time, one could obtain the appropriate porosity from a measurement of any intermediate travel time.

Figure 17-21 is an example from a chartbook which shows the solution to the time-average expression for the three common matrices.[7] However,

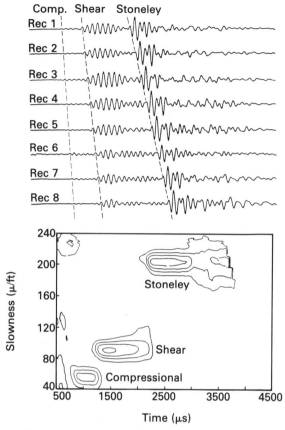

Figure 17-19. Waveforms from the eight receivers recorded in a formation showing distinct compressional, shear, and Stoneley arrivals. In the lower portion, the slowness-time coherence processing of the waveforms identifies the three arrivals and quantifies the interval transit time and relative amplitude of each. From Morris et al.[5]

note the cluttered appearance of the chart. The three principal lines correspond to a solution of:

$$\phi = \frac{\Delta t - \Delta t_{ma}}{\Delta t_f - \Delta t_{ma}},$$

where the fluid velocity has been fixed to 5300'/sec. In addition to the three linear relations expected for three matrices of differing matrix travel times, there are three slightly curved lines for the same three matrices. Raymer et al., established these additional transforms, which correspond to their judgments based on the observation of much field data.[8] They basically take into account that the sonic travel time seems to consistently underestimate the porosity in midrange. The additional lines to the right, carrying the notation B_{cp}, or compaction factor, correspond to an empirical method for correcting

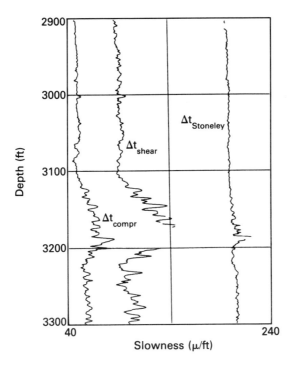

Figure 17-20. A log of compressional, shear, and Stoneley interval transit time from an array tool and signal processing. From Morris et al.[5]

the transit time measurements for formations which are not sufficiently compacted or which do not have sufficient effective stress ($\Delta p \approx 4000$ psi).

Despite the very empirical approach to the common interpretation of sonic logs, they do yield useful porosity estimates under many circumstances. For the technique to work well, the type of rock and its appropriate matrix travel time must be known, or a local transform between travel time and porosity must be established. Often, in cases of extreme hole rugosity or washout, when porosity readings from the density or neutron devices are useless, the sonic measurement will still be reliable.

Determination of other interesting formation properties can be made by using the sonic measurement in conjunction with other logging tools. One example of the sonic measurement in conjunction with two other porosity measurements is shown in Fig. 17–22. In the middle of the log, both the neutron and density readings indicate an increase in porosity, to a maximum of about 3 PU. The rest of the zone appears to be a zero porosity dolomite. However, in this same zone, where the dolomite appears to have 3% porosity, notice that the Δt curve has not shifted. It still indicates zero porosity.

The explanation for this type of occurrence, especially in carbonate rocks, is the presence of so-called secondary porosity. This is thought to be porosity which is unconnected with the majority of the pore systems in the rock. For

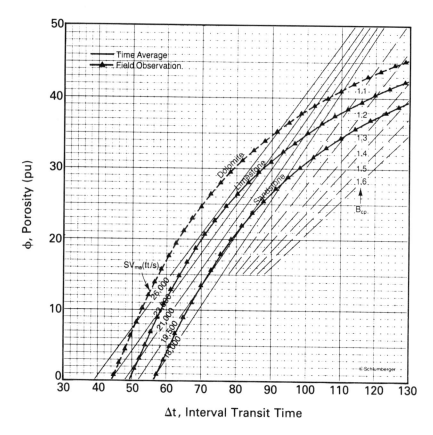

Figure 17-21. Chart for estimating porosity from compressional interval transit time. In addition to the Wyllie time-average solution and compaction corrections, another empirical solution by Raymer et al. is shown.[8] From Schlumberger.[7]

inclusions of porosity, unconnected to the rest of the rock, the acoustic wave energy follows the faster path around the inclusion in the rock matrix. In this case, there is no alteration of the travel time from that of the zero porosity matrix. In this sense, the sonic measurement does not "see" the secondary porosity.

Lithology and Pore Fluid Identification

The determination of lithology from borehole acoustic measurements is based on the variations of elastic parameters between rock types. These variations are reflected in the shear and compressional velocities. One convenient method of classifying lithologies is to compare the compressional velocity to

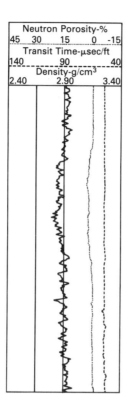

Figure 17-22. A log example of a carbonate section, showing indications of the presence of "secondary" porosity. This is inferred from the increase of density porosity and neutron porosity with no corresponding change in the Δt. From Timur.[3]

the shear velocity, as suggested by Pickett.[9] His laboratory and field data for many different formations showed that measurements corresponding to limestone and dolomites were found along lines of constant but different ratios: $v_c/v_s \approx$ 1.9 for limestone and $\approx$ 1.8 for dolomite. Sandstones showed a variation of velocity ratio from about 1.6 to 1.75, with the upper limit corresponding to high porosity sands under low effective stress.

Pickett's compilation of field points was made from painstaking manual analysis of recorded wavetrains. With the availability of routine shear and compressional velocities this technique becomes useful for lithology determination and the identification of gas. Figure 17–23 is a cross plot of the interval transit time for shear and compressional waves in the format of Pickett's original work. It is a composite of logging data from four different wells which contained, dolomite, limestone, halite, and sand formations. Some of the latter were gas-bearing. As found earlier, the limestone and dolomites fall on lines of constant ratios. The water-filled sandstone ratios

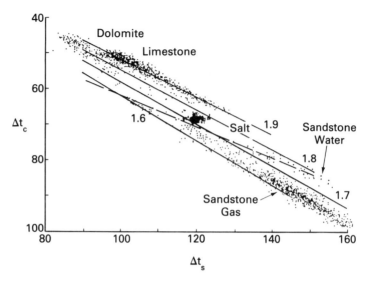

Figure 17-23. Composite log data of shear and compressional interval transit time in the presentation proposed by Pickett for lithology identification.[9] Identification of dolomite, limestone, and salt is indicated. The trends observed for water- and gas-saturated sandstones are predicted by Biot theory. From Leslie and Mons.[11]

vary from 1.6 to 1.8. However, the gas-filled points lie along a constant ratio of 1.6. This behavior has been shown to be consistent with predictions of the Biot theory.[10] In order to exploit this possibility, the logging tool must have the capability of separating the shear from the compressional arrivals.

In the case of simple compressional sonic logging, the effect expected in a gas zone is a large reduction in the compressional velocity. This corresponds to a large increase in Δt determined from the compressional wave arrival. An example from such a zone is seen in the log of Fig. 17–24. There is clearly a long zone over which the Δt is constant and much higher than elsewhere in the section. In a portion of the zone, the induction tool resistivity reads correspondingly high. But there is a curious phenomenon to explain. In the bottom portion of the zone, where the resistivity is much lower and the consequent gas saturation determination is much lower, the sonic log continues to indicate gas. An explanation for this effect was given earlier: The compressional velocity is not sensitive to saturation beyond the initial reduction seen at very small values of gas saturation.

Formation Fluid Pressure

One of the most practical applications of sonic compressional data is in the detection or prediction of overpressured shale zones. The procedure is schematically illustrated in Fig. 17–25. Normally the Δt measured in shales

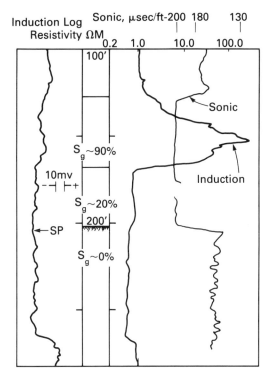

Figure 17-24. Gas effect observed in simple compressional wave logging. As expected, there is no sensitivity to saturation. From Timur.[3]

increases in a regular, if not logarithmic, fashion with depth or compaction. This may be related to changes in density as well as confining pressure as a function of depth. However in overcompacted shales, where the pore pressure is greater than hydrostatic pressure, a deviation from the normal trend is observed. According to Hottman and Johnson, the overpressure can be estimated from the difference in the observed transit time Δt_{ob} and the normal transit time Δt_n expected for that depth.[12]

Mechanical Properties and Fractures

From the relationships reviewed in Chapter 15, it is relatively easy to show that the elastic parameters of a rock formation can be obtained from a knowledge of its compressional and shear wave velocity, and its density. For example, the bulk modulus is given by:

$$B = \rho(v_c^2 - \frac{4}{3}v_s^2),$$

and the shear modulus by:

$$\mu = \rho v_s^2.$$

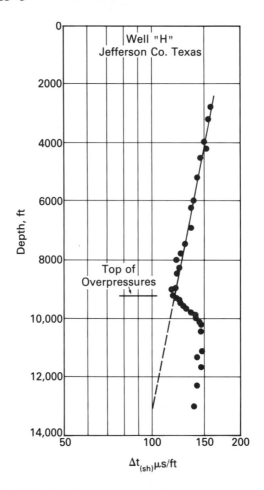

Figure 17-25. Determination of overpressured shale zones from a departure from the Δt-depth trend. From Hottman and Johnson.[12]

Until the development in the late 1970s of full waveform acoustic logging, the determination of shear velocities was a tedious exercise, and it was not routine to apply it in determining the mechanical properties of well bore rocks. However, one potential application received considerable attention: the prediction of fracture initiation.

The compressive strength of a rock is the stress required to produce a rupture or fracture. It depends on the shear modulus of the rock and substantially increases with applied pressure that results from internal friction. In porous rocks, the compressive strength is reduced by pore pressure. The difference between overburden pressure and pore pressure, or the effective confining pressure, plays an important role in the mechanical properties of rocks. The tensile strength of rocks, on the other hand, is only a fraction of the compressive strength. This is especially the case in poorly cemented sandstones. Fractures can be induced by producing a pressure in the well bore which will cause rock failure to occur. At any point on the borehole wall three forces are acting: the overburden pressure, the effective tangential stress which is a result of the overburden and pore pressure, and the pressure which is the result of the mud column. If the mud column pressure, P_{wf}, increases, it can effectively cancel the compressive tangential stress and place the rock forming the borehole wall in tension. Once the tensile strength is exceeded, the fracture is initiated. Hubbard gives an estimate of the fracture pressure P_{wf} to be:

$$P_{wf} = 2(P_o - P_f) \frac{\nu}{1-\nu} - P_f ,$$

where P_o and P_f are the overburden and pore pressures.[13] This expression involves a number of simplifications. The first is that the rock can be considered as an elastic material, so that the ratio of horizontal to equal vertical stresses is simply given by:

$$\frac{\nu}{1-\nu}$$

and the tensile strength of the rock is negligible. Although not the complete story,* a continuous measurement of the Poisson ratio is useful in predicting the relative ease of fracturing a formation.

There are two practical aspects to the fracture pressure of a formation. The hydrostatic pressure of the mud column must not exceed this value or unwanted fractures will be induced in the formation, causing caving of the borehole walls or worse. On the other hand, knowledge of the elastic parameters allows prediction of zones suitable for hydrofracturing. This technique consists of hydraulically isolating a zone which is pressurized by surface pumps and increasing the borehole pressure until fracturing of the rock occurs. This method is often used to stimulate production of low porosity and low permeability hydrocarbon reservoirs.

Identification of naturally occurring fracture zones from acoustic logs has received much attention in the past and is extensively reviewed by Timur[3] and by Jordan and Campbell.[14] Most of these methods, which use some portion of the acoustic wavetrain displayed in an analog form, are qualitative. One approach to the determination of the presence of fractures is to analyze the amplitude of the shear arrivals; the presence of fractures should decrease the ability of shear waves to be transmitted. Another method is a crude type of acoustic imaging. The presence of fractures intersecting the borehole wall is inferred from a characteristic chevron pattern in the displayed analog representation of the acoustic signal. The development of these chevron patterns arises from the conversion of shear and compressional energies at the

---

* A thorough treatment can be found in "Rock Mechanics," *The Technical Review*, Vol. 34, No. 3, 1986.

boundary of the fracture. These novelties will be rapidly displaced with the routine measurement of the important features of the detected wavetrain from the long spacing array devices mentioned earlier.

The presence of fractures can also be deduced from the Stoneley wave amplitude. In zones containing fractures, the Stoneley wave amplitude has been seen to change considerably, producing a noticeable effect on the VDL presentation.[15] Figure 17-26 illustrates the notion that, in passing a fracture (in this case depicted as horizontal and open), the low frequency tube wave actually forces fluid into the fracture, thereby reducing its amplitude. Note also that a given fracture will be apparent in the measurement over a distance equivalent to the transmitter-receiver spacing.

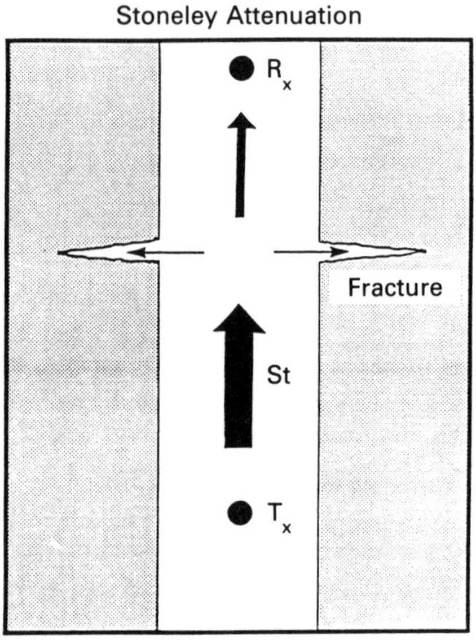

Figure 17-26. Schematic illustration of the amplitude reduction of a Stoneley wave as it passes an open fracture.

Permeability

One of the most important rock properties for the evaluation of hydrocarbon reservoirs is permeability. This parameter, related to the ease of flow of fluids through a porous medium, is obtained routinely through core analysis, well testing, or correlation schemes with other more easily measured rock properties such as porosity. These latter schemes have varied success. There is some evidence that improvements in log-derived permeability may come from acoustic logging.[16,17,18]

This exciting possibility for acoustic logging is a correlation between the normalized energy of the Stoneley wave and core-derived permeability. In

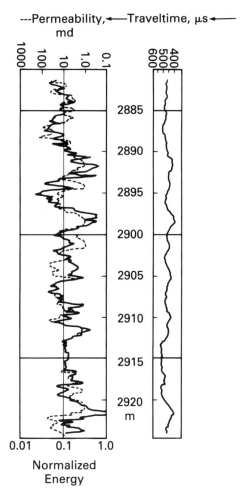

Figure 17-27. Correlation between the normalized Stoneley wave energy and the core-derived permeability. From Staal and Robinson.[19]

this particular example (Fig. 17–27) the correlation is seen to be very good over most of the zone. Another log example, in Fig. 17–28, shows good correlation using the ratio of Stoneley wave amplitudes at two receivers with the core permeability. The third track indicates a similar correlation with the Stoneley wave velocity. This example is intriguing since it contains a thick mudcake.

Various explanations for this effect have been proposed and relate to the loss of acoustic energy to formation permeability through the physical fluid motion in the porous medium induced by the passing pressure wave. If the formation is porous, then the fluid should move easily under this pressure wave: otherwise it does not. Why does this seem to work when there is a mudcake present?

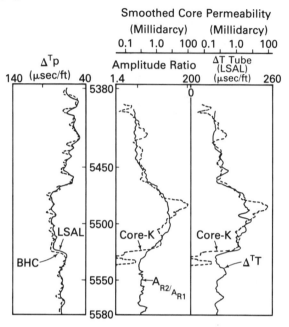

Figure 17-28. The Stoneley wave amplitude and its interval transit are compared to smoothed core-derived permeability. The zone of high permeability was reported to have a thick 3/4" mudcake present. From Williams et al.[18]

ULTRASONIC DEVICES

Unlike the conventional sonic tools which operate at frequencies in the range of 10–20 kHz, ultrasonic devices employ transducers capable operating from several hundred kHz to the MHz region. At these frequencies the wavelength can be as small as the millimeter range. This opens the possibility of performing a number of different types of measurements. One of the most widely used is that of acoustic imaging. The use of ultrasonic measurements also holds some promise for investigations of formation permeability. Biot theory indicates that at some critical frequency, dependent on permeability and viscosity, there should be a maximum in the attenuation of acoustic energy at frequencies between 10^4 and 10^5 Hz. This is the result of viscous losses between the fluid and the frame of rock. A second possibility is the measurement of grain size, since at very high frequencies ($\approx 10^7$ Hz) grain size becomes important as a loss mechanism. However, these are currently the subject of laboratory investigations. We turn instead to two logging tool applications.

One imaging device which has been used extensively for fracture identification is a specialized tool known as the BHTV, or televiewer, log. In this device, the centralized source and receiver rotates rapidly (see Fig. 17-29) to obtain a finely wound spiral image of the reflected signal from the borehole wall. Images can be made from either the amplitude of the reflected signal or its transit time. The presentation represents the unwrapping of the cylindrical borehole wall image. An example of the amplitude-derived image is shown in Fig. 17-30, which indicates the presence of vertical and dipping fracture planes. The dipping fracture appears as a sinusoid in this unwrapped display.

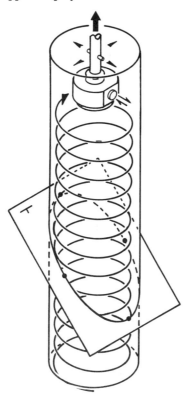

Figure 17-29. The principle of BHTV operation. A rapidly rotating high frequency transducer transmits and receives pulses of acoustic energy as it moves in the borehole. Bedding features which intersect the well will produce a characteristic sinusoidal pattern on the unfolded borehole wall image. Courtesy of Schlumberger.

Another device which operates in the reflection mode is used to evaluate the condition of cemented casing. Its operation is illustrated in Fig. 17-31. An ultrasonic transducer is used both as transmitter and receiver. In front of a steel casing, bonded to cement, the received signal contains several pieces of information. The internal reflections of acoustic energy in the casing can be used to determine its thickness and thus monitor its wear. The decay rate of the signal following the casing arrival is determined by the coupling between the casing and cement. The primary objective of this tool is to determine zones where the cement may be imperfect. In the logging tool an array of transducers is present to provide complete circumferential coverage.

Figure 17-30. An image obtained from the BHTV, showing both vertical and dipping fractures. Courtesy of Schlumberger.

Acoustic Logging Methods and Applications   **411**

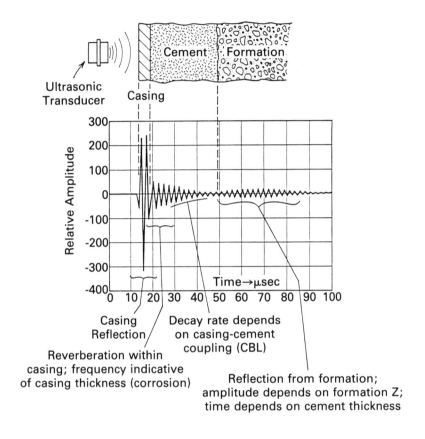

Figure 17-31. Use of an ultrasonic device for the inspection cement behind a steel casing. From Havira.[20]

References

1. Tittman, J., *Geophysical Well Logging,* Academic Press, Orlando, 1986.

2. Chemali, R., Gianzero, S., and Su, S. M., "The Depth of Investigation of Compressional Wave Logging for the Standard and the Long Spacing Sonde," Ninth S.A.I.D. Colloquium, Paper 13, 1984.

3. Timur, A., "Acoustic Logging," in SPE *Petroleum Production Handbook,* edited by H. Bradley, SPE, Dallas (in press).

4. Goetz, J. F., Dupal, L., and Bowler, J., "An Investigation into Discrepancies Between Sonic Log and Seismic Check Shot Velocities," *APEA Journal,* Vol. 19, 1979.

5. Morris, C. F., Little, T. M., and Letton, W., "A New Sonic Array Tool for Full Waveform Logging," Paper SPE 13285, Fifty-ninth Annual Technical Conference, Houston, 1984.

6. Kimball, C. V., and Marzetta, T. M., "Semblance Processing of Borehole Acoustic Array Data," *Geophysics,* Vol. 49, 1984.

7. *Schlumberger Log Interpretation Charts,* Schlumberger, New York, 1985.

8. Raymer, L. L., Hunt, E. R., and Gardner, J. S., "An Improved Sonic Transit Time-to-Porosity Transform," Trans. SPWLA, 1980.

9. Pickett, G. R., "Acoustic Character Logs and Their Applications in Formation Evaluation," *JPT,* Trans. AIME, June 1963.

10. Anderson, R. A., "Fluid and Frequency Effects on Sonic Velocities," Trans. SPWLA, Twenty-fifth Annual Logging Symposium, 1984.

11. Leslie, H. D., and Mons, F., "Sonic Waveform Analysis: Applications," Paper GG, SPWLA Twenty-third Annual Logging Symposium, 1982.

12. Hottman, C. E., and Johnson, R. K., "Estimation of Formation Pressures from Log-Derived Shale Properties," JPT, June 1965.

13. Hubbert, M. K., and Willis, D. G., "Mechanics of Hydraulic Fracturing," Trans. AIME, Vol. 210, 1957, pp. 153–166.

14. Jordan, J. R., and Campbell, F., *Well Logging II: Resistivity and Acoustic Logging,* Monograph Series, SPE, Dallas (in press).

15. Paillet, F. L., "Acoustic Propagation in the Vicinity of Fractures Which Intersect a Fluid-Filled Borehole," Trans. SPWLA, Twenty-first Annual Logging Symposium, 1980.

16. Burns, D. R., and Cheng, C. H., "Determination of In-situ Permeability from Tube Wave Velocity and Attenuation," Trans. SPWLA, Twenty-seventh Annual Logging Symposium, 1986.

17. Hsui, A. T., Jinzhong, Z., Cheng, C. H., and Toksoz, M. N., "Tube Wave Attenuation and In-situ Permeability," Trans. SPWLA, Twenty-sixth Annual Logging Symposium, 1985.

18. Williams, D. M., Zemanek, J., Angona, F. A., Dennis, C. L., and Caldwell, R. L., "The Long Spaced Acoustic Logging Tool," Trans. SPWLA, Twenty-fifth Annual Logging Symposium, 1984.

19. Staal, J. J., and Robinson, J. D., "Permeability Profiles from Acoustic Logging," Paper SPE 6821, SPE of AIME, 1977.

20. Havira, R. M., "Ultrasonic Cement Bond Evaluation," Trans. SPWLA, Twenty-third Annual Logging Symposium, 1982.

Problems

1. From the log of Fig. 17–4, determine the two sets of constants, a and b, which best relate Δt to porosity, i.e., $\Delta t = a + b\phi$.

2. If the full waveform is acquired with an appropriate sonic logging tool, it is possible to extract the formation compressional velocity v_p and shear velocity v_s. With an additional knowledge of the formation bulk density ρ_b, the elastic properties of the formation can be specified.

   Suppose you do not have a measurement of the formation density but know instead the mud properties (ρ_{mud}, v_{mud}, and its bulk modulus). How would you go about obtaining, for example, Young's modulus and Poisson's ratio of the formation? Write explicit expressions for the two moduli.

3. The log of Fig. 17–32 shows the results of an induction-sonic run in a shaly sand. Using the sonic log and the interpretation charts, answer the following questions.

   a. Estimate the average porosity in the reservoir zone. What is the range of porosity in this zone?

   b. How can the sonic trace, in conjunction with the resistivities, immediately indicate the presence of hydrocarbons?

   c. By comparing the neutron and density curves of the same zone, shown on Fig. 17–33, you can identify the gas zone. Why does the sonic trace not indicate the presence of gas?

   d. To what is the sudden decrease in the sonic transit time at 12678' due? Is it a real feature? Explain by comparison to the other logs.

e. What is the composition of the zones above and below the reservoir? Why does the density increase while the sonic transit time remains constant?

f. In the zone from 12660' to 12680', determine the values of Δt_{ma} and Δt_{fl} which best match the porosity estimated from the density.

4. Data obtained from three consecutive reservoir zones is given in the table below, where the average Δt is compared with core porosity measurements. Your job is to identify the appropriate parameters to be used in the time-average equation for future use in determining porosity from sonic log measurements. Using the interpretation chart of Fig. 17–21, do you obtain a consistent result for the three zones? What conclusion can you reach about the nature of the three zones?

	Δt	ϕ_{core}
upper zone	95	21
	97	25
	93	23
middle zone	75	15
	87	23
	85	21
lower zone	75	15
	65	10

5. The depth of investigation of an acoustic tool was estimated to be on the order of 6". How does this compare to the wavelength of the transmitted pulse in the case of a 20% porous, water-saturated sandstone?

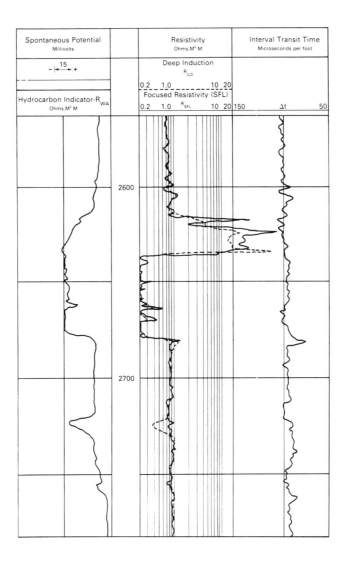

Figure 17-32. A typical low budget logging suite, the induction-sonic.

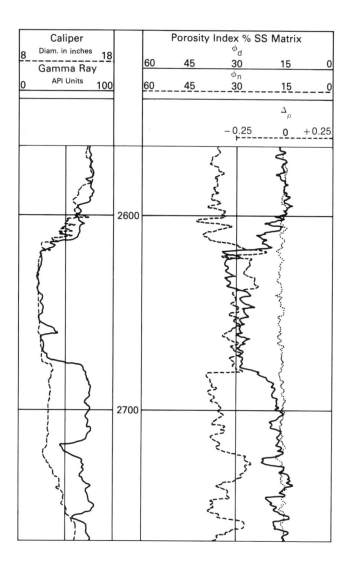

Figure 17-33. A companion density, neutron, gamma ray, and caliper for the log of Fig. 17-32.

18
LITHOLOGY IDENTIFICATION FROM POROSITY LOGS

INTRODUCTION

Porosity determination using the various logging devices presented in earlier chapters relies on a knowledge of the parameters related to the type of rock being investigated. In the case of the density tool, the density of the rock matrix must be known. The matrix travel time is used in interpreting the compressional wave interval transit time. In order to reflect porosity accurately, the matrix setting for the neutron tool must correspond to the rock type for the value of ϕ_n. Determining these parameters is not much of a problem if one has good geological knowledge of the formation and if the lithologies encountered are simple, such as a clean sandstone or limestone reservoir. However, what do you do when you are uncertain of the lithology, or if it is known to vary considerably in its composition, as in the case of limestone formations with variable inclusion of dolomite and anhydrite, or a sandstone with substantial calcite cementing?

To address this uncertainty, this chapter considers the question of lithology identification from log responses. The logs used are those which are primarily responsive to porosity yet retain some residual sensitivity to the rock matrix type. The techniques employed are simple graphical analyses developed in the 1960s and still useful today for quick evaluations.

For the graphical techniques considered, the term *matrix* is used to designate the principal rock types: sandstone, limestone, and dolomite. The lithological description often includes, in addition to the principal rock type, the presence of several minerals often encountered in logging. These include anhydrite, halite, gypsum, and others. To geologists, this list is limited but, for the most part, it usually is sufficient to solve the question of determining porosity in a previously unknown matrix.

In more complex lithologies there can be mixtures of many different minerals. For these cases one would like to have the use of a large number of logging measurements, each with a slightly different sensitivity to the various minerals, in order to make a complete mineralogical analysis. Graphical techniques for this type of analysis are inadequate. However, the problem can be solved numerically, and several approaches are considered. The ultimate approach to this problem may be based on nuclear spectroscopy and geochemistry, which is beyond the scope of this book but is touched on in the next chapter. This chapter is confined to the complex lithology analysis which can be unscrambled using conventional logging devices. It may, at some points, resemble a ramble through the chartbook.

GRAPHICAL APPROACH FOR BINARY MIXTURES

If we consider, for a moment, the response of the three porosity tools— density, neutron and sonic— we can idealize them as follows:

$$\rho_b = f(\phi, \text{lithology}, \cdots)$$

$$\phi_n = f(\phi, \text{lithology}, \cdots)$$

$$\Delta t = f(\phi, \text{lithology}, \cdots).$$

All three contain a dependence on porosity and a perturbation due to lithology. It seems natural to use these three measurements, two at a time, to eliminate porosity and thereby to obtain the lithology. This is precisely what is done in a number of well-known cross plotting techniques which are presented next, in order of their increasing usefulness.

The first is the density-sonic cross plot shown in Fig. 18–1. Because of the differing matrix densities and travel times for the three principal matrices, three distinct loci are traced out as water-filled porosity increases. As can be seen from the figure, there is not a great deal of contrast between the matrix endpoints. Since all the lines must join at 100% porosity, a bit of uncertainty in the measured pair (ρ_b, Δt) could cause considerable confusion in the ascribed lithology.

This is partially overcome in the next combination considered, neutron-sonic, which is shown in Fig. 18–2. In this case, the travel times are plotted as a function of the apparent limestone porosity for a thermal neutron porosity device. Due to the matrix effect of the neutron device, there is considerably more apparent separation between the three principal matrices which are shown.

Lithology Identification from Porosity Logs 419

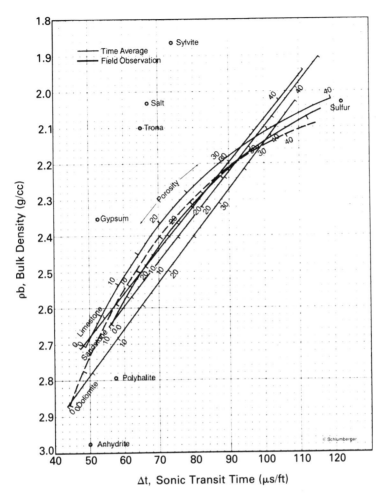

Figure 18–1. A density-sonic cross plot. Porosity variations of three major minerals produce trends of compressional interval transit time and bulk density. Location of measured pairs of values can help to identify the matrix mineral. Deviations from the trends can sometimes be attributed to significant portions of other minerals, some of which are also shown on the plot. From Schlumberger.[1]

The standard interpretation cross plot for binary mixtures, however, is the neutron-density, shown in Fig. 18–3. In this case the bulk density is plotted as a function of the apparent limestone porosity. The scale on the right is the conversion of the density to the equivalent limestone porosity for fresh-water pore fluid. To adjust this chart for other fluid densities, the density values are usually rescaled in accordance with the relation:

$$\rho_b = \phi \rho_{fl} + (1-\phi)\rho_{ma},$$

where ρ_{fl} is the density of the fluid filling the pores.

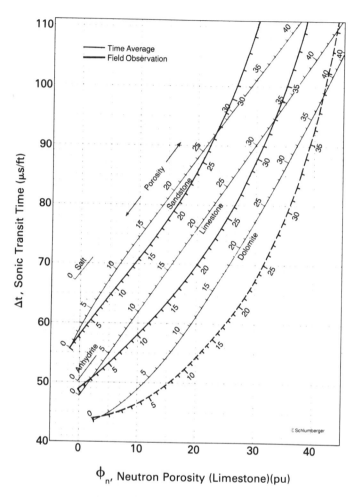

Figure 18–2. A neutron-sonic porosity cross plot showing an apparently larger resolving power for lithology discrimination. From Schlumberger.[1]

As an example of the use of the neutron-density cross plot, refer to the log of Fig. 18–4. It shows two apparent porosity traces from the density and neutron devices, scaled in limestone units. At the depth indicated, 15335', the density porosity reads about 2 PU, and the neutron 14 PU. To determine the lithology, we need only to find the intersection of these two points on Fig. 18–3. The surest way is to use the porosity scaling on the matrix curve for which the log was run. In this case it is limestone. By locating the 2-PU point on the Limestone curve, we see that the corresponding density value is about 2.68 g/cm^3. Dropping a vertical line from the 14-PU point for the neutron value (to the intersection with the horizontal density value), we see

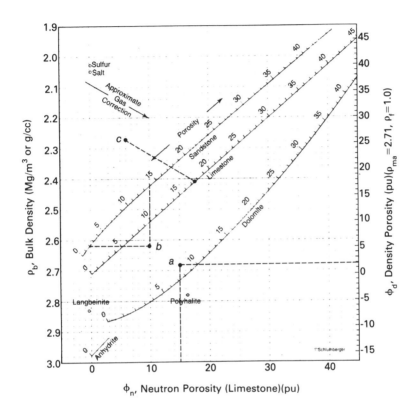

Figure 18–3. A neutron-density cross plot which is routinely used for lithology and porosity determination in simple lithologies. From Schlumberger.[1]

that the pair of points corresponds to a dolomite of about 8 PU, which is marked as point *a* in Fig. 18–3.

If the log of Fig. 18–4 had been run on a sandstone matrix and yielded the same apparent porosity values, the interpretation would be quite different. This can be seen by finding the 2-PU point on the sandstone curve in Fig. 18–3 which corresponds to a bulk density of about 2.62 g/cm^3. The 14-PU sandstone porosity for the neutron is equivalent to the reading expected in a 10-PU limestone. The intersection of these two points is marked at *b*. This corresponds to a formation which seems to be mostly limestone but could be a mixture of dolomite and sandstone, an unlikely possibility.

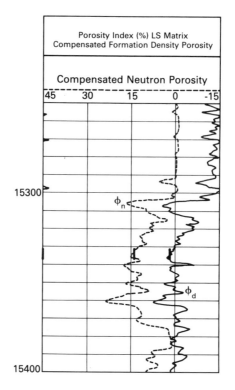

Figure 18-4. Log of apparent limestone porosity from a neutron and density device. The logged interval includes anhydrite, dolomite, and a streak of limestone. From Dewan.[2]

We have already seen the effect of gas on the neutron and density log presentation. Fig. 18-5 is another example which shows the evident separation in a 25' zone centered at about 1900'. The neutron reading is about 6 PU, and the density 24 PU. Both are recorded on an apparent limestone porosity scale. The location of this zone is shown as point C on the cross plot of Fig. 18-3. It is seen to be well to the left of the sandstone line. The trend of the gas effect is shown in the figure. Following this trend, the estimated porosity is found to be about 17.5 PU if the matrix is assumed to be limestone.

In the preceding example, two things are to be noted. The first is that even in the case of simple lithology mixtures, if the presence of gas is admitted, there is not enough information available from the neutron and density readings alone to decide on the matrix. From the available information, one could equally conclude that the formation was a gas-bearing sandstone as well. The second is the contrast between this visual method of identifying gas and the slowing-down length approach discussed in Chapter 12.

Lithology Identification from Porosity Logs 423

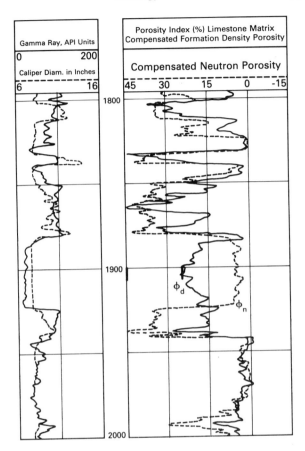

Figure 18–5. Gas separation between the neutron and density complicate the lithological analysis. From Dewan.[2]

The popularity of the neutron-density log combination for gas detection must be tempered with the need for knowing the matrix. Fig. 18–6 is a good example of a false gas indication running over nearly the entire section. What we have here is a tight sandstone formation (with some gas, perhaps) which has been presented in limestone units. The crossover is purely an artifact of the presentation in limestone units. Plotting a few of the points taken from the log onto Fig. 18–3 will be convincing.

It should be apparent from the previous few examples that in the case of gas, at least, additional information concerning the lithology is necessary. It could come from the addition of the sonic measurement, for example. Another possibility is use of the P_e, which can be obtained simultaneously with the density measurement. Fig. 18–7 shows an example of its use in a sequence of alternating limestones and dolomites. The first track contains the P_e with sections of dolomite and limestone clearly indicated. The density and

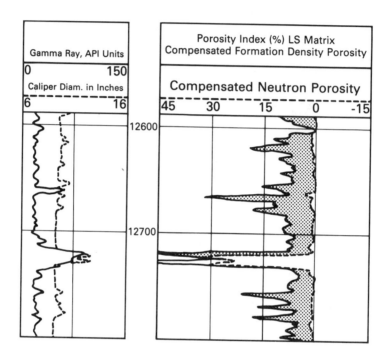

Figure 18–6. A neutron and density log in a carbonate. An inappropriate matrix setting has been used which indicates the presence of gas through most of the logged interval. From Dewan.[2]

neutron values which are presented in tracks 1 and 2 are on a scale for which water-filled limestone will show nearly perfect tracking. In the limestone zone, there is a clear separation indicating gas. However, in a short interval just below 10,000′, the density and neutron readings are nearly indistinguishable. Without additional information, this would be taken as a water-filled limestone. This can be verified by plotting the peak value of 21 PU for the neutron and 2.48 g/cm^3 for the density on the cross plot of Fig. 18–3. However, with the additional knowledge of the P_e, it can be seen to be a gas-bearing dolomite.

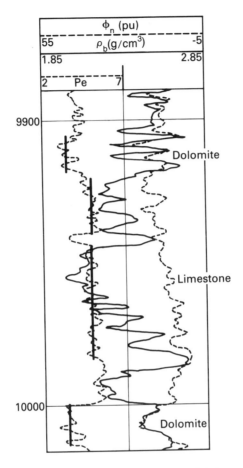

Figure 18–7. The companion P_e curve simplifies lithology determination in this sequence of alternating limestone and dolomites. The neutron and density information alone would not indicate gas in the lower interval. From Dewan.[2]

COMBINING THREE POROSITY LOGS

Before the availability of the P_e measurement, several methods were devised to combine the lithology information from the three porosity tools. The first approach was called the M–N cross plot. It attempts to remove the gross effect of porosity from the three measurements to deduce the matrix constants.

As indicated in Fig. 18–8 a combination of the sonic and density measurements is used to define the parameter M, which is nothing more than the slope of the Δt-ρ_b curve which varies slightly between the three major

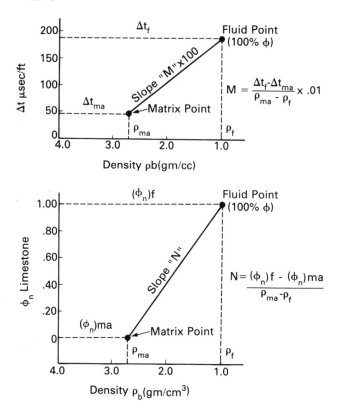

Figure 18–8. Idealized representations of density-sonic and neutron-density cross plots which define the M and N variables. Use of M and N allow two-dimensional representation of the three simultaneous logging measurements. Their definition effectively eliminates porosity from the response of each pair of measurements.

lithologies due to the matrix endpoints. The neutron-density cross plot yields a similar slope, designated as N. Once again the three matrix types produce slightly different values of N. The end product is indicated in Fig. 18–9 where the two slopes are plotted against one another. It corresponds to simultaneously viewing the two responses of Fig. 18–8 from the 100% porosity point. A number of frequently encountered minerals are shown in this manner.

M and N values can be obtained from log readings by replacing the matrix values in their definitions (see Fig. 18–8) by the appropriate log readings. If these pairs of values are plotted on the overlay of Fig. 18–9, it is possible to determine lithology, in the best of circumstances. The figure shows some spread in the matrix coordinates depending on the fluid density. Part of this is due to the nonlinearity of the neutron response, whereas the apparent values of M are determined by passing a straight line in the

Lithology Identification from Porosity Logs

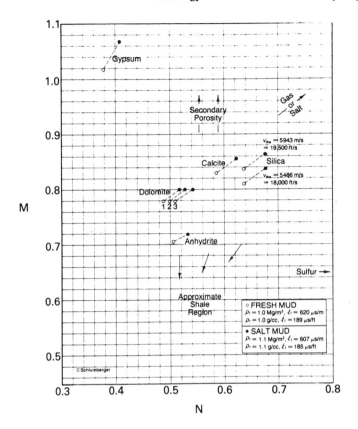

Figure 18-9. The M-N plot used for mineral identification. Porosity variation has nearly been eliminated, but there remains some sensitivity to the fluid density. This type of plot is frequently used to identify secondary porosity. From Schlumberger.[1]

appropriate space to the fluid-filled case. The location of the pivot point will change the values of M and N.

One of the best uses made of this type of presentation is to highlight the presence of secondary porosity, which causes M to change without any effect on N. This is because Δt remains constant for the inclusion of secondary porosity while, in the numerator of M, the density decreases. This will result in an apparent increase in M. For the case of N, both the density and neutron will change by about the same amount for the presence of secondary porosity, and thus there will be no change in its value.

The M-N plot was a first order attempt to get rid of the effects of porosity. Its successor, the MID (Matrix Identification) plot, goes one step further and tries, in a simplified way, to obtain the values of the matrix constants actually sought. The lower portion of Fig. 18-10 shows how this is done for the neutron and density values. Interpolating between the family of

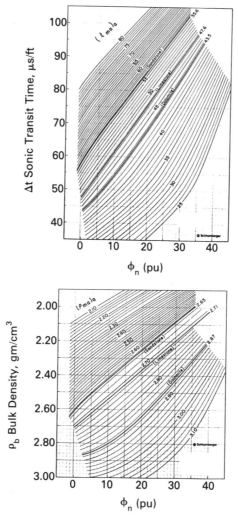

Figure 18-10. A alternative to the use of M–N variables is the use of apparent matrix values for Δt and ρ_b. These are obtained from the interpolated pair of cross plots using the same measurement combinations. From Schlumberger.[1]

curves of apparent matrix density $(\rho_{ma})_a$, the location of a point defined by (ρ_b, ϕ_n) determines the appropriate apparent matrix density. The upper half of Fig. 18-10 indicates how the apparent travel time of the matrix is determined from the neutron-sonic cross plot. Armed with these apparent matrix constants, we can enter the MID plot of Fig. 18-11. This diagram shows a considerably smaller spread of points than in the M–N plot, and the coordinates have some relationship to known physical parameters, rather than being abstract values.

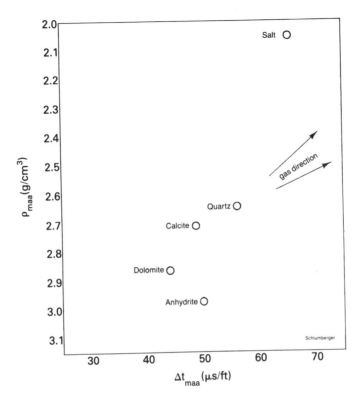

Figure 18-11. The matrix identification chart obtained using the defined values of apparent matrix density and Δt. From Schlumberger.[1]

One can debate the legitimacy of the linear interpolation of the matrix values used in the preceding figures; however, this approach does represent a large improvement over the M–N plot, for two reasons. First, it gets rid of a conversion to some rather meaningless parameters and attempts instead to find the more familiar values of matrix density and travel time. The second important point is that in this procedure the actual nonlinear tool response is taken into account rather than the linear slope determination of the M–N plot.* One weakness, however, is that this still is a method for presenting only three pieces of information simultaneously. It cannot be used if you want to consider ten simultaneous measurements.

---

* Concerning the nonlinearity of the neutron-density cross plot, the reader should consult Ellis and Case, who show that the curved "dolomite" line of the neutron-density cross plot represents the results of field observations, some in conditions of extreme water salinity and others in possible matrix mixtures of dolomite and anhydrite.[3] A clean dolomite formation should have a linear response.

LITHOLOGY LOGGING: INCORPORATING P_e

Before leaving the realm of two-dimensional graphical interpretation, let us consider a final example. In the previous techniques, measurements primarily sensitive to porosity were combined to eliminate their mutual porosity dependence and to emphasize their residual lithology sensitivity. However, one common measurement, the photoelectric factor, or P_e, is primarily sensitive to lithology and only mildly affected by porosity. An interesting aspect of the P_e measurement is illustrated in Fig. 18–12. On the left side is the conventional neutron-density cross plot. The ordering of the three major lithologies, from top to bottom, is sandstone, limestone, and dolomite, as is the case for the other possible combinations with the sonic measurement. The adjacent figure of P_e vs. ρ_b shows a startling difference. In this figure, the upper line is for limestone, and the dolomite matrix line is between the sand and limestone lines. Because of the reordering of the lithological groups, P_e adds a new dimension to the neutron-density cross plot. For this reason, use of the P_e easily resolves questions of binary lithology mixtures.

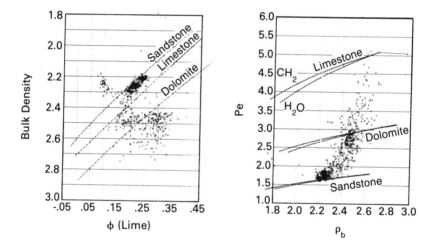

Figure 18–12. Comparison of log data on a neutron-density cross plot and a cross plot of P_e and density. Trend lines of three matrices from 0 to 50% porosity are shown in the plot of P_e. The upper lines correspond to hydrocarbon in the pores and the lower lines to water-filled porosity. The data points are from a shaly sand.

On the neutron-density cross plot, a data point falling near the limestone line might, in some extreme case, correspond to a dolomitic sand. If the P_e is available, this can be immediately confirmed or disproved. If, in fact, it is a lime/dolomite mixture, then it will lie above the dolomite line on the P_e plot and not below.

Recall that to obtain the P_e of a mixture involves computing the U value:

$$U_{total} = P_{e,1}\rho_{e,1}V_1 + P_{e,2}\rho_{e,2}V_2 + \cdots,$$

where $\rho_{e,i}$, is the electron density of material i, $P_{e,i}$ is the photoelectric factor of material i, and V_i is the volume fraction of that material. The final value of the average $\overline{P_e}$ is obtained from:

$$\overline{P_e} = \frac{U_{total}}{\overline{\rho_e}},$$

where the average electron density index $\overline{\rho_e}$ is given by:

$$\overline{\rho_e} = \rho_{e,1}V_1 + \rho_{e,2}V_2 + \cdots.$$

An interesting application of this calculation can be seen by referring to Fig. 18–12 and noting that two sets of lines have been drawn for the three matrices: one for oil-filled and the other water-filled porosity. The P_e value of water is 0.36 and that of oil 0.12. One might suspect that the upper curve of the three sets would be associated with water. However, the details of the calculation (see Problem 3) show that this is not the case.

The most obvious use of the volumetric cross section U is in combination with the grain density to clearly delineate lithologies. The approach is to eliminate porosity from the density measurement and from the P_e measurement to obtain the apparent grain density and the apparent value of U_{ma}. This latter parameter is conveniently found by use of the chart in Fig. 18–13. The measured bulk density and P_e values are entered into the left

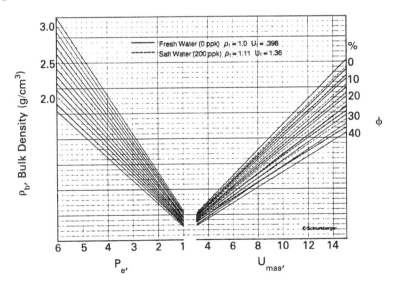

Figure 18–13. Chart for determining the apparent matrix cross section U_{maa} from a knowledge of P_e, density, and apparent porosity. From Schlumberger.[1]

of the figure, and U_{ma} is finally extracted on the right, based on knowledge of a cross plot porosity (from a neutron-density cross plot or other source). This type of data is then plotted on the overlay of Fig. 18–14, which shows a clear triangular separation between the three principal matrices. On this plot, the volumetric analysis of complex mixtures can be done very simply, since everything scales linearly.

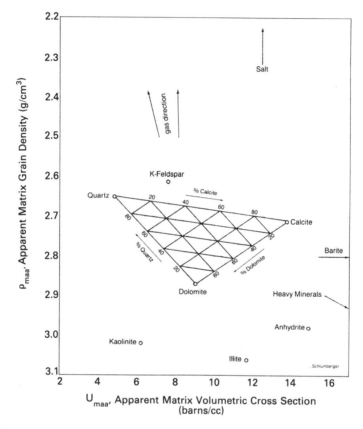

Figure 18–14. A matrix identification chart which uses the combination of P_e, density, and another porosity device. Both density and U scale volumetrically so that mixture proportions can be easily determined. From Schlumberger.[1]

NUMERICAL APPROACHES TO LITHOLOGY DETERMINATION

Two-dimensional cross plots have been very useful for interpretation and will continue to provide a simple method for obtaining quick estimates of volumes of major minerals. However, other methods must be considered for cases of more complicated lithologies and for the simultaneous inclusion of multiple logging information at each depth. One way to begin is to express the log response of various tools as equations which relate the response to the volume of each of the minerals present.

Consider the simple example of two logging measurements, density and P_e. First we begin with a model of the formation. We suppose the formation consists of a mixture of two minerals of relative volumes V_1 and V_2 and with densities of ρ_1 and ρ_2. The photoelectric absorption properties of the two minerals are given by U_1 and U_2. The porosity is given by ϕ, and it is assumed to be filled with a fluid characterized by U_{fl} and ρ_{fl}.

The tool response equations, or mixing laws, relate the measured parameters to the formation model. The complete set for this measurement is given by:

$$\rho_b = \rho_{fl}\phi + \rho_1 V_1 + \rho_2 V_2$$

and

$$U = U_{fl}\phi + U_1 V_1 + U_2 V_2 .$$

The final relation necessary to solve for the three unknowns is the closure relation of the partial volumes:

$$1 = \phi + V_1 + V_2 .$$

The solution can most easily be seen in terms of the matrix representation of the set of simultaneous equations:

$$M = R V ,$$

where M is the vector of measurements, R is the matrix of response coefficients, and V is the vector of unknown volumes. For the balanced case of N unknowns and N−1 logging values, the solution is the inverse of the response matrix, R^{-1}. Doveton shows a computer program for obtaining this inverse simply.[4]

Obviously this approach can be extended to as many measurements as are available, under the condition that their responses be written as linear combinations of the volumes. Two practical problems arise. The first is relatively straightforward and concerns the problem of overdetermination, which occurs when the number of logging measurements exceeds the number of minerals in the model. One solution to this problem is to find a least-squares solution to the set of equations. In this case, a weighting matrix is used to express the confidence level associated with each measurement. The programs necessary for implementing such an approach are described by Bevington.[5] A commercial processing package which utilizes such an approach for computerized log interpretation, not just lithology identification, is described by Mayer and Sibbit.[6]

However, the more serious problem concerns the more probable occurrence that the number of minerals in the formation will vastly outnumber the set of logging measurements. This problem has been addressed from the geological point of view by Abbot and Serra, who defined the concept of *electrofacies*.[7] This term refers to the collective set of logging tool responses which allow discrimination between one bed and another. The set of N logging measurements at a particular level can be viewed as a single point in N-dimensional space. Clustering of points will indicate zones of a similar facies. A statistical approach to the problem of lithology identification uses an extensive data base for more than a hundred facies incorporating up to nine different logging tool responses.[8] Instead of using a two-dimensional cross plot, each mineral or electrofacies is defined in the N-dimensional space of log responses.

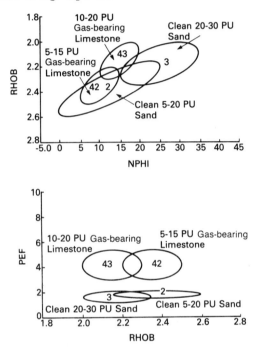

Figure 18–15. An approach to the N-dimensional space of multiple logging measurements, considering them two at a time. From Delfiner et al.[8]

To implement this approach, the characteristic response of the tools is established on a two-by-two basis, as shown in Fig. 18–15. By combining all the various cross plot combinations, ellipses in N-dimensional space can be defined for the tool response to a particular type of lithology and porosity range. To identify the particular type of matrix, a given set of logging measurements is tested statistically to see in which of the many volumes it may be contained. An example of the type of display which can be produced is shown in Fig. 18–16. The results strongly resemble a core description made by a trained geologist.

These various approaches involve a practical issue of determining the model parameters as well as defining the appropriate model for a given formation and set of logging measurements. Raymer and Edmundson give many valuable parameters for most of the logging tools,[9] but often one is faced with unusual minerals for which the logging tool response is unknown but can be estimated from the logging measurements. Quirein et al. describe a generalized program for lithological interpretation which permits solution of the problem in various steps.[10] It can be used, in a first stage, to generate the electrofacies from the given set of logging measurements. If the appropriate tool responses are known, it can evaluate simultaneously up to five different models of the formation. The choice of the final model is left to the user.

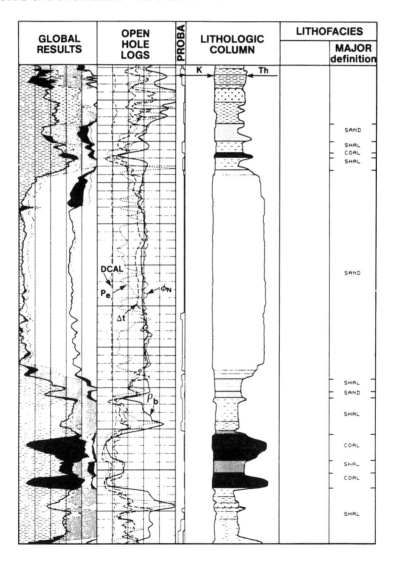

Figure 18–16. Output of a program for automatically determining lithology from a number of predefined geological models. From Delfiner et al.[8]

References

1. *Schlumberger Log Interpretation Charts*, Schlumberger, New York, 1985.

2. Dewan, J. T., *Essentials of Modern Open-hole Log Interpretation*, PennWell Publishing Co., Tulsa, 1983.

3. Ellis. D. V., and Case, C. R., "CNT-A Dolomite Response," Paper S, SPWLA Twenty-Fourth Annual Logging Symposium, 1983.

4. Doveton, J. H., *Log Analysis of Subsurface Geology, Concepts and Computer Methods*, John Wiley, New York, 1986.

5. Bevington, P. R., *Data Reduction and Error Analysis for the Physical Sciences*, McGraw-Hill, New York, 1969.

6. Mayer, C., and Sibbit, A., "Global: A New Approach to Computer-Processsed Log Interpretation," Paper SPE 9341, SPE Fifty-fifth Annual Technical Conference, 1980.

7. Serra, O., and Abbott, H., "The Contribution of Logging Data to Sedimentology and Stratigraphy," Paper SPE 9270, SPE Fifty-fifth Annual Technical Conference, 1980.

8. Delfiner, P. C., Peyrat, O., and Serra, O., "Automatic Determination of Lithology from Well Logs," Paper SPE 13290, SPE Fifty-ninth Annual Technical Conference, 1984.

9. Edmundson, H., and Raymer, L. L., "Radioactive Logging Parameters for Common Minerals," SPWLA Twenty-third Annual Logging Symposium, 1979.

10. Quirein, J., Kimminau, J., Lavigne, J., Singer, J., and Wendel, F., "A Coherent Framework for Developing and Applying Multiple Formation Evaluation Models," Paper DD, SPWLA Twenty-seventh Annual Logging Symposium, 1986.

Problems

1. Fig. 18–4 is a log of a tight (low porosity) carbonate section. Using the neutron-density cross plot (Fig. 18–3), identify zones of the different matrix types present in this section of the well.

2. To gain some practice with the manual determination of lithology and matrix values, consider the following set of data taken from sections of a clean sandstone reservoir. Although the sandstone is free of clay, it does contain some pyrite. The question to answer is how close does the manual cross plot technique get you to the true porosity?

Lithology Identification from Porosity Logs   437

The following table, which lists the values painstakingly read off the logs, is in a format which will help you to complete the task. Note the column of matrix density values, which have been determined from core analysis. Make a plot of porosity obtained from the density tool alone under two different conditions: using the core-measured grain density, and using the cross plot grain density ρ_{ma}n–d. For both calculations, assume that the formation fluid has a density of 1.20 g/cm^3.

Δt	ρ_b	ϕ_n	ϕ_{n-d}	ρ_{ma} n–d	ϕ_{n-s}	Δt_{ma} n–s	ρ_{ma}
95	2.35	21					2.74
90	2.40	19					2.74
94	2.41	18					2.84
87	2.52	26					2.70
97	2.36	24					2.84
95	2.38	20					2.70
90	2.38	19					2.76

3. To verify the identification of the sets of matrix lines in the P_e plot of Fig. 18–12, compute the ρ_b and P_e values for 50%-porosity limestone, dolomite, and sandstone. Consider two cases of pore fluid, water and CH_2. The appropriate values for the computations may be found in Table 10–1.

4. With reference to Figs. 18–1 through 18–3, which tool combination would you prefer to use for lithology definition in a carbonate reservoir containing limestone and dolomite? Specifically what are the maximum errors tolerable in a 5%-porous limestone so that it is not misidentified as a dolomite? For each pair of cross plots, you can evaluate the maximum tolerable error by either of the measurements or assume a simultaneous error of the two.

5. Consider a barite-loaded mud with a density of 14 lb/gal which is known to be 46% $BaSO_4$ by weight. What is the P_e of the mud? If the mud infiltrates a 20%-porous sandstone what, P_e do you expect to see?

19
CLAY TYPING AND QUANTIFICATION FROM LOGS

INTRODUCTION

The presence of shale in hydrocarbon reservoirs has a large impact on estimates of reserves and producibility. The clay minerals present in the shale complicate the determination of saturation and porosity. Permeability is often controlled by very low levels of clay minerals in the pore space. Without specific knowledge of the clay minerals present, there is a risk of impairing the permeability of a reservoir by introducing improper fluids.

Log examples in earlier chapters show that clay in rocks affects all of the log readings that have been considered. How is clay different from any other mineral? The distinction lies, perhaps, in the magnitude of the effects which are observed. Clays, unlike previously considered minerals, increase the conductivity of the formation so that a straightforward application of the Archie relation yields water saturations which are too large. For the neutron log response, the presence of hydrogen associated with the clay can increase the apparent porosity by up to 40 porosity units.

In traditional well log analysis, clay has been treated as a "fourth" mineral exactly as was done for lithology determination in the previous chapter. A large body of successful techniques is based on crossplotting log measurements to determine the average properties of the clay or shale. These methods produce an estimate of the volume fraction present with little regard for the details of its composition or distribution. The traditional cross plotting techniques determine the clay properties from indicators over large depth intervals, often including massive shales. In zones where the evaluation is to be performed, the shale or clay volume is usually at a considerably reduced level. The absolute magnitude may be crucial in determining the permeability of the formation. Traditional shale indicators are known to be quite inadequate for this type of refined shale volume estimation. In addition, there is the question of whether the shale in the zone of interest corresponds to the type of shale used to determine the average properties.

Rather than rely on the "fourth" mineral approach, a more sensible alternative would be to use measurements specifically designed to be sensitive to clay minerals and their distribution in rocks. At present, although we fall short of this ideal, there are a number of new methods for clay typing and quantification. If we look at the characteristics of clay minerals, a number of them are found to be responsible for the perturbation of conventional logging measurements. It is their chemical composition, however, which is most accessible for their quantification by logging measurements, in particular those based on nuclear spectroscopy. Some of the methods require a reinterpretation of familiar measurements, whereas others, using new measurements, require a geochemical framework to describe clay mineralogy.

WHAT IS CLAY/SHALE?

Although we have used the terms shale and clay interchangeably throughout the text, it is appropriate to bear in mind that shale is a fine-grained rock containing a sizable fraction of clay minerals. It is the presence and distribution of the clay minerals which affect the log readings. Rather than attempt to completely answer the question posed in the title above, it is more useful to concentrate on the properties of clay minerals which are important for well logging. In assessing the important parameters, only two aspects are considered here: perturbations of traditional logging measurements, and properties of the clay which may be detectable by newer logging measurements.

The hundreds of clay minerals which have been studied by crystallographers, chemists, soil scientists, and clay mineralogists can be lumped, on the basis of structure and composition, into five groups: kaolinite, mica (illite), smectite (montmorillonite), chlorite, and vermiculite. In common, they are all aluminosilicates, which gives a clue as to the dominant elements present. The clay mineral structures are sheetlike as a result of the geometric nature of the fundamental structural units which contain the aluminum and silicon.* One sheet type consists of octahedral units of oxygen or hydroxyl around a central atom (usually aluminum but sometimes magnesium or iron). The other is a tetrahedral unit consisting of a central silicon atom surrounded by oxygen. The five groups listed above are stacking

---

* See Grim for detailed discussion of clay crystal structure.[1]

combinations of these two types of sheet structure. Figure 19–1 indicates the layer structure and composition of members of the five groups formed by various stackings of the tetrahedral and octahedral sheets which result in characteristic lattice dimensions or spacings.

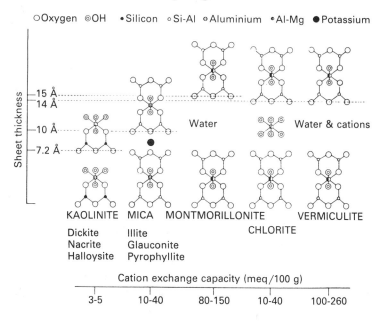

Figure 19–1. Schematic representation of the structure and composition of members of the five clay mineral groups. Adapted from Brindley.[16]

The simplest structure belongs to kaolinite and its 1:1 stacking of single octahedral and tetrahedral sheets. The other clay minerals consist of a 2:1 layer stacking of two tetrahedral sheets sharing oxygen atoms with an octahedral sheet between them. The difference between groups is related to the amount of layer charge in the lattice, created by isomorphous substitution (e.g., Mg^{+2} for Al^{+3} or Al^{+3} for Si^{+4}) and the type of interlayer complex present to balance the charge deficiency. Smectite has the smallest amount of substitution, hence the least layer charge deficiency. This small negative charge is balanced by hydrated cations in the interlayer space, often causing the smectite to swell. Vermiculites and chlorites both have intermediate charge deficiencies created by substitutions. In vermiculites the negative lattice charge is also balanced by hydrated cations, though the swelling is reduced by the increased attractive forces between lattice and cation. Chlorites balance their charge by interlayer complexes of Al and Mg hydroxide sheets, much like the octahedral sheet in the lattice. Micas or illites have the greatest substitution in the lattice, and the charge is balanced by dehydrated cations, usually K^+, in the interlayer space. The absence of water in illites results in nonswelling behavior and a smaller lattice spacing.

Since the structure of clay minerals, normally determined by X-ray diffraction methods, is not available in logging measurements, the immediate interest of the information presented in Fig. 19–1 is to identify elements which are susceptible to identification by logging techniques. Unfortunately this task is complicated by the lack of standard chemical formulas for most of the members of each family. This lack is the result of the previously mentioned substitution in the lattice and interlayer sites.

Hydrogen is a prominent member of all the clay minerals listed. It has a large impact on the neutron porosity response. Identification of some of the other elements can be made using gamma ray spectroscopy, since this has proven to be a practical, although indirect, means of formation chemical analysis in the borehole. We have already seen an example of the identification of clays by determining the range of U, Th, and K concentrations. From the abbreviated set of clay minerals presented in Fig. 19–1, we can see that only K is associated with the mica group. The other abundant characteristic element which is easily detectable is Al. This detection can be done through neutron activation, a proven technique which is becoming a routine logging procedure.

By itself, the Al content of a formation will not thoroughly quantify the presence of clay minerals, since feldspars contain Al and the concentration of Al in clay minerals is variable. In a number of clay mineral families, there is substitution of Al for Si and Fe for Al; this latter substitution is frequently observed in illite and chlorite, and to a much lesser degree in kaolinite. Iron is easily detectable either through capture gamma ray spectroscopy or the attenuation of low energy gamma rays in the determination of P_e.

A second unique feature of clay minerals for log response is associated with their platy nature, which results from their structure. Water trapped between the plates contributes to conductivity and to porosity measurements, although it not considered a part of "effective" porosity. Layers of water can be trapped in the inner layers during sedimentation. Some of this water is released during compaction or as a result of mineralogical reactions. If the permeability of the surrounding material is not sufficiently high, overpressured shales can result.

Another aspect of the platy nature of clay minerals is the presence of a substantial negative surface charge, which is responsible for their ability to adsorb ions. In some instances these ions are radioactive and account for the GR activity frequently associated clay minerals. An important property of clay minerals is their ability to adsorb ions, primarily cations. This results from the negative surface charge created by isomorphous substitution within the lattice and is responsible for the classification of the different clay mineral groups. In the presence of an electrolyte this surface charge is also responsible for creating a thin layer of altered fluid composition. Polar water molecules, and sodium or potassium ions are adsorbed whereas chlorine ions are repelled. This double layer, considered in more detail in the next chapter, has large implications for the resistivity measurements.

The ability of a clay mineral to form the electrical double layer is called the cation exchange capacity (CEC). It corresponds to the excess of cations over anions on the surface of a solid. One method of measuring this property is to first saturate the clay-bearing sample with salt water and then to pass a barium solution through it. The barium will replace the Na, and the quantity that does so can be measured. The units for CEC often are *milli-equivalents* per 100 g of material. An *equivalent* corresponds to a number of cations required to neutralize the charge of 6.023×10^{23} electrons. Typical ranges of CEC values of the five groups of clays are shown along the bottom of Fig. 19–1.

Although Berner attributes the trends of cation exchange capacity to the substitution (or lack of it) of Al for Si in the clay mineral structure,[2] there are other sources. One important source of the negative surface charge is the presence of broken bonds in the sheet structure. Yariv and Cross suggest that the cation exchange capacity should depend on the surface area; they cite a study of kaolinites which demonstrated a linear relationship between CEC and particle size.[3] Patchett has taken this a step further, suggesting that there is a direct link between specific surface area and CEC.[4] A portion of his data is shown in Fig. 19–2.

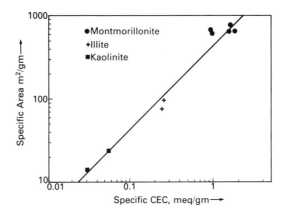

Figure 19–2. Correlation between specific surface area and bulk cation exchange capacity. From Patchett.[4]

This brings us to the final aspect of clay: its physical size. These thin, sheetlike particles can have very large surface areas, based on simple calculations from lattice dimensions. Particle size is typically less than 5 microns. The actual specific surface area depends on the clay mineral. Table 19–1 lists the specific surface area ranges for a number of clay minerals.

SPECIFIC SURFACE AREAS OF CLAY MINERALS	
	m²/g
Smectite	700-800
Illite	113
Chlorite	42
Kaolinite	15-40

After Almon & Davies[5] and Yariv and Cross[3]

Table 19-1. Specific surface areas of several clay minerals. After Almon and Davies,[5] and Yariv and Cross.[3]

Distribution of Clay

The distribution of shale or clay in formations has different impacts on some of the log measurements. For this reason, log analysts have identified three types of distributions: laminar, structural, and dispersed shale. These are illustrated in Figure 19-3. Laminar shale is shown to be discrete interspersed layers of shale in an otherwise clean sandstone. In this case the unit volume of formation is considered to share both matrix and porosity volume with this type of shale. Structural shale is considered to be composed of grains consisting of a rock matrix and clay particles. None of the pore volume is occupied by the shale. The third variety is dispersed shale, in which the clay is present throughout the system, dispersed in the pore space.

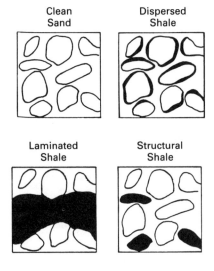

Figure 19-3. Classification of shale by distribution. From Poupon et al.[7]

Dispersed shale or clay minerals occurring at small volume concentrations are of great interest for hydrocarbon reservoir evaluation. As a result of the use of the scanning electron microscope, several types of distributions have been identified: pore lining, pore filling, and pore bridging. Dramatic photomicrographs of these three types are shown in Fig. 19–4.

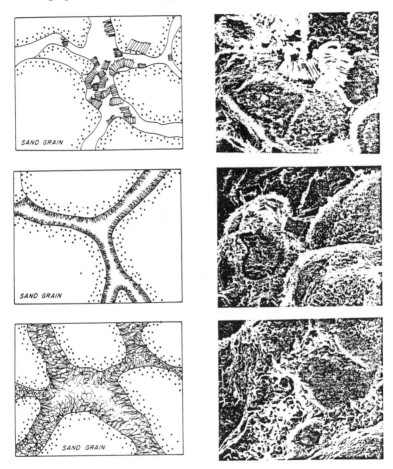

Figure 19–4. Photomicrographs of dispersed clay in sandstone reservoir rocks. Adapted from Almon.[17]

Almon and Davies have studied the impact of some dispersed clay minerals on hydrocarbon production.[5] The problem resulting from the occurrence of kaolinite is related to its structure. Stacks of kaolinite, seen in the upper photograph of Fig. 19–4, are characterized by a booklike shape. These stacks are loosely attached to sand grains and can be detached by high flow rates. Because of the large size of these crystals, they can block large pore throats, resulting in a permanent, irreversible damage to the permeability.

The members of the smectite family pose problems because of their large specific surface area and their swelling capability. Due to the affinity of the clay particles and water, the production of water-free oil in the presence of high water saturation has been attributed to montmorillonite. Unless the presence of montmorillonite is recognized, productive zones may be abandoned because of derived S_w values which appear too high. The swelling of the clays (produced by fresh water) also can cause the particles to dislodge and plug pore throats. In this case, oil-based mud or KCl additives must be used to prevent formation damage.

The large specific surface area of illite can create large volumes of microporosity that result in high values of S_w. In the lower photo of Fig. 19–4, long filaments of illite are shown to have completely bridged the pore volumes. In this case, the porosity reduction may not be very great, but certainly the impact on permeability will be significant.

Members of the chlorite family, particularly the iron-rich types, are sensitive to acid, which is sometimes injected to stimulate production by dissolving carbonate cements. Exposing chlorite to acid precipitates a gelatinous iron compound, which may permanently halt production.

Influence on Logging Measurements

For a summary of the influence of shale on a few of the log measurements considered thus far, refer to Table 19–2. Implicit in this table is that all the measurements are affected by the quantity or volume fraction of clay present in the formation. In addition, there are perturbations due to any of three clay descriptors: the chemistry, surface area, or distribution of clay in the formation.

Log Parameter	CLAY DESCRIPTOR		
	Chemistry/ Structure	Surface Area	Clay Distribution
R_T		CEC	X
ρ_b	X		X
P_e	Fe		
ϕ_n	$(OH)_x$	B?	
Σ	K, Fe	B, Gd?	
Δt			X
GR	K	Th, U?	

Table 19–2. Influence of clay type and distribution on logging responses.

The chemical composition has its largest impact on P_e, primarily through the presence of iron. The concentration of hydroxyls will strongly influence the neutron porosity response. Highly absorptive constituents such as iron and potassium will also affect the capture cross section Σ, and the presence of potassium in the clay mineral will influence the gamma ray.

Clay Typing and Quantification from Logs   447

Logging Tool Response in Sedimentary Minerals

Name	Formula	ρ_{PDC} g/cc	ϕ_{SNP} p.u.	ϕ_{CNL} p.u.	t_c μs/ft	t_s μs/ft	P_e barn/elect	U barn/cc	ϵ farads/m	t_p nsec/m	GR API units	Σ c.u.
SILICATES												
Quartz	SiO_2	2.64	-1	-2	56.0	88.0	1.81	4.79	4.65	7.2	–	4.26
d-Cristobalite	SiO_2	2.15	-2	-3			1.81	3.89			–	3.52
Opal (3.5% H_2O)	$SiO_2(H_2O)_{.1209}$	2.13	4	2	58		1.75	3.72			–	5.03
Garnet	$Fe_3Al_2(SiO_4)_3$	4.31	3	7			11.09	47.80			–	44.91
Hornblende	$Ca_2NaMg_2Fe_2AlSi_8O_{22}(O,OH)_2$	3.20	4	8	43.8	81.5	5.99	19.17			–	18.12
Tourmaline	$NaMg_3Al_6B_3Si_6O_7(OH)_4$	3.02	16	22			2.14	6.46			–	7449.82
Zircon	$ZrSiO_4$	4.50	-1	-3			69.10	311			–	6.92
CARBONATES												
Calcite	$CaCO_3$	2.71	0	-1	49.0	88.4	5.08	13.77	7.5	9.1	–	7.08
Dolomite	$CaCO_3MgCO_3$	2.88	2	1	44.0	72	3.14	9.00	6.8	8.7	–	4.70
Ankerite	$Ca(Mg,Fe)(CO_3)_2$	2.86	0	1			9.32	26.65			–	22.18
Siderite	$FeCO_3$	3.89	5	12	47		14.69	57.14	6.8 – 7.5	8.8 – 9.1	–	52.31
OXIDATES												
Hematite	Fe_2O_3	5.18	4	11	42.9	79.3	21.48	111.27			–	101.37
Magnetite	Fe_3O_4	5.08	3	9	73		22.24	112.98			–	103.08
Geothite	$FeO(OH)$	4.34	50+	60+			19.02	82.55			–	85.37
Limonite	$FeO(OH)(H_2O)_{2.05}$	3.59	50+	60+	56.9	102.6	13.00	46.67	9.9 – 10.9	10.5 – 11.0	–	71.12
Gibbsite	$Al(OH)_3$	2.49	50+	60+			1.10				–	23.11
PHOSPHATES												
Hydroxyapatite	$Ca_5(PO_4)_3OH$	3.17	5	8	42		5.81	18.4			–	9.60
Chlorapatite	$Ca_5(PO_4)_3Cl$	3.18	-1	-1	42		6.06	19.27			–	130.21
Fluorapatite	$Ca_5(PO_4)_3F$	3.21	-1	-2	42		5.82	18.68			–	8.48
Carbonapatite	$(Ca_5(PO_4)_3)_2CO_3H_2O$	3.13	5	8			5.58	17.47			–	9.09
FELDSPARS-Alkili												
Orthoclase	$KAlSi_3O_8$	2.52	-2	-3	69		2.86	7.21	4.4 – 6.0	7.0 – 8.2	~220	15.51
Anorthoclase	$KAlSi_3O_8$	2.59	-2	-2			2.86	7.41	4.4 – 6.0	7.0 – 8.2	~220	15.91
Microcline	$KAlSi_3O_8$	2.53	-2	-3			2.86	7.24	4.4 – 6.0	7.0 – 8.2	~220	15.58
FELDSPARS-Plagioclase												
Albite	$NaAlSi_3O_8$	2.59	-1	-2	49	85	1.68	4.35	4.4 – 6.0	7.0 – 8.2	–	7.47
Anorthite	$CaAl_2Si_2O_8$	2.74	-1	-2	45		3.13	8.58	4.4 – 6.0	7.0 – 8.2	–	7.24
MICAS												
Muscovite	$KAl_2(Si_3AlO_{10})(OH)_2$	2.82	12	20	49	149	2.40	6.74	6.2 – 7.9	8.3 – 9.4	~270	16.85
Glauconite	$K_2(Mg,Fe)_2Al_6(Si_4O_{10})_3(OH)_{12}$	~2.54	~23	~38			6.37	16.24				24.79
Biotite	$K(Mg,Fe)_3(AlSi_3O_{10})(OH)_2$	~2.99	~11	~21	50.8	224	6.27	18.75	4.8 – 6.0	7.2 – 8.1	~275	29.83
Phlogopite	$KMg_3(AlSi_3O_{10})(OH)_2$				50	207						33.3
CLAYS												
Kaolinite	$Al_4Si_4O_{10}(OH)_8$	2.41	34	37			1.83	4.44	~5.8	~8.0	80 – 130	14.12
Chlorite	$(Mg,Fe,Al)_6(Si,Al)_4O_{10}(OH)_8$	2.76	37	52			6.30	17.38	~5.8	~8.0	180 – 250	24.87
Illite	$K_{1-1.5}Al_4(Si_{6.5-7}Al_{1-1.5}O_{20})(OH)_4$	2.52	20	30			3.45	8.73	~5.8	~8.0	250 – 300	17.58
Montmorillonite	$(Ca,Na)_7(Al,Mg,Fe)_4(Si,Al)_8O_{20}(OH)_4(H_2O)_n$	2.12	40	44			2.04	4.04	~5.8	~8.0	150 – 200	14.12
EVAPORITES												
Halite	$NaCl$	2.04	-2	-3	67.0	120	4.65	9.45	5.6 – 6.3	7.9 – 8.4	–	754.2
Anhydrite	$CaSO_4$	2.98	-1	-2	50		5.05	14.93	6.3	8.4	–	12.45
Gypsum	$CaSO_4(H_2O)_2$	2.35	50+	60+	52		3.99	9.37	4.1	6.8	–	18.5
Trona	$Na_2CO_3NaHCO_3H_2O$	2.08	24	35	65		0.71	1.48			–	15.92
Tachydrite	$CaCl_2(MgCl_2)_2(H_2O)_{12}$	1.66	50+	60+	92		3.84	6.37			–	406.02
Sylvite	KCl	1.86	-2	-3			8.51	15.83	4.6 – 4.8	7.2 – 7.3	500+	564.57
Carnalite	$KClMgCl_2(H_2O)_6$	1.57	41	60+			4.09	6.42			~220	368.99
Langbenite	$K_2SO_4(MgSO_4)_2$	2.82	-1	-2			3.56	10.04			~290	24.19
Polyhalite	$K_2SO_4MgSO_4(CaSO_4)_2(H_2O)_2$	2.79	14	25			4.32	12.05			~200	23.70
Kainite	$MgSO_4KCl(H_2O)_3$	2.12	40	60+			3.50	7.42			~245	195.14
Kieserite	$MgSO_4H_2O$	2.59	38	43			1.83	4.74			–	13.96
Epsomite	$MgSO_4(H_2O)_7$	1.71	50+	60+			1.15	1.97			–	21.48
Bischofite	$MgCl_2(H_2O)_6$	1.54	50+	60+	100		2.59	3.99			–	323.44
Barite	$BaSO_4$	4.09	-1	-2			266.82	1091			–	6.77
Celestite	$SrSO_4$	3.79	-1	-1			55.19	209			–	7.90
SULFIDES												
Pyrite	FeS_2	4.99	-2	-3	39.2	62.1	16.97	84.68			–	90.10
Marcasite	FeS_2	4.87	-2	-3			16.97	82.64			–	88.12
Pyrrhotite	Fe_7S_8	4.53	-2	-3			20.55	93.09			–	94.18
Sphaterite	ZnS	3.85	-3	-3			35.93	138.33	7.8 – 8.1	9.3 – 9.5	–	25.34
Chalcopyrite	$CuFeS_2$	4.07	-2	-3			26.72	108.75			–	102.13
Galena	PbS	6.39	-3	-3			1631.37	10424			–	13.36
Sulfur	S	2.02	-2	-3	122		5.43	10.97			–	20.22
COALS												
Anthracite	$CH_{.358}N_{.009}O_{.022}$	1.47	37	38	105		0.16	0.23			–	8.65
Bituminous	$CH_{.793}N_{.015}O_{.078}$	1.24	50+	60+	120		0.17	0.21			–	14.30
Lignite	$CH_{.849}N_{.015}O_{.211}$	1.19	47	52	160		0.20	0.24			–	12.79

Table 19–3. Logging parameters for sedimentary minerals. From Schlumberger.[18]

The surface area of the clay will have the largest impact on the resistivity distortion, because of the associated cation exchange capacity, which determines, in part, the inherent resistivity of the clay. The surface area may contribute to the adsorption of other ions, of which some are radioactive and others have a large capture cross section. Consequently it will play a role in the value of Σ and GR measured in a shale.

The clay distribution will largely affect the resistivity measurement: A laminated shale will produce quite different results for the same volume of a dispersed clay, because of anisotropy or accessible surface. To some extent the distribution may play a role in the effect on the density reading. It is possible that the appropriate grain density for a laminated shale will be significantly different than for a dispersed clay. There are similar effects for the magnitude of perturbation on the Δt measurement.

A slightly more quantitative list of logging tool responses to a variety of minerals can be found in Table 19–3. The portion concerning clay minerals will be useful in later discussions.

SOME TRADITIONAL INDICATORS

We have already seen that the gamma ray can furnish a shale volume estimation if we use a linear interpolation between minimum and maximum readings observed over a lengthy zone of log. The results of this linear interpolation can be converted to shale volume through the use of one of the curves of Fig. 9–7, depending upon the rock type. Although such an estimate of V_{shale} can be misleading, it is the mainstay of traditional log interpretation. Commercial interpretation programs developed in the 1970s use a wide variety of shale indicators and cross plots to determine shale parameters.[7,8] Reference 6 lists nine types of programs which combine pairs of log readings, usually using the gamma ray as a discriminator in the third axis.

As an example of an improvement over the simple use of the GR reading alone, consider the use of M (from Chapter 18), which combines the sonic and density measurements. As the frequency cross plot of Fig. 19–5 shows, the value of GR corresponding to clean formations is found to be about 12 API units. From there, a fairly definite trend exists to the somewhat arbitrarily chosen shale point. Along this trend line, an estimate of V_{shale} can be obtained from linear interpolation of the GR reading between the minimum and maximum values. The horizontal trend of increasing GR values at constant M is attributed to radioactive dolomite. In these zones another estimate of V_{shale} must be used.

Two questions can be raised from this example. Suppose the "clean" value of GR should be chosen as 10 rather than 12 API units? How reliably can estimates of clay volume at the level of several percent be made? Another approach is needed for this type of refined estimate. A second point concerns the use of multiple parameters for the estimation of V_{shale}. From this example one might imagine a regression of the type:

$$V_{shale} = a\,M + b\,GR\;,$$

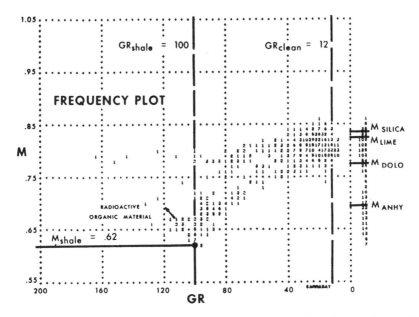

Figure 19-5. Use of three logging parameters for estimating shale volume. The parameter M combines the sonic and density measurement. From Schlumberger.[8]

as a substitute for the use of a single indicator. However, regardless of the parameters used, there is a tacit assumption of linearity with shale volume. Considering the variability of clay mineral mixtures in shale, is there any reason to believe that this would be the case?

The neutron-density cross plot is of fundamental importance for many log interpretation procedures. To be useful for porosity determination and gas detection, corrections must be made for the effects of clay. To see the traditional approach to the correction of the neutron log for the effects of clay, refer to Fig. 19-6 which shows a cross plot of neutron porosity and density values. Clean water-bearing sandstones will fall on the straight line of equal density and neutron porosity estimates. To the left are points from clean gas-bearing sandstones. Because of the additional hydrogen (in the form of hydroxyls), the apparent neutron porosity is usually higher than the porosity estimate from the density tool. From the cross plot in a very shaly zone (determined usually by reference to the GR), one can establish a 100% clay point to the right of the straight line defining the clean water-bearing sandstone line. If we connect this point to the 0-PU and 100-PU points, we can then establish a linear grid which can (if we believe this model) give the shale fraction and porosity corresponding to any pair of density and porosity readings as indicated in Fig. 19-7.

This leads to a graphical method for obtaining formation porosity, as well as an estimate of shale volume. Consider the points near the region marker A

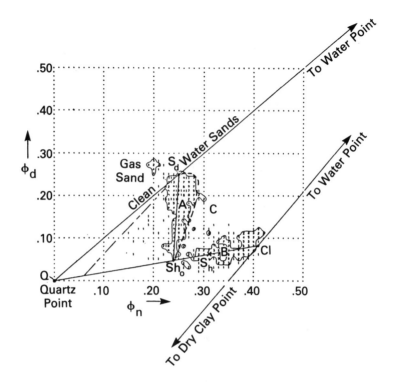

Figure 19-6. A neutron-density cross plot showing groups of points from a sand-shale sequence. Region A is identified as a "shale trend" in one interpretation. The "clay" point is consistent with an eight-hydroxyl clay mineral. From Schlumberger.[8]

in Fig. 19-6. According to the scaling procedure for Fig. 19-7, in a water-filled formation they would be interpreted as containing about 25% shale. The porosity would be obtained by projecting a line parallel to the shale trend until its intersection with the sand line. These points would appear to be associated with a formation of about 15% porosity.

Suppose the points in region A were actually gas-bearing. In this case, the effect has been completely masked by the presence of shale. It could still be detected by using an estimate of clay from another measurement, such as the gamma ray. If another reliable shale indicator suggested that the shale volume of the region was 50% rather than 25% as first suspected, then the graphical technique would shift the points of zone A by an amount corresponding to the clay volume. They would be found to the left of the clean line in the gas-bearing region.

The next section gives an alternative explanation for this type of cross plot behavior. It is based on the response of the neutron porosity tool to the hydroxyl content of the clay minerals present.

Clay Typing and Quantification from Logs   451

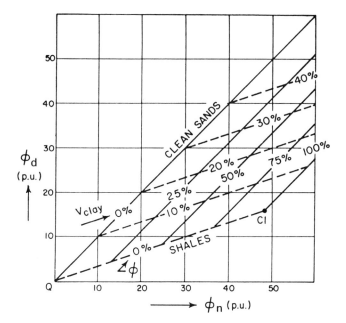

Figure 19–7. Determination of clay volume from the neutron density cross plot or, in the presence of another independent indicator of V_{clay}, the correction of the neutron-density reading for the clay. From Schlumberger.[8]

SOME NEW METHODS OF CLAY QUANTIFICATION

We have seen that almost any well logging device can be used to detect clays under specific circumstances. However, no single logging measurement discussed so far can be counted upon for providing an estimate of the clay content at all times. A measurement system which is primarily sensitive to a unique characteristic of clay minerals would represent a significant improvement in this situation. Before considering such a set of measurements, let us first review conventional logging measurements which, interpreted in a new light, can be made to yield more precise information about the nature of clay minerals present in the ubiquitous shale.

Interpretation of P_e in Shaly Sands

An interesting quantitative clay indicator for use in shaly sands can be constructed from the P_e measurement. It is based on the fact that P_e values measured in shales are primarily related to the iron content of the clay minerals. Without the presence of iron, the P_e of the aluminosilicates would reflect the silicon content and be indistinguishable from sand. A clay

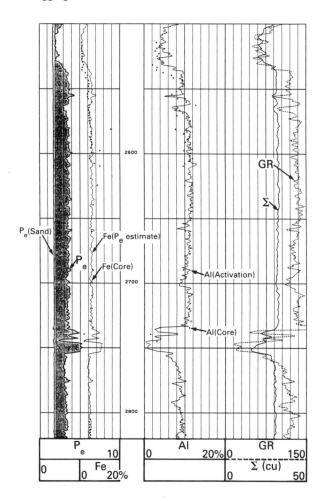

Figure 19–8. Log example for the determination of iron content from P_e. In track 1, the observed P_e is compared to the value of P_e expected in a sandstone of a porosity determined from a density log. The shading corresponds to the excess P_e. The Fe estimate is compared to core analysis. Track 2 shows a comparison between Al concentration data from an activation log and core measurements. Track 3 displays Σ and GR as indicators of shale.

indicator can be formed by calculating an artificial P_e curve based on the measured values of formation density, assuming that the matrix is sand and comparing it to the measured value of P_e. Figure 19–8 shows such an artificial P_e log with shading between the value expected for clean sand and the measured value. The shaded portion of excess P_e is primarily the result of the iron in the clay minerals.

The quantitative interpretation of the excess P_e can best be understood in terms of another method of calculating the P_e of mixtures. But first we use the parameter U to determine the base P_e curve which reflects the density variations. The tacit assumption here is that in a shaly sand, the matrix density of the sand and the shale are not all that different. Shale can be considered as quartz silt plus clay minerals which might have grain densities between 2.6 and 2.8 g/cm^3. Thus a rough estimate of porosity, which is sufficient to determine the small variation induced on P_e, can be obtained from ρ_b by assuming a grain density close to 2.65 g/cm^3. The expected sand value $P_{e,exp}$ can then be computed from the volumetric relation for U:

$$U_{exp} = U_{fl}\,\phi + U_{ma}(1-\phi),$$

and

$$P_{e,exp} = \frac{U_{exp}}{\rho_{e,log}}.$$

Consider the example of a 20%-porous sand filled with fresh water. From Table 10–1, the expected value of U is found to be:

$$U_{exp} = (0.4)(0.2) + (4.79)(0.8) = 3.91.$$

The corresponding density measured will be:

$$\rho_b = 0.2 + 2.65(0.8) = 2.34 \text{ g/cm}^3,$$

and the $P_{e,exp}$ is found to be 1.67, after converting ρ_b to ρ_e; however, this step can usually be omitted.

An alternative calculation which is useful in the interpretation of the excess P_e, considered later, uses the fact that the P_e of a composite mineral is the weighted average of the weight concentration of each of the elements. The weighting factors are related to the atomic number Z of each element raised to the 3.6 power (see Chapter 10 and Problem 3, Chapter 19). As an example, take the case of quartz (SiO$_2$). The molecular weight is 60.09 (= 28.09 + 2×16), and thus the weight fraction of Si is 46.7%, and that of O is 53.3%. The P_e of SiO$_2$ can then be written as:

$$P_e = \left[\frac{14}{10}\right]^{3.6} Wt_{Si} + \left[\frac{8}{10}\right]^{3.6} Wt_O$$

$$= 3.36 \times .467 + 0.44 \times .533 = 1.80,$$

which agrees with Table 10–1.

The computation of the expected P_e of a 20-PU water-bearing sandstone can be done in a similar manner. The weight percent of each component must first be determined. The density of the formation is 2.34 g/cm^3; thus it is 8.5% H$_2$O and 91.5% SiO$_2$. Since the mass fraction of hydrogen in water is 11%, it represents only 1% of the formation. Since quartz is 47% silicon, its mass fraction of the formation will be 42.7%. Combining the two sources of oxygen shows the formation to be 56.4% oxygen by weight. The expected value of the P_e is thus:

$$P_{e,exp} = \left[\frac{1}{10}\right]^{3.6} \times .01 + \left[\frac{8}{10}\right]^{3.6} \times .564 + \left[\frac{14}{10}\right]^{3.6} \times .427 = 1.68,$$

which agrees with the earlier computation.

From this exercise, it should be obvious that elements with large atomic numbers can have a large effect on the global value of P_e, even when they occur as very small weight concentrations. Equally obvious is the interpretation of the excess P_e: It is the value expected for the dominant host matrix plus a quantity due to the trace heavy mineral, or:

$$P_e \approx \sum_i \left[\frac{Z_i}{10}\right]^{3.6} Wt_i + \left[\frac{26}{10}\right]^{3.6} Wt_{Fe}, \qquad (1)$$

in the case of iron. The weight fraction of iron in the formation is simply the excess P_e ($P_{e,log} - P_{e,exp}$) divided by 31.2. For any other suspected high Z material, the factor of division is $(Z/10)^{3.6}$.

The approximation in Eq. (1) results from the fact that presumably the concentrations of the host elements sum to unity, but an additional element which was unaccounted for has been added. If 10% of the formation mass is due to iron, then the weight of H, O, and Si must be adjusted downward. However, if we perform the calculation we see that little difference will occur in the expected P_e. This approximation is obviously better for trace amounts of high Z materials.

To demonstrate the validity of this procedure, refer to the first track of Fig. 19–8, which shows the excess P_e and its conversion to the iron weight fraction using Eq. (1). The log-derived curve is compared to the analysis of the iron content of core samples, which are shown as small stars. There is very close agreement.

Clay Mineral Parameters and Neutron Porosity Response

The relationship between clay minerals and expected neutron porosity response has only recently been explored,[9,10] but it has not yet been exploited. It is determined primarily by the hydrogen content of the clay mineral (in the form of hydroxyls). The matrix composition is of little importance except at very low porosity. From this point of view, two large groups of clay types may be identified if intercalated water is ignored: those with low hydroxyl content $(OH)_4$, and those with high hydroxyl content $(OH)_8$.

The neutron log porosity estimate is largely determined by the slowing-down length of a formation, and this parameter mainly reflects the amount of hydrogen present. A porous medium with a given volume of clay can have a considerable range of slowing-down lengths, depending on the mix of high and low hydroxyl clays. In addition, a thermal neutron porosity estimate will be further influenced by the presence of common neutron absorbers associated with the clay: Fe, K, and B. Neither the thermal nor the epithermal porosity estimates are representative of the "true" porosity in a shaly sand. Both respond to the hydrogen in the pore space as well as to the hydrogen associated with the clay mineral. Accurate determination of the clay volume will depend on knowledge of the hydroxyl type.

To quantify the influence of these two different families of clay minerals on the neutron porosity response, refer to the information presented in Fig. 19–9. It shows the range of slowing-down length for the three primary rock types over the range of porosity. As indicated, the slowing-down length of a clay mineral containing four hydroxyl groups is seen to be about 17 cm. It is equivalent to a porosity of 10% in sandstone. The slowing-down length of the eight-hydroxyl clay mineral is even shorter, about 10 cm, which corresponds roughly to a 40%-porous water-bearing sandstone.

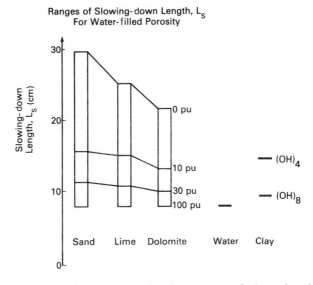

Figure 19–9. The slowing-down length of two types of clay minerals compared with the range in three lithologies over variations in porosity. The effect of the hydroxyls in the clay minerals is to shorten the slowing-down length. From Ellis.[9]

Since an adequate slowing-down length mixing law has not yet been developed, it is difficult to appreciate the effect on apparent porosity that small quantities of either of these two families of clay minerals would have. The effect was already demonstrated in Fig. 12–9 which shows the slowing-down length as a function of porosity for three rock compositions. The top curve is for a clean water-filled sandstone and represents the conversion of measured slowing-down length into sandstone porosity units. The second trace shows the slowing-down length as a function of porosity for a matrix which is composed of a 50/50 mixture of sand and illite. The third curve shows the relationship between L_s and porosity for a 50/50 mixture of sand and kaolinite. As the illite contains only four hydroxyl groups, its influence is smaller than that observed for the eight-hydroxyl kaolinite.

The important point to note is that an epithermal porosity estimate of a sand formation containing a clay mineral will exceed the density porosity (assuming $\rho_{ma} = 2.65$) by an amount related to the clay fraction and the concentration of hydrogen in the particular clay mineral. For a thermal porosity estimate, the same line of reasoning holds, but in addition the apparent porosity value will be somewhat higher because of additional thermal neutron absorption from associated Fe or K atoms or accessory elements such as B or Gd. At best, in an iron- or potassium-free clay mineral with no additional neutron absorbers, the thermal and epithermal values will be equal, but both will still be larger than a simple estimate made from the density measurement, assuming a grain density of about 2.65 g/cm^3. In the absence of a better mixing law for slowing-down length, the value of $\phi_{epi} - \phi_d$ can be roughly related to the volume fraction of four-hydroxyl or eight-hydroxyl clays by scaling the difference by 12 PU or 40 PU.

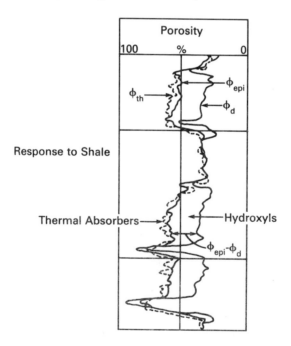

Figure 19-10. Log example showing the effect of hydroxyls and absorbers associated with the shale on the thermal and epithermal neutron log response. From Ellis.[9]

Without additional information, it is possible to interpret the cross plot of Fig. 19–6 in the following manner. Suppose there are two types of clay minerals present in the shale. The indicated "clay" point might correspond to the highest concentration of eight-hydroxyl minerals, whereas the "shale" point may be a mixture of four- and eight-hydroxyl minerals. This considerably complicates the determination of the shale volume from the neutron-density cross plot. In order to do it correctly, we need to know the mixture of clay types. Before pursuing the types of additional information available for such a distinction, we first look at a log example illustrating the points mentioned above.

Figure 19–10 illustrates the separation between density and neutron porosity estimates that result from the presence of clay. In the track shown, three porosity estimates on a scale of 0 to 100 PU are exhibited: ϕ_d, ϕ_{epi}, ϕ_{th}. In the middle zone, which is a clean sand, all three estimates agree at a value slightly in excess of 30 PU. On either side of this zone, both neutron porosity estimates exceed the density porosity. The separation of nearly 20 PU between the density and epithermal estimate is the result of the hydroxyls in this predominantly kaolinite/illite clay mixture. The additional several-PU difference between ϕ_{th} and ϕ_{epi} is the result of the iron and potassium content of the illite, as well as some boron which is associated with it.

Response of Σ to Clay Minerals

Although the conventional use of the measurement of the formation thermal capture cross section (Σ) is for the determination of water saturation in cased-hole applications, it can be used as a shale indicator. This is clearly seen in the log of Fig. 19–8 where Σ and GR are presented in the third track. In the shale zones, Σ is in the range of slightly more than 30 capture units and decreases significantly in the two clean zones. The observed correlation with shale content is due to the fact that some of the clay families discussed earlier contain elements with relatively large thermal absorption cross sections. Two examples are potassium and iron (see Table 13–1). Thus if the composition of a clay mineral is well-known, its corresponding Σ can be determined. Table 19–3 indicates some typical Σ values, which range from 15 to 30 capture units. The largest values are associated with clay minerals containing K and Fe.

One difficulty in exploiting Σ for quantitative clay volume estimates is that frequently trace elements with extremely large thermal absorption cross sections are associated with some clay minerals. These might include B, Gd, and Sm. Depending on the concentration of these and other rare earth elements, the observed Σ may be much larger than that expected from Table 19–3, which was computed for average chemical compositions of the dominant elements. In the well of Fig. 19–10 (which had extensive core analysis of trace elements), boron, in concentrations of up to 400 ppm, was found to be associated with the illite but not with kaolinite. This may reflect the large difference in specific surface areas between the two clay minerals.

Another difficulty with the use of Σ is that it responds to elements present in both the rock matrix and formation fluid. Chlorine is high on the list of common elements with a significant absorption cross section, and it is present in substantial quantities in formation fluids. Consequently there is some difficulty in extracting the Σ value for the rock matrix if there are uncertainties in porosity or the nature of the formation fluids. A more appealing measurement would be sensitive only to the rock matrix. One such measurement is discussed below.

Aluminum Activation

One relatively constant component of clay minerals is aluminum. The abundance of aluminum in sedimentary rocks is significant only in clays and shales, where it can be as large as 20% by weight. Thus the detection of aluminum is an attractive possibility for identifying and quantifying the presence of shale in a formation.

The aluminum concentrations of clay minerals and feldspars vary according to the type, as can be determined from a perusal of Table 19–3. Extensive core analysis in a number of different wells has found that the weight percent of Al in kaolinite is about 20%, in illite 15%, and in K-feldspar about 10%. Thus a simple measurement of Al is not sufficient to directly determine clay volume. Knowledge of the types of clays present is necessary. However, with additional information about the clay type or feldspar mixture, it is possible to compute a very accurate clay fraction, at low levels, from the Al content.

A technique which has been proposed is the activation of ^{27}Al by neutron absorption to produce ^{28}Al. This isotope decays with a half-life of 2.3 minutes by emitting a relatively easily detected 1.78 MeV gamma ray. The requirements for a continuous logging device to perform this measurement are a source of high flux, low energy neutrons, and a gamma ray detector able to distinguish the 1.78 MeV characteristic radiation from other capture gamma rays. The low energy source of neutrons is required to avoid conversion of ^{28}Si to ^{28}Al, which would obscure the distinction between sand and shales.

Fig. 19–11 shows a typical gamma ray spectrum obtained in a formation containing aluminum, after activation. The middle spectrum is the natural activity that results from the products of Th, U, and K in the formation. The bottom spectrum is the result of subtracting the natural radiation background from the activation spectrum to reveal the contribution due to aluminum.

Experimental devices incorporating these basic features have been demonstrated in borehole applications.[11,12] The results show that aluminum activation provides an important piece of information for evaluating the mineralogy of formations. Track 2 of the log in Fig. 19–8 shows the good correlation between such a measurement and the results of chemical analysis of the core. Log-derived aluminum concentrations, which are directly related to the presence of clay minerals, are seen to be in good agreement with the core analysis over the range of 0 to 12% by weight. Such a measurement is obviously useful when an SP is not available, when sands contain additional radioactive minerals, or in cased-hole applications when the casing becomes radioactive because of precipitation from formation water movement. Any serious attempt at clay evaluation would clearly benefit from a measurement of aluminum activation. Such a scheme is discussed in the next section.

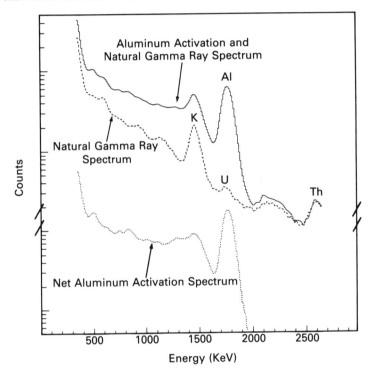

Figure 19-11. A typical gamma ray spectrum obtained after activation of a naturally radioactive aluminum-bearing formation. Courtesy of Schlumberger.

Clay Typing (Geochemical Logging)

We have seen the need to quantify low levels of clay content in reservoirs and to identify the types of clay minerals present. How can this be done without resorting to core analysis? Although we have discussed a number of measurements which are sensitive to clay types, it has been pointed out that additional information is necessary to refine the estimate. How much additional information is necessary and how far can it take us?

Consider the combination of the neutron response and Σ. Both contain information about clay minerals: the epithermal porosity ϕ_{epi} provides an indication of the hydroxyl content, and an estimate of Σ_{ma} indicates the clay type. The usefulness of such a breakdown can be seen in the table of Fig. 19–12. In this classification, the complex problem of clay description has been rather summarily treated; only four clay mineral groups represent the wide variety known to exist. In addition to the classification by OH type, a distinction with respect to capture cross section is also made. Additional characteristic elements associated with the four representative clay minerals, such as potassium or iron, are also noted. The presence of these two elements can be quantified by the use of auxiliary measurements, as discussed earlier.

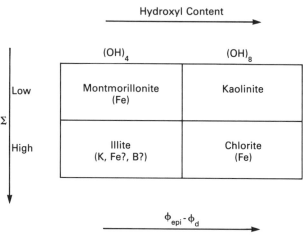

Figure 19–12. A simplified view of clay types with auxiliary logging parameters for distinguishing the groups.

An estimate of the hydroxyl content of a shaly sand can be obtained from the difference between the epithermal porosity and density porosity. For clean sands, this difference will be zero if sandstone units are used for the epithermal porosity and a grain density of 2.65 g/cm^3 is used for the density porosity ϕ_d. Thus the porosity estimate difference may be used to distinguish between the two major categories. The problem of identifying the hydroxyl type of the clay still remains before this separation can be converted to a meaningful clay volume. The Σ measurement can help to discriminate the clay type.

The principle of this technique is illustrated in Figure 19–13 which identifies the locations of the four principal clay types on a plot of Σ versus the epithermal/density porosity difference. To the right of the plot at a porosity separation of about 40 PU, chlorite and kaolinite are shown. The Σ of the chlorite is somewhat greater than that of kaolinite because of the iron in the clay mineral. At a porosity separation of about 10 PU, the two (OH)$_4$ clays are shown. Illite is shown at a larger value of Σ because of the presence of boron. The distinction between illite and montmorillonite can be made on the basis of potassium content from the spectral gamma ray log. If we use this graphical technique and assume that only two clay types are present, the Σ and $\phi_{epi} - \phi_d$ measured at each level in the well can lead to a clay type and volume estimate.

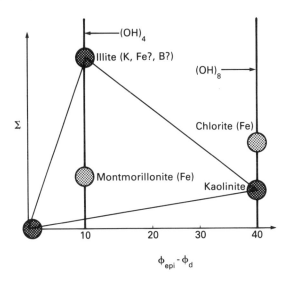

Figure 19–13. Schematic approach for combining Σ and a measure of hydroxyl content to determining the relative proportions of two clay types.

The preceding example has been presented in the traditional spirit of well log analysis: cross plotting two sets of measurements which are not quite related to the measurement desired. To the extent that ρ_{ma} is 2.65 g/cm^3, then the porosity estimate $\phi_{epi} - \phi_d$ reflects the hydroxyl content of the clay mineral. However, Σ may be responding to absorbers in the fluids rather than those associated with the clays. One can imagine a multitude of cross plots of this type which will separate, with varying degrees of success, the various clay minerals present.

An improvement in discrimination of clay types and their quantification would result from restricting the cross-plotted measurements to those which have a direct bearing on clay minerals. Chemical composition is one key to identifying clay minerals. We have already seen examples of logging measurements which respond to particular elements in the formation. Considering the information available from Al-activation, natural gamma ray spectroscopy, and capture neutron gamma spectroscopy, a large number of elements are detectable with logging measurements. This number can be reduced somewhat by concentrating on those which are associated only with the rock matrix rather than the formation and borehole fluid; this would eliminate H and Cl, for example.

An alternative to cross plotting the numerous elemental measurements would be to employ the matrix representation of the relationship between tool response and formation constituents of Chapter 18. The difference this time is that we would like to relate the measured elemental concentrations to the formation minerals, including clay minerals.

The first step of this process was demonstrated by Flaum and Pirie, who calibrated the gamma ray yields of a pulsed neutron spectroscopy logging tool to chemical analysis of the core, to do a detailed lithological analysis of the formation.[13] This approach has been extended much further, and no longer requires calibration to core measurements, because of the availability of Al activation logging data and the work of M. Herron and others.[12,14,15,19]

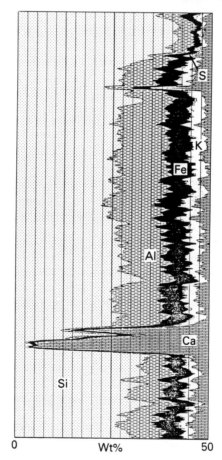

Figure 19–14. A 300' log-derived chemical analysis of the matrix portion of the formation for all major elements, excluding oxygen.

Because of the recent availability of Al concentration from activation logging, combined with natural gamma ray spectroscopy and neutron-induced spectroscopy, it is now possible to routinely obtain the elemental concentration of a number of geochemically important elements. Six of the important elements derived from such a logging combination are shown in Fig. 19–14 over a 300' interval. The track is calibrated in weight percent of the rock matrix and shows, from left to right, the weight concentrations of Si, Al, Fe, S, K, and Ca, which comprise about 50% of the formation (the other 50% being due to oxygen).

A new technique, which attempts to use logs of elemental concentration, is described next. Its basic premise is that one can perform an element-to-mineral transformation. This is done by constructing a series of simultaneous equations, as described in Chapter 18, relating the elemental concentrations to the mineral content of the formation. Because there are hundreds of possible minerals, the task may seem hopeless at first. However, it is frequently the case that formations of interest may contain as few as four to six minerals. The key to unraveling the lithologic variations of even such limited variety is the measurement of the appropriate significant elements. This is to be contrasted with the traditional approach, examined in Chapter 18, in which logging measurements, designed to be sensitive to entirely different parameters, have been combined to emphasize their residual dependence on the mineral composition of the formation. One example of an inappropriate set is the measurement of U, Th, and K by natural gamma ray spectroscopy, which has met with only limited success in the determination of clay mineralogy.

A second problem concerns the variability of chemical composition. This is particularly a problem for clays, although the composition of the usual suite of quartz, dolomite, and limestone are well-known. Herron suggests that compositional variability of clay minerals is determined by the clay type and lattice order.[14] The elemental concentrations derived to describe the disordered clay minerals in a South American shaly sand have been applicable in describing the same minerals at other geographically diverse sites.

As an example of the power of log-derived geochemical analysis, Herron and others have applied the method to a Venezuelan well which contains four major minerals: quartz, kaolinite, illite, and K-feldspar.[12,14,15] These minerals were estimated from a minimal set of log measurements, K, Al, and Fe. Potassium was determined from the natural radioactivity, and aluminum from the activation measurement described earlier; iron was derived from capture gamma ray spectroscopy. Although silicon was also measured, it was not necessary for this relatively simple lithology. Figure 19–15 shows the log-derived elemental concentrations in the well compared to numerous determinations from core analysis. The use of linear model relating the concentrations of K, Al, Fe, and Si to the four minerals results in the log-derived mineralogy of Fig. 19–16 along with the X-ray diffraction estimate of the mineralogy made on the core samples. This method clearly gives a more than satisfactory agreement with the core analysis and does it on a continuous basis.

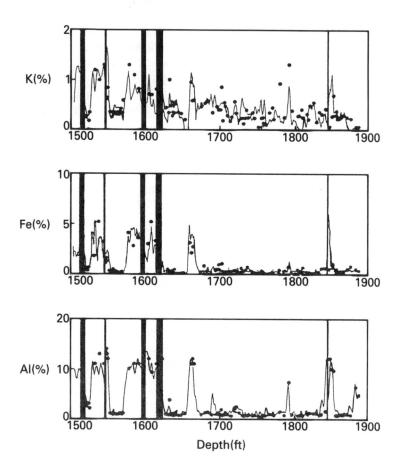

Figure 19–15. Comparison of log-derived Al, Fe, and K concentration with results of neutron activation of core samples. The black bands correspond to coal zones which are not appropriate to the model used to derive mineralogy. From Herron.[15]

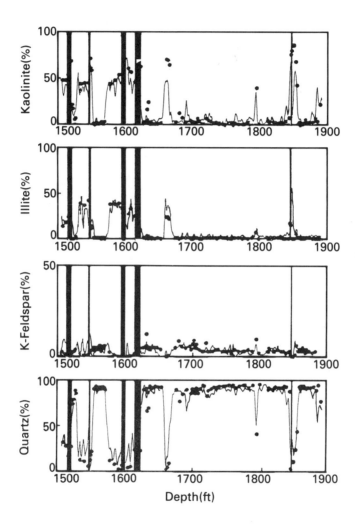

Figure 19-16. Mineral abundances determined from the log values of elemental concentrations are shown as the continuous logs. The mineral analysis from X-ray diffraction measurements on the side-wall core samples are shown as solid dots. From Herron.[15]

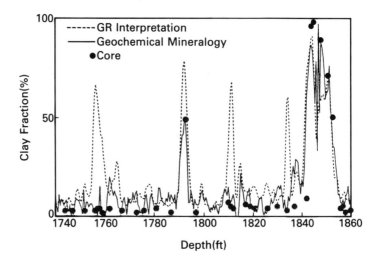

Figure 19–17. The formation clay fraction V_{clay} derived from the geochemical model. It is compared to X-ray diffraction clay analysis and an estimate made from the total GR log, shown as the dashed line. From Herron.[15]

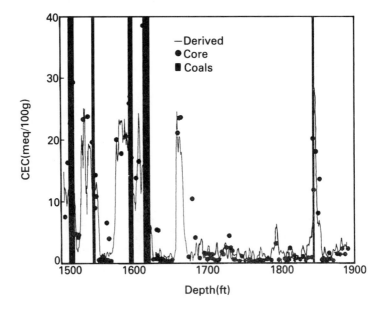

Figure 19–18. A log of the formation bulk cation exchange capacity derived from the illite and kaolinite concentrations. The solid points represent the values of CEC determined on the side-wall core samples. From Herron.[15]

Following the success in predicting the formation mineralogy, we can determine other formation parameters of interest. The clay fraction of the formation can easily be obtained by summing the derived clay minerals. The log of V_{clay} derived in this manner is compared in Fig. 19-17 to the traditional scaling of the total GR log. Although the GR estimate agrees frequently with the geochemically derived log,— and, incidentally, the X-ray diffraction analysis,— there are several zones of glaring disagreement. In these intervals there are sources of nonclay-associated thorium and uranium, so that the GR vastly overestimates V_{clay}.

As a final example of extracting auxiliary parameters from the geochemical mineralogy, consider the case of CEC. The earlier discussion notes that the various clay minerals are associated with typical ranges of CEC. If we know the relative proportion of the two clay minerals, in this example, we can derive the CEC of the entire interval in a straightforward manner as shown in Fig. 19-18. Note the comparison with measurement of CEC made on the core samples. Other derived properties of the formation considered by Herron include grain size estimates and permeability.[15]

Wells with more complex mineralogy will require a greater number of significant element inputs. However, at a minimum, ten or twelve elements are easily determined. In constructing the response matrix for expanded mineral models, the problem to be addressed will be that of compositional variability. Comparisons with core analysis will continue to refine these models, which will become more useful in time.

References

1. Grim, R. E., *Clay Mineralogy*, McGraw-Hill, New York, 1968.

2. Berner, R. A., *Principles of Chemical Sedimentology*, McGraw-Hill, New York, 1971.

3. Yariv, S., and Cross, H., *Geochemistry of Colloid Systems for Earth Scientists*, Springer-Verlag, Berlin, 1979.

4. Patchett, J. G., "An Investigation of Shale Conductivity," Paper U, SPWLA Symposium, 1975.

5. Almon, W. R., and Davies, D. K., "Formation Damage and the Crystal Chemistry of Clays," in *Clays and the Resource Geologist*, edited by F. J. Longstaffe, Mineralogical Assn. of Canada, Toronto, 1981.

6. *Schlumberger Log Interpretation*, Vol. 2: *Applications*, Schlumberger, New York, 1974.

7. Poupon, A., Clavier, C., Dumanoir, J., Gaymard, R., and Misk, A., "Log Analysis of Sand-Shale Sequences: A Systematic Approach," JPT, July 1970.

8. Poupon, A., Hoyle, W. R., and Schmidt, A. W., "Log Analysis in Formations with Complex Lithologies," JPT, Aug. 1971.

9. Ellis, D. V., "Neutron Porosity Logs: What Do They Measure?" *First Break*, Vol. 4, No. 3, 1986.

10. Ellis, D. V., "Nuclear Logging Techniques," *SPE Petroleum Production Handbook*, edited by H. Bradley, SPE, Dallas (in press).

11. Scott, H., and Smith, M. P., "The Aluminum Activation Log," Paper F, SPWLA Fourteenth Annual Logging Symposium, 1973.

12. Everett, R., Herron, M., and Pirie, G., "Log Responses and Core Evaluation Case Study Techniques: Field and Laboratory Procedures," Paper OO, SPWLA, 1983.

13. Flaum, C., and Pirie, G., "Determination of Lithology from Induced Gamma-Ray Spectroscopy," Paper H, SPWLA Twenty-second Annual Logging Symposium, 1981.

14. Herron, M., "Mineralogy from Geochemical Well Logging," *Clay and Clay Minerals*, Vol. 34, No. 2, 1986, pp. 204–213.

15. Herron, M. "Subsurface Geochemistry: 1. Future Applications of Geochemical Data," Presented at IAEA Consultant's Meeting, "Nuclear Data for Applied Nuclear Geophysics," Vienna, April, 1986.

16. Brindley, G. W., "Structure and Chemical Composition of Clay Minerals," in *Clays and the Resource Geologist*, edited by F. J. Longstaffe, Mineralogical Assn. of Canada, Toronto, 1981.

17. Almon, W. R., "A Geologic Appreciation of Shaly Sands," Paper WW, SPWLA Symposium, 1979.

18. *Schlumberger Log Interpretation Charts*, Schlumberger, New York, 1985.

19. Colson, L., Ellis, D. V., Grau, J., Herron, M., Hertzog, R., O'Brien, M., Seeman, B., Schweitzer, J., and Wraight, P., "Geochemical Logging with Spectrometry Tools," Paper SPE 16792, Sixty-seventh Ann. Technical Conf., 1987.

Problems

1. Compute the specific surface area (m^2/g) for a dry 35%-porous sandstone composed of spherical grains with negligible contact area. The diameter of the grains is 250 μ. What the is surface area/cm^3?

2. Suppose 1/4 of the available porosity of the sandstone of Problem 1 is occupied by kaolinite. What is the surface area/cm^3?

3. From the volumetric mixing law for U:

$$U \approx P_e \rho_b = U_1 V_1 + U_2 V_2 + \cdots + U_n V_n,$$

show that the mixing law for any material can be written in terms of the atomic numbers of its constituents as:

$$P_e = (\frac{Z}{10})^{3.6} Wt_1 + \cdots + (\frac{Z_i}{10})^{3.6} Wt_i.$$

Wt_i is the weight fraction of element i in the mixture.

4. What is the percent by weight of Al contained in the form of kaolinite, as shown in Fig. 19–1? What is the weight percent of potassium in illite?

5. In the log of Fig. 19–8 the zone between 2550′ and 2740′ is known to be predominantly illite. The standard chemical formula for illite can be found in Table 19–3 which shows no iron, although it is known that Fe can substitute for Al. In view of the Al and Fe traces on the log, what is a more reasonable formula for illite in this well?

6. What is the apparent Fe concentration, as derived from P_e, of a sandstone formation which contains 5% calcite cement?

20
SATURATION ESTIMATION

INTRODUCTION

The cornerstone of the saturation interpretation of resistivity measurements is the evaluation of the Archie relationship, which was presented in Chapter 3. Because of its simplicity, it has many shortcomings, and it is not directly applicable to shaly formations. And, despite its simplicity, its application to practical interpretation problems is not always straightforward; constants appropriate to the formation must be determined. In clean formations, this is relatively easy to do. Two graphical solutions of the Archie relation for determining the water resistivity and saturation are examined in this chapter. One of them also permits the determination of the cementation exponent appropriate to the given zone.

There are no such clear-cut methods for determining saturation in shaly formations. Dozens of different prescriptions exist. As a background to this bewildering array, this chapter considers in detail the effect of clay on resistivity measurements, alluded to in earlier chapters. The observational results of the influence of clay on rock resistivity is discussed, along with models to explain the behavior.

Many of the saturation equations for shaly formations are based on empirical observations and are of limited validity, despite good predictive success in certain applications. Two current models which are based on an important property of clays (the cation exchange capacity) are also discussed.

CLEAN FORMATIONS

The basic interpretation problem, given the corrected resistivity of the uninvaded formation R_t and the porosity ϕ, is in the evaluation of the Archie relation. In its simplest form, it can be written as:

$$S_w = \sqrt{\frac{R_w}{R_t} \frac{1}{\phi}} . \qquad (1)$$

The first question, in a practical application, concerns the value to use for R_w. This may present a problem if there is no proper SP development, as is often the case. In addition, if the matrix values for the formation are not known, there may even be some considerable doubt about the porosity values to be associated with the resistivity values measured. Finally, there can be uncertainty about the cementation exponent to be used. In the simple case above, it was taken to be 2 (i.e., $F = \frac{a}{\phi^m} = \frac{1}{\phi^2}$). However, this need not always be the case.

There are two graphical methods available for interpreting the water saturation of a zone when R_w is assumed to be constant but unknown. The basic measurements necessary are R_t, corrected for environmental effects, and a porosity log (usually density or sonic). A further requirement is the presence of a few water-bearing zones of different porosity in the logged interval, and, of course, the formations of interest must be clean (shale-free).

The first cross plot technique to be considered is the Hingle plot.[1] In this case, assuming that a density or sonic measurement is available, even if the matrix values are unknown, a plot can be constructed which will give porosity and water saturation directly. This expedience, coupled with the ease with which sonic and resistivity logs can be run in a single pass, has contributed much to the success of the induction/sonic logging combination.

To see the logic behind the Hingle plot, note that the simplified saturation expression of Eq. (1) indicates that ϕ will vary as $\frac{1}{\sqrt{R_t}}$ at a fixed value of water saturation, assuming, of course, that the water resistivity is constant. This leads to the construction of a plot, shown in Fig. 20–1, of inverse square root of resistivity versus porosity. It is obvious that formations of constant water saturation will lie on straight lines. Since we can rewrite Eq. (1) as:

$$\frac{1}{\sqrt{R_t}} = S_w \frac{1}{\sqrt{R_w}} \phi , \qquad (2)$$

it is clear that the 100%-water-saturated points will fall on a straight line of maximum slope. Less-saturated points, at any fixed porosity, must have a larger resistivity and thus fall below this line. Once these points have been identified, the line corresponding to $S_w = 100\%$ can be drawn, as shown in Fig. 20–1. It is relatively easy to construct lines of the appropriate slopes corresponding to partial water saturations.

Saturation Estimation   473

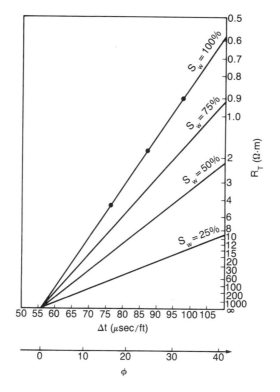

Figure 20–1. The Hingle plot, which combines resistivity and porosity (in this case the Δt measurement) to estimate water saturation. Implicit use is made of the square-root saturation relation in constructing the chart.

The value of R_w can be determined immediately from inspection of the graph. In the construction of Fig. 20–1, the uppermost line corresponds to R_o, since it is fully water-saturated and satisfies the relationship:

$$F = \frac{R_o}{R_w} = \frac{1}{\phi^2}.$$

The implication is that at a porosity of 10% the indicated value of R_o will be 100 times the value of R_w. For the example given, the value of R_o at 10 PU is 12 Ω·m, which indicates that the water resistivity is 0.12 Ω·m.

In the case of unknown porosity values, the horizontal axis may be scaled in the raw log reading: Δt or ρ_b. The intersection of the R_o line with the horizontal axis (corresponding to an infinite resistivity) will give the matrix value for constructing a porosity scale.

The second useful graphical technique is the result of work by Pickett.[2] A knowledge of porosity is required, but the values of m (the appropriate cementation exponent), R_w, and S_w can be obtained. In this method, the power law expression for saturation is exploited by using log-log graph paper.

Starting with the general saturation expression:

$$S_w^n = \frac{a}{\phi^m} \frac{R_w}{R_t}$$

and taking the log of both sides of the equation results in:

$$m \log(\phi) + \log(R_t) = \log(a) + \log(R_w) - n \log(S_w) .$$

This can be rearranged to:

$$\log(\phi) = -\frac{1}{m} \log(R_t) + \frac{1}{m} \left[\log(a) + \log(R_w) - n \log(S_w) \right] . \qquad (3)$$

Thus at a constant water saturation, a log-log plot of porosity versus R_t should result in a straight line (see Fig. 20–2) of negative slope whose value is the cementation exponent, and should be in the neighborhood of 2.

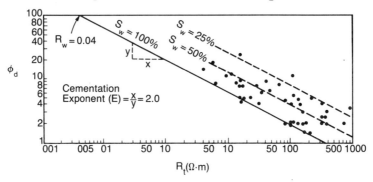

Figure 20–2. A log-log representation of resistivity and porosity attributed to Pickett. It is useful for determining the saturation exponent which best describes a given formation.

If we consider the value a to be unity, then we can write:

$$\log(\phi) = -\frac{1}{m} \log(R_o) + \frac{1}{m} \log(R_w) ,$$

which represents the line of 100% water saturation. In this case the intercept at the 100%-porosity point gives the value of R_w directly. For values of S_w less than 100%, the relationship between ϕ and R_t will be represented by lines parallel to the 100%-saturation case but displaced to the right. Using the simplified expression of Eq. (3), with a saturation exponent of 2, the displacement is easily quantified:

$$\log(R_t) = -2 \log(S_w) + \log(R_w) - m \log\phi .$$

Lines representing different values of water saturation can be placed in relation to the $S_w = 1$ line by recognizing that at a fixed porosity a saturation decrease by a factor of 2 (i.e., from 100% to S_w of 0.5) corresponds to a resistivity increase by a factor of four. Thus one obtains the 50%-saturation line by shifting a line parallel to the $S_w = 1$ line by a factor of four in resistivity, and the 25%-saturation case by shifting another factor of four, and so forth. An example of the result of this procedure is shown in Fig. 20–2.

As an exercise in applying these two techniques, let us compare them in a simple interpretation. The logs to be used for the evaluation are shown in Figs. 20–3 and 20–4. In the bottom section, four clean zones are indicated. With reference to the R_{xo}/R_t and SP overlay in track 1 of Fig. 20–3, the lower three zones seem to indicate the presence of hydrocarbon. Zone 4 appears to contain water. How can this be confirmed?

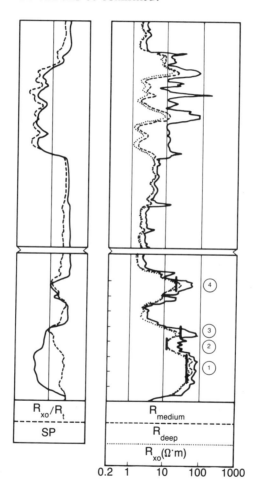

Figure 20–3. A resistivity log with zones for evaluation indicated. Adapted from Hilchie.[3]

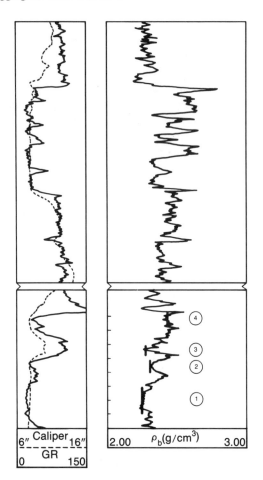

Figure 20-4. A companion density log for use in the Pickett plot and Hingle plot. Adapted from Hilchie.[3]

Since the formation rock type is not given, the porosity scaling for the density log (Fig. 20-4) is not obvious. However, the Hingle plot can be used to determine the saturation of the four zones in question. First, a number of points of resistivity and ρ_b are selected from the upper section of the log in the clean section, where the R_{xo}/R_t and SP indicate water-bearing formations. These establish the 100%-saturation line.

Fig. 20-5 shows the selected points, in the inverse $\sqrt{R_t}$ format, with density along the linear scale. (Note that the resistivity on this figure may be by any multiplicative factor desired to accommodate the log readings.) The 100%-saturation line is quite easy to identify, and with appropriate scaling the saturations of the four zones of interest can be determined. Zone 4 seems to have a water saturation of about 40%, or the same as zone 2, which appears less promising on the resistivity log.

Saturation Estimation 477

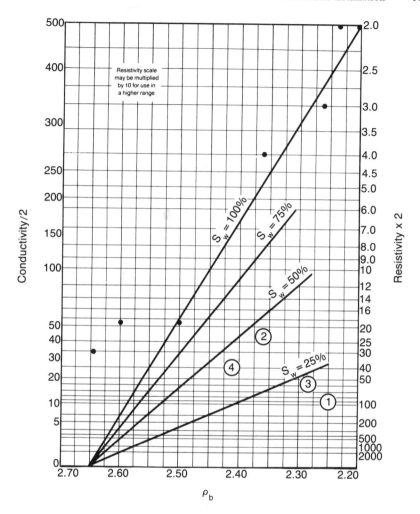

Figure 20–5. Hingle plot for selected data from the logs of Figs. 20–3 and 20–4. Points selected as water-bearing have been used to define the $S_w = 100\%$ line. Points representing the four suspected hydrocarbon zones are indicated.

Using the porosity scale, which can be derived from the Hingle plot, a similar analysis can be made using the Pickett technique, which is left as an exercise. The points corresponding to the water zones will nearly fall on a straight line, but with a slope nearer 1.4 than the value of 2 which was implicit in the graphical construction of the preceding analysis. Constructing the saturation lines parallel to the line defined by the water zones will indicate that the water saturations of zones 2 and 4 are very similar and close to a value of 40%.

SHALY FORMATIONS

Before reviewing the variety of saturation equations used in the analysis of shaly formations, let us consider why they are necessary. The reason is that frequently real rocks do not follow the simple conductivity behavior described by the Archie equation. Before considering the complication of partial saturation, it is useful to see how complicated the behavior of fully water-saturated rock samples is.

The experimental data which confirm that there is an additional complication in the interpretation of the resistivity of clay-bearing or shaly rocks is given in Chapter 7 (Figs. 7–13 and 7–14). Fig. 20–6 shows the conductivity of the fully saturated rock as a function of the saturating water conductivity. The clean sand response, shown as a dashed line, represents the Archie relation; its slope is the reciprocal of the formation factor F. At large values of water conductivity, the response of the shaly formation is seen to be simply displaced with respect to the Archie-type behavior. This additional conductivity associated with the clay can be put into the Archie relation as:

$$C_o = \frac{C_w}{F} + X,$$

where X, the additional term that results from shale, must decrease to zero as the clay content vanishes. Above some value of water conductivity, it simply appears as a linear shift. The slope of the line beyond this region yields the same formation factor F as would be obtained for the rock without the presence of clay. However, it is seen that at very low values of water salinity there is a nonlinear region in which the additional clay conductivity appears to be a function of C_w.

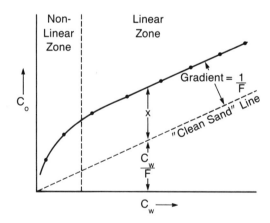

Figure 20–6. Schematic behavior of the conductivity of water-saturated shaly rocks showing the nonlinear behavior at low values of water conductivity and offset at higher values. From Worthington.[4]

To further illustrate this problem, Fig. 20–7 shows the results of some measurements of the formation factor as a function of porosity for some sandstone core samples. In the top figure, the samples have been saturated with very saline water (C_w is large). The behavior for these shaly samples is as expected for clean cores. There is a definite porosity relationship for the formation factor. Shown in the lower portion of the figure are results of measurements on the same cores, this time with quite fresh saturating water. In this instance, the apparent formation factors are seemingly quite random. The intrinsic conductivity of the shale portion dominates the conductivity in this case.

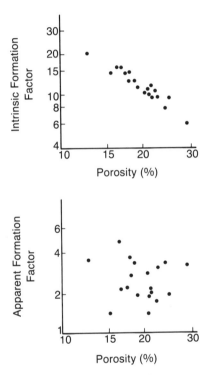

Figure 20–7. Data from Worthington which demonstrate the influence of clay content at low values of water conductivity.[4] The formation factors of the cores in the upper figure were determined with highly conductive saturating water, the lower set with very fresh water. The spread in the lower set is a reflection of the clay content and cation exchange capacity of the rock.

From this we can conclude that the abnormal electrical behavior of shaly sands is of minor importance when the resistivity of the formation is low. However, it is much more important when the sand is saturated with a dilute brine or when the saturated sample contains a large fraction of nonconductive hydrocarbon. Because of this last point, a number of empirical techniques to cope with the clay-affected resistivity measurements have been developed over the years.

V_{sh} Models

The first model for attempting to quantify this shale conduction term, and the only one which we will consider in any detail, applies the simplified Archie relation:

$$C_o = \frac{C_w}{F} = \phi^2 C_w ,$$

to the shale. When all of the interstitial electrolyte is replaced by a wetted shale which completely fills the porosity, the volume of shale, V_{sh} will be equal to ϕ. It will have, by analogy with the equation above, an additional conductivity of magnitude $V_{sh}^2 C_{sh}$, where C_{sh} is the conductivity of the wetted shale. Thus the completely water-saturated conductivity of a shaly formation can be written as:

$$C_o = \frac{C_w}{F} + X = \frac{C_w}{F} + V_{sh}^2 C_{sh} .$$

The interpretation task is to evaluate the shale conductivity and the volume fraction contained in a given formation in order to correct the resistivity reading for the perturbation due to the shale.

$$C_o = \frac{C_w}{F} + V_{sh}^2 C_{sh}$$

$$C_o = \frac{C_w}{F} + V_{sh} C_{sh}$$

$$\sqrt{C_o} = \sqrt{\frac{C_w}{F}} + V_{sh} \sqrt{C_{sh}}$$

$$\sqrt{C_o} = \sqrt{\frac{C_w}{F}} + V_{sh}^{1-V_{sh}/2} \sqrt{C_{sh}}$$

Table 20–1. Four empirical conductivity relationships for fully water-saturated rock. The shale content of the rock is described by a single bulk parameter V_{sh}. From Worthington.[4]

Table 20–1 presents four of the general types of V_{sh} models which have been developed over the years to cope with local situations. Of interest is the third expression, attributed to an unpublished work by H. G. Doll. It seems to be obtained from the first expression by simply taking the square root of each of the terms. If the Doll expression is then squared, the result is:

$$C_o = \frac{C_w}{F} + V_{sh}^2 C_{sh} + 2 V_{sh} \sqrt{\frac{C_w}{F}} \sqrt{C_{sh}} .$$

Despite the fact that there seems to be no logic for the procedure, the cross-term provides the ability to match the behavior of the conductivity at low salinities in the nonlinear region of Fig. 20–6.

The problem with the four V_{sh} models presented here is that C_{sh}, the shale conductivity, needs to be altered to fit the linear and nonlinear regions. From this sampling it should be noted that there is no universal V_{sh} model to fit all interpretation needs. Different approaches work in different circumstances. Most importantly, these V_{sh} models do not take into account the mode of distribution of the shale or any other physical attribute of the shale. In fact, the models may be partially driven by the method employed to determine V_{sh}.

Effect of Clay Minerals on Resistivity

As briefly mentioned in Chapters 7 and 19, one of the most important properties of shale in terms of electrical effects is the cation exchange capacity, or CEC. The CEC of a shale is related to the ability of the shale to adsorb electrolytic cations, such as Na, onto or near the surface. Another factor which must be related to this adsorption is the amount of surface accessible to the electrolyte. This is a combination of the specific surface area of the clay minerals and their distribution.

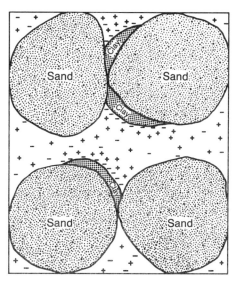

Figure 20–8. A conceptualization of the distribution of clay in a sand which affects electrical conductivity. From Winsauer et al.[5]

The mechanism for the excess conductivity was first proposed by Winsauer et al. in 1953 and is illustrated in Fig. 20–8.[5] In this representation, the clay is seen to coat some of the sand grains. The negative surface charge of the clay platelets causes the attraction of the positive Na cations. Thus in the regions of fluid close to the clay there will be a much larger concentration of positive charge carriers than in the rest of the solution. The conductance of this "near" fluid will be higher than that of the undisturbed portion of the fluid. This visualization is usually referred to as a double layer model and gives rise naturally to the image of two conduction paths for electrical current: one through the unbound "far" fluid and another through the more conductive "near", or bound, layer.

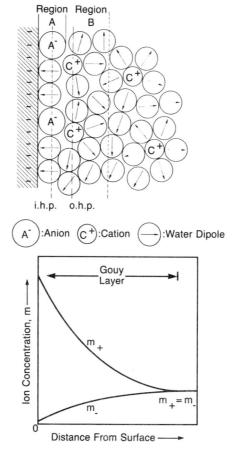

Figure 20–9. The distribution of water molecules and electrolytes near the surface of a clay crystal with excess negative surface charge. The lower portion of the figure indicates the distribution of ions as a function of distance from the surface. Increased conductivity is expected in the layer next to the clay as a result of the increased cation density. The distance at which equilibrium between cation and anion densities (n_- and n_+) is reached depends on temperature and water salinity. Adapted from Yariv and Cross,[6] and Berner.[7]

Electrical double layer theory, known to physical chemists long before its application to shaly formations, attempts to describe the distribution of ions in solution near the interface of a solid. For our purposes, the solid is a clay crystal with excess negative surface charge. Such a region is sketched in the upper portion of Fig. 20–9. Several distinct regions of the electrolyte have been identified. In region A, there are preferentially oriented water molecules and some unhydrated exchangeable counter ions. The next region, referred to as the outer Helmholtz layer, is the limit of closest approach of fully hydrated cations. Beyond this layer is a diffuse region where the concentration of cations decreases until reaching equilibrium with the bulk of the electrolyte. The lower portion of Fig. 20–9 schematically shows the ion concentration near the surface, indicating an excess of cations which tails off to the equilibrium value. It is this region, whose dimensions are temperature and salinity dependent, which is considered to provide an additional path of conduction in shaly formations.

Double Layer Models

The recognition of the importance of the CEC spurred the development of another set of models for dealing with clay effects on conductivity which are a bit more sophisticated than the V_{sh} models. The first of these, proposed by Waxman and Smits of Shell, used the concept of CEC directly to explain the increased conductivity.[8] Rather than use CEC, it is more convenient to define a new quantity, Q_v, which is the CEC normalized to the pore volume. The definition of Q_v is given by:

$$Q_v \equiv \frac{CEC\, \rho_b (1-\phi)}{\phi}.$$

Since the dimensions of CEC are in meq/g, the dimensions of Q_v are in meq/cm^3.

The Waxman–Smits model then expresses the conductivity relation as:

$$C_o = \frac{C_w}{F} + \frac{BQ_v}{F},$$

where B is the conductance of the Na cations. This quantity has been measured in the laboratory, and is known as a function of temperature and concentration of the NaCl solution.

In this model, B is not a constant and must change at low values of C_w in order to fit the nonlinear region of the conductivity data. Waxman and Thomas find that it can be related simply to an exponential function of the water conductivity.[9] The most important shortcoming is that, at present, there is no direct way to obtain CEC from log measurements aside from induced polarization measurements and the geochemical inference discussed in the last chapter.[10] For the present, CEC values must be painstakenly obtained from measurements on cores or estimated from other logging parameters.

Another model, the "dual water" model, has been developed to rectify some of the deficiencies of the Waxman–Smits model.[11] In this representation, the authors view the clay as attracting not only the Na cations, but also a layer of polar water molecules, as illustrated in Fig. 20–9. In this manner, the porosity is viewed as having two components, a bound water component which is directly in contact with the clay, and a free component which is not associated with the clay particles.

This model also holds the shale-bound water to have a certain invariable conductivity (although there is a temperature dependence) which is not dependent on the type of clay. In contrast to the Waxman–Smits approach, the conductivity of the enriched layer of Na ions near the surface is diluted by the presence of the bound water. A crucial parameter in the dual water model is the fractional portion of the porosity which is bound to the clay. To compute this volume, one considers the thickness of the cation-enriched layer and multiplies by the specific surface area of the clay. Drawing upon the correlation between specific surface area and CEC, the fractional volume of bound water is directly proportional to Q_v. The dual water model representation of conductivity is given by:

$$C_o = \frac{C_w}{F_o} + \frac{(C_{bw} - C_w) v_q Q_v}{F_o},$$

where C_{bw} is the conductivity of the bound water, and v_q is the constant of proportionality between the fractional pore volume of bound water and Q_v. Dewan details the steps for applying the "dual water" model to a log example.[12]

Saturation Equations

So far the discussion has been centered on the description of the electrical behavior of fully water-saturated rock. What will be the effect of changing water saturation by introducing nonconductive hydrocarbon? The result most certainly will depend on the details of the fluid distribution. If the hydrocarbon is distributed in irregular drops surrounded by water-wet pore walls, it seems difficult to imagine a generalized relationship between the resistivity index (R_t/R_o) and saturation which works in all ranges of porosity and pore distribution systems. However, empirical observations seem to indicate that the saturation index is independent of porosity and has a value of about 2. In the absence of definitive physical models for the understanding of saturation relations in clean rocks, we have but one recourse for clay-bearing rocks: We must resort to the empirical Archie relation:

$$C_c = \frac{C_w}{F} S_w^n,$$

where n is the saturation exponent. If only the clean portion of the conductivity is expected to change by reducing the saturation, we might expect the general saturation equation to be given by:

$$C_t = \frac{C_w}{R} S_w^n + X .$$

However, experimental evidence indicates that the shale conductivity term can be considered to vary with saturation and is generally expressed as:

$$C_t = \frac{C_w}{F} S_w^n + X S_w^s ,$$

where s is the shale saturation exponent. It has also been recognized that a decrease in the water saturation increases the importance of the electrical double layer. The enhanced shale conductivity is often approximated by $\frac{X}{S_w}$. Consequently the generalized saturation equation above usually has a shale conductivity term which is linear in S_w.

Worthington has made a comprehensive survey of the more than thirty saturation equations which have been used over the years.[4] A summary of the four basic types of saturation equations is given in Table 20-2. Of the four types listed, only one, that based on the double layer model, is less empirical than the others. Why are there so many approaches? Certainly one reason is related to the good predictive performance in localized applications. One wonders, also, if the number of models proliferated through some basic misunderstanding of the porosity derived from the neutron-density combination. We have seen how different clay types can affect the neutron porosity, and thus the cross-plot-derived porosity. Another reason has been the drive to put some saturation models on a sound scientific basis. But for the moment this has not entirely been achieved.

$$C_t = \frac{C_w}{F} S_w^n + X$$

$$C_t = \frac{C_w}{F} S_w^n + X S_w^s$$

$$\sqrt{C_t} = \sqrt{\frac{C_w}{F} S_w^{n/2}} + \sqrt{X}$$

$$\sqrt{C_t} = \sqrt{\frac{C_w}{F} S_w^{n/2}} + \sqrt{X} S_w^{s/2}$$

Table 20-2. Four basic saturation equations written in terms of the excess conductivity due to the contained shale X. The Waxman–Smits and "dual water" model use an equation of the second form.

It is clear that the CEC is one reasonable way to approach the correction of shaly resistivity measurements, despite the difficulty in obtaining reasonable values to use for CEC. It is probably through the use of such a parameter that a unified approach can be obtained. As an example of this,

Worthington has contributed the following observation.[4] If we rearrange the basic conductivity equation:

$$C_o = \frac{C_w}{F} + X,$$

we can obtain the following:

$$\frac{C_o - X}{C_o} = \frac{C_w}{F} \frac{1}{C_o},$$

where the term $\frac{C_w}{C_o}$ is recognized as the apparent formation factor F_{app}. Thus the ratio of F_{app}/F represents the fraction of total conductivity which cannot be attributed to shale. Plots of the ratio as a function of the water conductivity for four sets of data with widely varying clay content (and thus widely varying Q_v) are shown in Fig. 20–10. It is obvious that the four curves resemble one another almost as if each were shifted according to the CEC value. Casting the available data into this framework may provide a unifying basis for comparing further work. However, the definitive work on saturation determination from electrical measurements has not yet appeared.

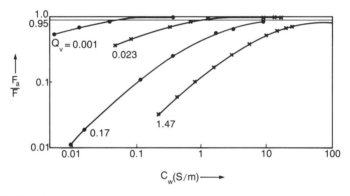

Figure 20–10. An attempt to demonstrate the generally observed trends of conductivity in clay-bearing rocks. The ratio of apparent formation factor to F is the fraction of the total conductivity not attributed to shale conductivity. It is plotted as a function of saturating water conductivity for four core samples of varying clay content, quantified by a particular value of Q_v. The curves have a similar shape and seem to be offset by a value related to Q_v. From Worthington.[4]

References

1. Hingle, A. T., "The Use of Logs in Exploration Problems," Twenty-ninth Annual International Meeting of SEG, Los Angeles, 1959.

2. Pickett, G. R., "Acoustic Character Logs and Their Application," *JPT*, June 1963.

3. Hilchie, D. W., *Applied Openhole Log Interpretation*, D. W. Hilchie, Golden Colorado, 1978.

4. Worthington, P. F., "The Evolution of Shaly-Sand Concepts in Reservoir Evaluation," *The Log Analyst*, Jan.–Feb. 1985.

5. Winsauer, W. O., and McCardell, W. M., "Ionic Double-Layer Conductivity in Reservoir Rock," Petroleum Transactions, AIME, Vol. 198, 1953.

6. Yariv, S., and Cross, H., *Geochemistry of Colloid Systems for Earth Scientists*, Springer-Verlag, Berlin, 1979.

7. Berner, R. A., *Principles of Chemical Sedimentology*, McGraw-Hill, New York, 1971.

8. Waxman, M. H., and Smits, L. J. M., "Electrical Conductivities in Oil-Bearing Sands," SPEJ, June 1968.

9. Waxman, M. H., and Thomas, E. C., "Electrical Conductivity in Shaly Sands, I. The Relation Between Hydrocarbon Saturation and Resistivity Index, II. The Temperature Coefficient of Electrical Conductivity," Paper SPE 4094, SPE-AIME Forty-seventh Annual Meeting, 1972.

10. Vinegar, H. J., Waxman, M. H., Best, M. H., and Reddy, I. K., "Induced Polarization Logging–Borehole Modeling, Tool Design and Field Tests," SPWLA Twenty-sixth Annual Logging Symposium, 1985.

11. Clavier, C., Coates, G., and Dumanoir, J., "The Theoretical and Experimental Basis for the 'Dual Water' Model for the Interpretation of Shaly Sands," SPE Paper 6859, 1977.

12. Dewan, J. T., *Essentials of Modern Open-hole Log Interpretation*, PennWell Publishing Co., Tulsa, 1983.

Problems

1. The Hingle plot of Fig. 20–5 shows the fully water-saturated line corresponding to sandstone (since the intercept is at 2.65 g/cm^3).

    a. Estimate the value of R_w.

    b. Two of the data points at low apparent porosity do not lie on the trend line for $S_w = 0$. Give several plausible reasons to explain the positions of these two points on the plot.

    c. Redraw the Hingle plot assuming that the matrix is limestone. How different are the saturation estimates for zones 1 and 3?

2. Using estimates of porosity from Fig. 20–5, use the Pickett plot technique of analyzing the resistivity data. Graphically determine the following:
   a. water resistivity
   b. cementation exponent
   c. water saturation in zones 1–4

3. The table below lists the values of R_t and Δt observed in a number of clean zones in a well with zones of hydrocarbons, as well as some water zones of different porosities.
   a. Using the Hingle plot technique on the graph paper of Fig. 20–3, determine the value of R_w to be used in the analysis of the suspected hydrocarbon zones.
   b. From the plot of part a, what value should be used for the matrix travel time Δt_{ma}, in order to convert ρ_b to porosity?
   c. Which zones have an oil saturation greater than 50%?
   d. What is the porosity of level 16?

Zone	Δt	R_t	Zone	Δt	R_t
1	107	0.95			
2	87	1.9	11	75	5.8
3	107	4.1	12	89	5.9
4	76	5.0	13	92	1.9
5	102	6.1	14	71	12
6	96	12	15	97	1.9
7	97	0.88	16	95	6
8	96	1.2	17	82	10
9	65	23	18	102	1.8
10	85	3.2	19	89	15

4. Using your knowledge of porosity from the preceding analysis, use the Pickett plot technique of analyzing the resistivity data. Graphically determine the following:
   a. water resistivity
   b. cementation exponent
   c. water saturation in zones 1–4

21
EXTENDING MEASUREMENTS AWAY FROM THE BOREHOLE

INTRODUCTION

In this chapter the depth of investigation of the various measurements at our disposal in well logging is reviewed. The depth of investigation is controlled by the geometry of the measurement sonde and by loss mechanisms associated with the particular measurement under consideration. Most of the conventional borehole measurements are very localized. Extrapolating measurements made in the vicinity of the borehole to distances on the scale of reservoir dimensions assures many surprises. Low frequency acoustic waves offer promise for the extension of measurements away from the borehole.

Before discussing the use of low frequency acoustic waves in borehole logging, a brief summary of surface seismic principles is given. The development of a borehole seismic measurement, known as a vertical seismic profile (VSP), is shown to have been a natural outgrowth of using conventional sonic and density logs to provide support for the interpretation of surface seismic sections.

The vertical seismic profile combines relatively simple down-hole acoustic measurements to produce an image which can easily be compared to surface seismic sections or logs. Some of the simple processing techniques for obtaining the images are reviewed, along with the relation of the vertical seismic profile (VSP) to surface seismic measurements. An additional feature of the VSP, which already has better vertical resolution than surface seismic measurements, is the possibility of extending its lateral coverage. Some illustrations of this technique, known as offset VSP, are given.

DEPTH OF INVESTIGATION AND RESOLUTION OF LOGGING MEASUREMENTS

The well logging measurements discussed to this point have all been of rather limited depth of investigation. These limits are based on a combination of attenuation mechanisms which reduce the signal strength and the geometry imposed by operating in a borehole. The borehole geometry restricts the manner in which experiments can be performed; for example, it is not possible to do a straightforward transmission measurement. In other cases, operating in the borehole environment places restrictions on the transmitter or source strength. This may be the result either of power requirements or limitations on physical dimensions. The parameter which limits the depth of investigation is called the skin depth or mean free path. We will consider the three domains of physical measurements employed (electromagnetic, nuclear, and acoustic) to reach some broad conclusions on the depth of investigation to be expected from each.

Nuclear

There are several elements involved in the determination of the depth of investigation of a nuclear device. One of the most important controlling parameters is the mean free path of the radiation used in the measurement. This quantity is related to the cross section for the radiation employed and thus depends on the energy of the radiation and the nature of the material under investigation. Neutrons and gamma rays emitted from sources used in logging devices have mean free paths on the order of 10 to 20 cm, at most.

The angular distribution of the source radiation can be altered somewhat through the use of shielding to produce some influence on the overall depth of investigation. It seems reasonable to assume that radiation directed into the formation may penetrate further than radiation allowed to stream at shallow angles close to the sonde. However, the energy of the radiation detected in the sonde, at some distance from the source, is usually much reduced compared to the source. Thus the mean free path of the detected radiation is likely to be much smaller than the source radiation. The combined effects of source shielding and source-to-detector spacing (limited to distances usually much less than a meter because of source strength and counting rate statistics) contribute to determining the depth of investigation. A representative depth of investigation for nuclear tools is in the range of tens of centimeters.

Electrical

Electrical measurements can be grouped by operating frequency range. For very low frequency measurements, the depth of investigation is controlled by the spacing between the current emitter and the voltage monitor. The potential varies simply with inverse distance from the current source, and one can imagine achieving any depth of investigation desired by simply changing the spacing. This may have some practical limits, associated with the detection of minute potential differences at very long spacings.

Variations on the normal device design have been developed with some extremely long electrode configurations. One such device, the ultra-long-spaced electric log (ULSEL), is used to detect conductive and nonconductive anomalies far from the borehole.[1] With a range of electrode spacings between 600' and 4000', detection and mapping of salt domes have been performed at distances of several thousand feet from the well. For a special application to blow-out control, a shorter configuration (75' and 600') has been used to detect the distances from a relief well to a cased well out of control.

Although the long spacing measurements can provide a considerable lateral depth of investigation, the vertical resolution is on the same order of magnitude. Induction tools provide some sort of compromise. In Chapter 6 the simple two-coil induction device is described in terms of the geometric factor. The vertical and radial responses are found to be closely related to the coil spacing. In theory, an appropriate combination of coils could alter the position of maximum radial sensitivity curve to any depth desired. However, there are the practical engineering limitations of signal strength, cancellation of the mutually induced signal, and skin depth. This latter factor is related to operating frequency and the conductivity of the formation. In the best of cases, the induction tool, which operates at 25 kHz with a basic coil spacing of 40", derives 90% of its signal from within 1.5 meters of the sonde. However, the skin depth may reduce this region of investigation to the order of a meter. For the electromagnetic propagation tool, which operates at much higher frequencies, the depth of investigation, which is controlled by the skin depth, is only a few centimeters. For this reason, the spacing of the antennae on this tool is of the same order of magnitude.

At intermediate frequencies, the skin depth is somewhere within these extremes. In this range, measurements can be extended away from the borehole. This requires abandoning the concept of a measurement device which contains both the source and receiver. Some studies of cross well tomography have been made at frequencies between 1 and 30 MHz.[2,3] In this technique, a transmitter is placed in one well and a receiver in an adjacent well. Even though the skin depth is smaller than for the induction tool, this type of transmission geometry permits measurements over large distances. The signal attenuation is measured for a variety of depths for both transmitter and receiver. The attenuation is the result of variations in the conductivity along the electromagnetic wave travel path. Depths of penetration up to 100′ have been attained with vertical resolution on the order of 2–3′. However, the practicality of this operation is limited because of the requirement of adjacent wells with such close spacing.

Acoustic

The depth of investigation of the conventional sonic logging device was considered in Chapter 17. Because of the rather crude nature of the measurement technique, which simply detects the first arrival above a given signal threshold, the depth of investigation is somewhat difficult to define. It is found to be a function of the velocity contrast between the altered zone and the virgin formation, and the source-to-detector spacing. For the conventional sonic logging tool with a pair of receivers at 3′ and 5′ from the transmitter, a typical value may be on the order of 6″.

Low frequency seismic sources produce a pressure wave which, in a uniform homogeneous medium, radiates spherically from the source of initial energy. The energy density falls off with a $\frac{1}{r^2}$ dependence. The quantity which is generally measured, the formation velocity or displacement, is proportional to the square root of the acoustic energy. In addition to the spreading loss of amplitude, there is additional loss to frictional dissipation. Thus the acoustic amplitude at a distance r from the source can be written as:

$$I = I_0 \frac{e^{-\alpha r}}{r} ,$$

where I_0 is the initial amplitude of the disturbance. The absorption coefficient α depends upon the particular rock material. One observation that has been made at frequencies below several hundred Hz is that α is proportional to frequency. Table 21–1 indicates the value of α in cm^{-1}

Attenuation Coefficients for 50-Hz Waves		
Material	Velocity, km/s	Attenuation ×10⁻⁵
Granite	5.0-5.1	0.21-0.38
Basalt	5.5	0.41
Diorite	5.8	0.2
Limestone	5.9-6.0	0.04-0.37
Sandstone	4.0-4.3	0.7-1.8
Shale	2.15-3.3	2.32-0.68

Table 21–1. Velocities and attenuation coefficients for 50 Hz seismic waves. Adapted from Dobrin.[6]

evaluated at 50 Hz for a number of rock formations. As can be seen, there is considerable variation in this quantity over the small sample of formations presented in the table. Considering only the highest value of attenuation (Pierre shale), we find a value of $2.32 \times 10^{-5} \text{cm}^{-1}$. This is equivalent to a mean free path or skin depth of 400 m (at 50 Hz). Thus attenuation losses of low frequency acoustic waves are mainly dominated by the $\frac{1}{r}$ spreading loss. For this reason extremely large depths of penetration are possible employing this type of energy source. However, there is a price to be paid; the low frequencies which permit large penetration distances also imply wavelengths which are large compared to length scales of interest to underground geological structures.

Some idea of the vertical resolution of the acoustic measurements can be obtained from a consideration of the wavelengths involved. The frequency range over which measurements are made covers three orders of magnitude: 10–20 kHz for continuous borehole sonic logging, to around 10 Hz for seismic reflection surveys. To put the spatial resolution of acoustic measurements into perspective, it is instructive to make a simple plot of the resolution vs. frequency. For the seismic region where beds are detected by reflection, the resolution is taken to be equal to one quarter of a wavelength. For the higher frequencies used in transmission, the estimate is based on a resolution of half a wavelength. These estimates are plotted in Fig. 21–1 for three values of formation velocity ranging from 5,000 to 20,000'/sec.

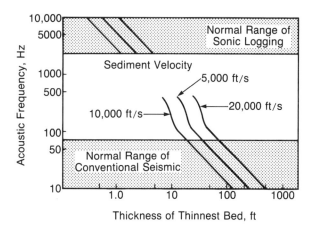

Figure 21–1. Bed thickness resolution as a function of acoustic frequency. Courtesy of Schlumberger.

The upper band corresponds to frequencies used in borehole logging tools and shows a capability of resolving beds of thickness on the order of 1'. The lower band shows the normal range of seismic frequencies where the thinnest resolvable beds are on the order of 100', depending on formation velocities and frequency content.

Due to the frequency dependence of acoustic attenuation, the earth is effectively opaque for frequencies roughly greater than 100 Hz. For large distances of penetration, lower frequencies must be used. With the exception of electrical and acoustic measurements between closely spaced wells, the only real hope for extending measurements away from the borehole is the use of seismiclike experiments.

SURFACE SEISMICS

In order to introduce the subject of well seismics, we first review the rudiments of surface seismics. This technique can furnish an image of the subsurface by using a pulse of low frequency acoustic energy at the surface and a number of geophones (also at the surface) to record, as a function of time, the resultant echos from subsurface reflectors.

Two simple principles describe the seismic reflection process. The most useful simplification is the so-called ray theory. It represents the propagation of energy of an acoustic wave by a single ray in the direction of interest. The usefulness of this representation is the prediction of the refraction of the acoustic energy when crossing boundaries of rock formation with different seismic velocities. When we keep to a flat earth model with virtually no separation between source and detector, this predictive extravagance is not necessary, but it is useful for the case of dipping (nonhorizontal) reflectors or when there is considerable separation between source and detector.

Acoustic reflection occurs at boundaries between layers of differing acoustic impedance. For a material of density ρ and compressional velocity v, the acoustic impedance for compressional waves z is simply the product ρv. A derivation of this relationship is given in the appendix to this chapter. The reflection coefficient for the amplitude of particle motion for acoustic waves at normal incidence to the boundary of two layers characterized by z_1 and z_2 is given by:

$$\frac{z_2 - z_1}{z_2 + z_1}.$$

The values may be either positive or negative, depending on the impedance contrast.

To illustrate the imaging capabilities of the seismic technique, consider the upper portion of Fig. 21–2, which shows the location of a succession of shot points and associated geophones. The subsurface is shown to have a single acoustic impedance contrast, and thus a single reflection is seen at each geophone location in the lower portion of the figure. Normally the trace of geophone amplitude vs. time is displayed vertically, with time running downward. Each trace corresponds to a particular shot point and receiver position.

When all of the seismic traces are suspended from a representation of the earth's surface, what we have is a rudimentary seismic section. The reflection features will follow (to some extent) the geometric outline of the

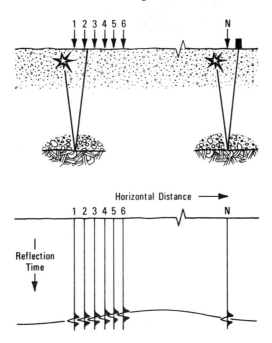

Figure 21-2. Simplified seismic section generated by a succession of shot-point and receiver pairs. The lower portion of the diagram shows the conventional vertical display of the received signal as a function of time, so that a reflector appears as a continuous line through coherent maxima. Courtesy of Schlumberger.

underlying structure. However, it must be kept in mind that the vertical axis on such a plot is not depth but time. An example of a seismic section is shown in Fig. 21-3. What information can such a graphical representation bring?

First of all, it can assist in large-scale structural interpretation by delineating the times to seismic reflectors. The signal amplitude at these reflectors has some significance. Reflections are generated by impedance contrasts, which may signify an important change in lithology or the presence of gas. The shape of the reflector surface, especially if convex, may have important consequences for the placement of wells. The dip of the structure may also be determined. If sets of orthogonal seismic lines are used, then some reasonable three-dimensional description can be obtained about portions of the subsurface.

Because of the way seismic shots are actually made, it is possible to derive the formation velocities without actually knowing the depth of the reflector. A rudimentary example of this is illustrated in Fig. 21-4. Two shots are used. The base time to the reflector is determined with the source and geophone placed close together. A second shot, with source and receiver each displaced from the original position, will provide a second time to the

496   Well Logging for Earth Scientists

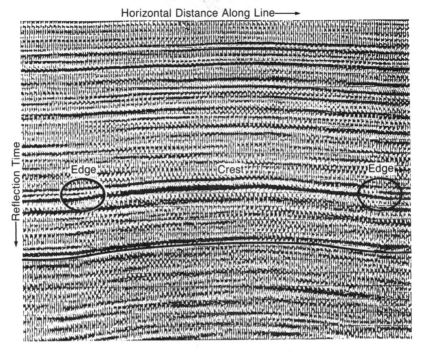

Figure 21-3.   A more complicated seismic section, showing some distinct reflecting horizons. Courtesy of Schlumberger.

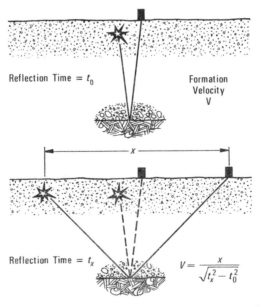

Figure 21-4.   Determination of seismic velocity by comparison of the travel time with an offset source and the receiver with the perpendicular travel time. Courtesy of Schlumberger.

reflector. The time difference between the two measurements and the offset distance allows calculation of the formation velocity. This velocity information may provide clues as to the formation lithology or the presence of overpressured zones. Another piece of information that can be extracted from the seismic section is the amplitude of the reflections, which is related to the impedance of the formation.

Another aspect of the information from the seismic traces is the lateral zone of investigation. Because of the fact that the waves are not planar but spherical, the reflected signal comes from a roughly circular Fresnel zone whose diameter can be in the range of 200 to 700 m. This is in great contrast with the lateral extent of logging measurements, which is generally less than several feet.

It must be admitted that, short of drilling a well, the surface seismic method gives more information concerning the subsurface than any other exploration technique. Despite this power, it has some shortcomings. One of the most obvious concerns its vertical resolution. As indicated earlier, the resolution is linked to the frequency. Generally seismic sources radiate most of their energy at low frequencies. The amplitude spectrum of one seismic source is shown as the upper curve of Fig. 21–5. It indicates an effective

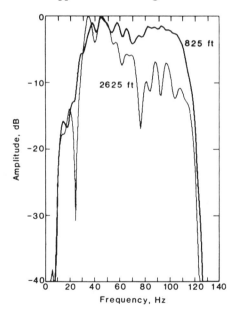

Figure 21–5. The frequency content of a surface source recorded at two depths, showing the attenuation at higher frequencies. From Balch and Lee.[5]

output in the range of 30 to 120 Hz. Even if it were practical to increase the high frequency content of seismic sources, resolution would not be much improved. This is because the higher frequencies are strongly attenuated. The lower trace in Fig. 21–5 shows the amplitude spectrum of the source after attenuation through 1800′ of formation. A significant reduction is indicated for frequencies above 70 Hz. Thus the frequency content will be related to the surface source characteristics and the depth. The average frequency will be lower at greater depths.

For an order of magnitude calculation of resolution, a representative frequency can be taken to be about 50 Hz. In a formation with a transit time of 100 μsec/ft or a compressional velocity of 10,000′/sec, the acoustic wavelength is 200′. Layered formations of a quarter wavelength in thickness can be resolved giving 50′ as the vertical resolution. This is quite gross compared to the normal 6″ sampling of most logging data.

Leaving aside other seismic problems, such as multiple reflections and dipping beds, all of which can more or less be overcome with appropriate data collection and data processing, there is the problem of associating the proper seismic wiggles with precise geologic features. This practical question can begun to be answered when a well is available along the seismic line.

BOREHOLE SEISMICS

The discipline of acquiring seismic data and knowledge in the vicinity of the borehole is called borehole seismics. The data may consist of surface seismic data, check shot surveys, density and acoustic logs, dipmeter results, and VSP surveys.

Check shot surveys are mentioned in Chapter 15 as a precursor to sonic logging. They provide a correlation between transit times and the depth of specific reference formations. These are of crucial importance for establishing depths on seismic sections.

The equipment involved in such a survey is a surface source and a down-hole tool equipped with one or more geophones. The surface source is often an air gun (especially useful offshore) or, on land, a vibrator. The down-hole equipment makes a measurement of the particle motion induced by the seismic energy reaching the position of the sonde in the borehole. The sonde is required to be firmly coupled to the formation. The earliest sensors were geophones which consist of an electric coil suspended by springs in a permanent magnet field. The coil movement produces a current in the coil. This type of device is primarily sensitive to movement in the vertical direction. More modern VSP equipment is equipped with triaxial sensors which are based on ceramic accelerometers.

As we have already seen, borehole acoustic logging was developed to supplant the need for the check shot. Rather than supplant this need, borehole acoustic logging, or sonic logging, has taken on a life of its own. One of the reasons the sonic log has not become the absolute standard for time-to-depth correlation is that the travel time to a given depth involves the integration of the interval travel times (Δt) at each depth from the surface to the point desired. Rarely do the sonic logs begin at the surface. The other problem is that sonic tools operate at roughly 20 kHz, whereas seismic frequencies are about 1000 times smaller. Over this large frequency range there is noticeable dispersion; the velocity varies with frequency. Consequently integrated travel times obtained from the sonic log will show some drift with respect to actual seismic travel time. The sonic logs thus need to be adjusted, and this is done through the check shot technique which was developed in the 1960s. It uses a source nearly equivalent in frequency content to the normal seismic source and records the total travel time to a given depth in the well, thus providing the necessary calibration of the integrated sonic transit times.

In an attempt to tie geologic horizons present in borehole measurements to surface seismic data, the synthetic seismogram was developed. This simplified modeling procedure uses as input the density and sonic logs. From these two, a log of acoustic impedance can be produced. At each of the major formation interfaces, a reflection coefficient can be calculated. Its amplitude is determined by the impedance contrast between layers, and its sign indicates the polarity of the reflected wave. Both of these attributes are illustrated in the left portion of Fig. 21-6.

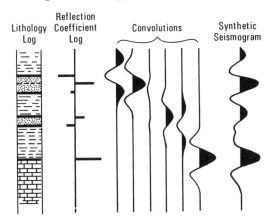

Figure 21-6. Generation of a synthetic seismogram through the combination of two logging measurements: compressional velocity and bulk density. Using log values, the layers traversed by the well can be zoned into lithologic sections of uniform acoustic impedance. The trace of acoustic impedances can then be computed on a layer-by-layer basis. Convolution of the reflection coefficient log with an idealized source signal produces the synthetic trace. Courtesy of Schlumberger.

The next step in the process is to convolve the form of surface signal with each of the reflection coefficients. This is achieved essentially by multiplying the representative waveform by the reflection coefficient and lining up the resultant at the proper depth associated with the waveform. This is continued for each of the reflection coefficients. The final result, shown on the right, is obtained from the summation of the various modulated signals. It represents an approximation to the up-going acoustic wavefield.

The synthetic seismogram can account satisfactorily for many of the events seen on the seismic section. If it does not, then the model has been too simple, or the primary data of the density and sonic log have been affected by environmental effects. This can particularly be true of the density log in the upper sections of wells that pass through very shaly or unconsolidated formation in which the borehole may be very washed out or rugose.

Rather then refine the model for the synthetic seismogram, another possibility exists, that of the vertical seismic profile. This procedure circumvents the synthesis of a seismic signal by actually measuring one down-hole.

THE VSP

A vertical seismic profile is an extension of the velocity check shot. Fig. 21–7 shows an idealized setup. A seismic source is stationary at the surface and quite close to the well head. At a number of receiver positions in the well, the signal is recorded. At each level the recorded signal, as a function of time, will contain first the primary down-going signal from the source and any multiples generated at the surface. At later times, the source

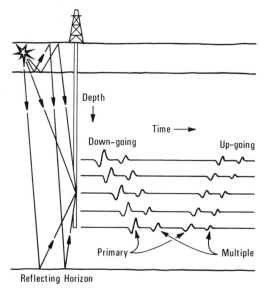

Figure 21–7. Schematic of a VSP recording showing typical traces recorded at five depths above a reflector. The prominent down-going and up-going waves for a single reflector are illustrated for the typical VSP recording. As depth increases, the first arrival occurs later and the reflected signal occurs earlier. Courtesy of Schlumberger.

signature (and its multiples) will reappear if a reflector exists below the tool position. In the case sketched, one such reflector is located below the greatest depth at which the wavetrains have been recorded. Comparison of successive recordings, as depth increases, shows items of interest. The first arrival (or first break) appears later in time than at the shallower depths. This is because of the increased transit time over the interval corresponding to the difference in depth between the two tool positions. Also, the reflected signal occurs earlier in time, because as the tool approaches the reflector, the time required for the return trip is less. The down-going and up-going waves in this sketch are seen to be approaching each other as the sonde depth increases. They will intersect or overlap when the tool is on depth with the reflector.

What is the lateral extent of the VSP measurement? A hint that the VSP data has some lateral extent can be seen in the sketch of Fig. 21–7 which shows the origin of the reflections at one depth. These reflections come from the so-called Fresnel zone, illustrated in Fig. 21–8. The Fresnel zone is that region of a spherically expanding constant phase signal which will arrive at an observation point within half a wave length of the direct signal. Thus the signal will originate from a nearly circular area of several hundred feet in diameter, depending on the distance from the detector to the reflector. Because of this fact, a synthetic seismogram which can be produced from the VSP data set will be much more representative of the area around the borehole than the synthetic seismogram produced from the sonic and density logs.

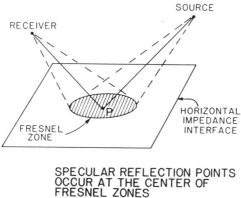

SPECULAR REFLECTION POINTS OCCUR AT THE CENTER OF FRESNEL ZONES

Figure 21–8. The lateral extent of the VSP measurement can be estimated from the size of the Fresnel zone. This corresponds to the surface of constant phase of the expanding spherical wave as it intersects a plane boundary. From Hardage.[4]

An estimate of the depth of investigation of the VSP can be made for the first arrivals of the down-going wave by considering the consequence of the spherical divergence of the wavefront. Consider the situation in Fig. 21–9: a geophone is at some position b below a reflector which is at depth a. Any point on the surface can be taken as a source of secondary wavefronts which will arrive at the geophone. Depending on the offset of the point, the wavefronts will arrive out of phase with the first arrival. Those which arrive at the geophone less than $\frac{\lambda}{2}$ out of phase will contribute to the amplitude of the first arrival. How far off-axis, x, can the secondary signal come from, in order to satisfy this criterion?

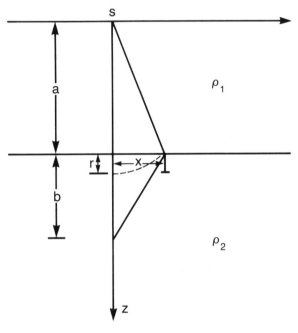

Figure 21-9. Estimation of the size of the Fresnel zone for the geometry of the VSP.

The spherical wave originating from point x is in phase with the spherical wave (of radius a+r) at a distance r below the interface, as indicated in the figure. The maximum path length condition can be written as:

$$b - r + \frac{\lambda}{2} = (b^2 + x^2)^{1/2}.$$

To eliminate the distance r from this expression, we note that:

$$a + r = (x^2 + a^2)^{1/2} = a\left(\frac{x^2}{a^2} + 1\right)^{1/2},$$

and consider the case where x<<a, which yields,

$$r \approx \frac{x^2}{2a}.$$

This may be put into the original path condition:

$$b - \frac{x^2}{2a} + \frac{\lambda}{2} = (b^2 + x^2)^{1/2}.$$

Using a similar criterion that $x << b$, we can write:

$$b - \frac{x^2}{2a} + \frac{\lambda}{2} = b + \frac{x^2}{2b},$$

which upon simplification gives:

$$x = \sqrt{\lambda \frac{ab}{a+b}}.$$

This zone of investigation attains a maximum for regions equidistant between source and receiver. For the case of the geophone anchored at 4000′ with an assumed wavelength of 200′, the zone of investigation is seen to have a radius of about 400′.

Now that we have examined the lateral resolution of a single WST measurement, what should the depth spacing of the multiple shots be? The answer to this will depend upon the application. For any type of imaging in which it will be important to separate up-going from down-going waves, the wavefield must be sampled with a resolution somewhat smaller than half a wavelength. Thus the sampling interval will depend on the highest frequencies present and on the local velocity. A rough estimate is that a shot is needed every 50′ for imaging work.

Now we examine a real VSP data set. Shown in Fig. 21–10 is a nearly 7000′ section where the seismic wavetrain has been recorded every 20′. The direct arrivals are seen clearly at the left edge of the trace. The slightly parabolic envelope running through these first arrivals indicates that the velocity is steadily increasing with depth. A number of near-surface down-going reflections can be perceived running parallel to the first arrival envelope but with reduced amplitude.

The up-going reflections, which have a mirror image slope compared to the down-going waves, can easily be traced to the levels at which they originate. Along the top of this display is a portion of the surface seismic data taken from a region of shots quite close to the well. It can be seen that there is a good correlation between the VSP trace and the seismic section; furthermore, the surface seismic events can be traced to their origin in the well.

It is apparent that VSP data will be of immediate use when tied into surface seismogram data. However, a VSP record as naturally recorded and presented is not particularly convenient to use. Although Fig. 21–10 is a particularly good example of a VSP, this method of raw data presentation is analogous to presenting the measured gamma ray spectra at each depth rather than the extracted concentrations of Th, U, and K. As in the case of other logging devices, a certain amount of signal processing must be done to provide a more convenient and useful result. For this reason, the next section briefly discusses basic imaging and provides a look at some simple methods for transforming the VSP display for correlation with subsurface logging data and for use in a splice with surface seismic data.

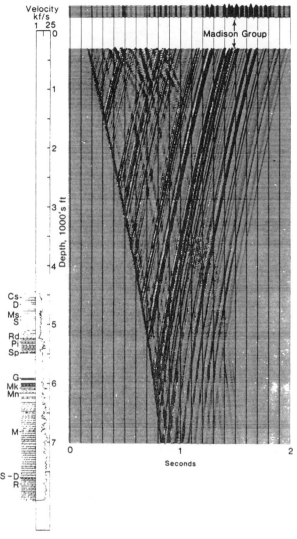

Figure 21-10. Raw data from a VSP over a 6000' zone. It shows a number of clear reflecting horizons which have been correlated with lithologic changes determined from other logs. From Balch and Lee.[5]

Rudimentary VSP Processing

The primary goal of a rudimentary processing is to separate the recorded wavefield into up-going and down-going waves: The up-going reflections have the information content for comparison with the surface seismic traces. The process involves determining the form of the down-going signal and then subtracting it from each trace, leaving an up-going wavefield. How can this most easily be accomplished? One way is to determine an average down-going waveform by aligning all of the WST traces starting with the first break. Assuming that the gain of each recording level has been adjusted so that the first break amplitudes are the same, then a summation over all the waveforms at a fixed time can provide an averaged down-going signal. The down-going waveform will be repeated in phase at each level, while reflected signals will not be aligned. This simple summation procedure will effectively cancel out any but the down-going waves.

To illustrate this process, refer to the upper portion of Fig. 21-11, which shows the raw WST waveform as initially recorded. In the upper trace, five sets of fairly strong down-going waves can be identified. Two sets of up-going waves, which are much fainter, can also be seen. On the lower left the original WST data has been aligned with respect to first break times. Now the down-going waves are clearly seen as vertical bands. It is not difficult to imagine that summing along each of the columns will produce a representative down-going waveform (after suitable normalization), which can then be subtracted (with the appropriate gain) level by level from the raw waveforms.

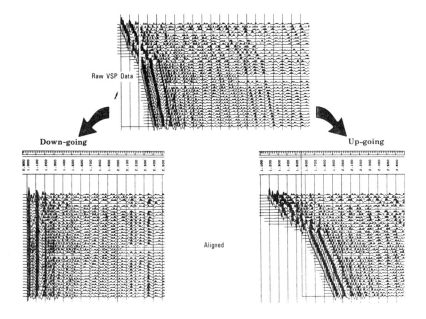

Figure 21-11. Raw VSP data with two simple types of processing for the emphasis of down-going and up-going waves. Courtesy of Schlumberger.

Another way to emphasize up-going events occurring on the VSP record and to make them immediately compatible with surface seismic signals is to shift each trace by an amount equal to the first arrival time. In this manner each reflection is shifted by the amount of time required to reach the surface (which is equal to the time for the source signal to travel from the surface to the level in question). This puts each trace on a true two-way travel time basis. In this case, up-going waves occur at the same time on each trace and thus present themselves as vertical bars. The VSP wavefield in the lower right of Fig. 21-11 illustrates the effect of the two-way travel shift on the original data. Although the down-going wavetrain is still present at the beginning of each wavetrain in this illustration, two sets of up-going reflections are indicated.

A convenient display for surface seismic comparison, referred to as a limited time window or "corridor stacking" illustrates the simple concepts just presented. For the case in question, see Fig. 21-12 for the raw VSP waveforms. Once again the first arrivals are clearly visible, as well as one prominent reflection originating at the very bottom trace. In order to emphasize the up-going waves, the right panel of Fig. 21-12 indicates the two-way transit time shift of the same data. Despite the presence of the down-going signal, the characteristic columns of up-going waves are clearly visible.

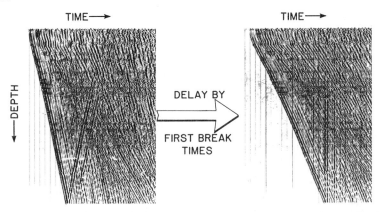

Figure 21-12. A recorded VSP wavefield and its first break time delay representation, which emphasizes up-going waves. From Hardage.[4]

It is clear from inspection of Fig. 21-12 that each trace, at later times, contains many multiples and other events which can be regarded on a first cut as noise. Thus it is convenient to deal with merely a portion of the early arrival part of the wavetrain. In Fig. 21-13, the arbitrarily chosen early portion of the two-way shifted data is presented after subtraction of the estimated down-going waveform. Comparison of Fig. 21-13 with the right panel of Fig. 21-12 shows the dramatic increase in visibility of the up-going waves.

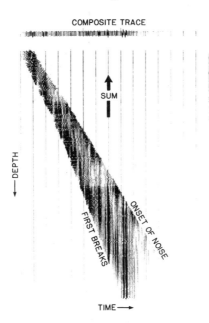

Figure 21-13. An example of corridor stacking. A selected envelope of the signals from Fig. 21-12 have had the average down-going signal subtracted. Horizontal summing of the traces produces a single composite trace at the top of the figure. If suspended vertically, the composite trace will resemble an ordinary surface seismic section near the region of the well. From Hardage.[4]

The final step in the production of a trace compatible with a surface seismic section is the summation, at constant times, over all depth intervals contained within the selected envelope. The result of this summation is shown at the top of Fig. 21-13. This artificially produced trace may then be spliced into a seismic section in the vicinity of the well. Fig. 21-14 shows such an example. Here the corridor-stacked trace has been duplicated a number of times for easy visual comparison with the surrounding surface seismic data.

More sophisticated forms of signal processing are usually employed on the VSP wavefield to separate the up-going and down-going waves. They are extensively discussed in Reference 4. For determining the signal of the down-going field, a nonlinear digital filter known as a median filter can be used. A second technique involves the use of a two-dimensional Fourier transform to convert the recorded wavefield to the space of frequency and wave number. In this representation, the down-going portion of the transformed data can be suppressed before the inverse transform. The result is a wavefield representative of up-going signals only. Predictive deconvolution is also used to remove unnecessary obscuring detail from the recorded trace. Due to the "ringing" of the source and possible multiple reflections at the surface, the down-going signal can extend over a considerable time interval, easily obscuring the reflection of two closely spaced beds. This type of deconvolution involves the generation of a pulse representative of the source waveform at depth. Correlation of this pulse with the recorded wavetrain effectively produces a trace similar to one obtained with a sharp source signature.

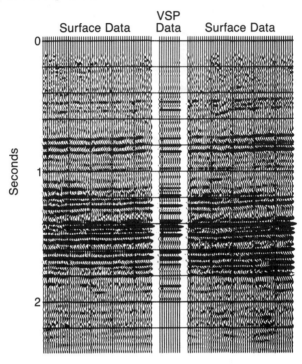

Figure 21-14. Another example of corridor stacking in which the final result has been spliced into a conventional surface seismic section. From Balch and Lee.[5]

Information from the VSP

What practical information can be extracted from the VSP? First, from the down-going waves the direct travel time can be established. The second major use is in the elimination of reflected multiples which can corrupt both VSP-processed data and surface data. An example of the type of event which can be identified and suppressed is shown in Fig. 21–15. In the layer between the two indicated reflectors, there can be multiple reflections which, seen from the surface, will simply appear as a later (and therefore deeper) reflection. However, in this case the arrival does not correspond to an actual reflector. In the figure, the up-going and down-going waves are drawn to two positions of the tool. In the upper location, above the reflecting layer, one down-going and three up-going waves will be recorded. Below the layer, the ray diagram indicates that there will be only two down-going events observed. The simple two-way alignment procedure is sufficient to identify those multiple up-going reflections, since they disappear from the traces at depths below the reflector level.

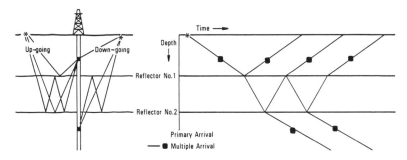

Figure 21-15. Idealized waveform of multiples generated between two subsurface boundaries. Courtesy of Schlumberger.

The direct arrivals also allow, through comparison between records, an analysis of the variation in amplitude changes. This allows determination of the absorption of the various layers which can be used in gain control and reconstitution of surface traces.

The most interesting use of the zero offset VSP is to identify reflectors below the final depth of the well. These may be layers invisible on surface seismic sections because of the attenuation over the two-way travel path. In some cases one is able to identify the existence of overpressured zones ahead of the drill bit. Fig. 21-16 shows an example of how this might be used. In

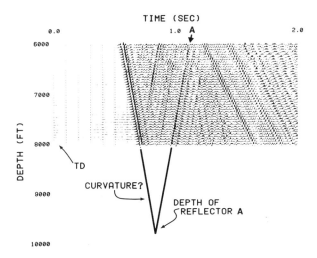

Figure 21-16. Extrapolating VSP data to identify reflectors at depth below the total well depth. From Hardage.[4]

this raw waveform presentation, two strong up-going waves are noticed. One of them comes from a depth just above 8000', and the other, which occurs somewhat later in time, seems to originate from a depth which can be estimated to be near 10,000'. The uncertainty in this type of extrapolation will come from variation in the velocity in the layers below the recorded data.

Increasing the Lateral Investigation: Offset VSP

There are several applications for the use of lateral extension of VSP data. However, they all basically attempt to define the geological structure at some distance from the borehole. The questions may concern the structure of a layer traversed by the borehole, to determine if a fracture occurs in it, or if it is the flank of a producing structure, the location of the next well. Another application is for the localization of a salt dome or a possible producing zone which has not been traversed by the borehole.

Although we have seen that there is some vertical extent to the signals detected by the VSP, this is mainly an accident related to the spherical spreading of the wavefront and the wavelengths involved. It is possible, however, to extend the vertical resolution of the VSP by the simple expedient of removing the surface source to some distance from the well head. This will emphasize reflections from points between the well and the source. Operationally there are two basic approaches to increasing the lateral coverage from a vertical well; one uses a single source position and various receiver locations, and the other a fixed measurement depth with various source positions. The second case can be more practical in offshore work than on land. Both cases are illustrated in Fig. 21–17 along with a schematic indication of the lateral zone covered by the measurement.

To understand this coverage quantitatively, we must use a model of the subsurface. For simplicity, we consider only parallel nondipping beds. The question to be answered is the following: Assuming a source at some substantial distance from the well head, where does the ensuing seismic trace, measured down-hole, come from as a function of time?

Fig. 21–18 sketches the geometry for determining the locus of all possible reflection events as a function of time. At each depth x, the distance y from the well bore is determined to yield the minimum travel time from the source to the point y and back to the geophone. The simple result obtained for the radial position of the reflector y for any depth x is:

$$y = \frac{Y_o x - X_o y_o}{(2x - X_o)},$$

where Y_o is the source offset distance and X_o is the depth of the measurement point. This is seen to behave correctly in the two limits. At depth x, equal to the location of the geophone, the offset reflection comes at the borehole.

Extending Measurements from the Borehole   **511**

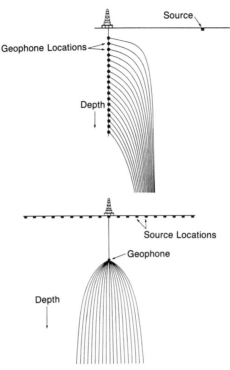

Figure 21–17. Lateral coverage from two types of offset VSP configurations. In the top figure a single receiver station is used with a multiple offset source position. In the lower figure, a single offset source position is used for a number of geophone locations in the well. Adapted from Dillon and Thomson.[7]

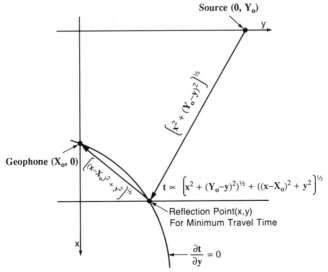

Figure 21–18. Geometry for locating the locus of points of equal travel time for the offset VSP.

For very great depths, the reflection point is seen to come from halfway between the source and detector.

Fig. 21–19 illustrates the coverages possible for a single geophone station at a depth of 500', and a multitude of source positions ranging between 500'

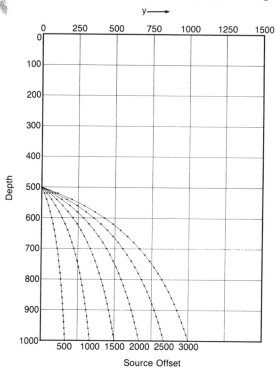

Figure 21–19. Offset coverage for a geophone station at a depth of 500', with source locations ranging from 500' to 3000' from the vertical position.

and 3000' from the well head. Fig. 21–20 illustrates the coverage for a fixed source at 3000' from the well head for a number of measurement stations between 500' and 1500'. The coverage lines of the previous figures suggest a method of imaging the data. One simply plots, at a given station depth, the wavetrain as a function of time along the appropriate curve, as indicated in the previous figures. An example of such a display is shown in Fig. 21–21, in which a number of apparently horizontal reflectors are immediately obvious.

Forward modeling, using ray tracing, provides a means of interpreting the VSP wavetrains. The following simple example of a fault with a 200' throw, shown in the left portion of Fig. 21–22, illustrates the combination of modeling and waveform interpretation. From the raypath analysis for the fixed source, the predicted wavetrain is computed and displayed alongside. The fault is clearly visible in the normal VSP presentation. In the two-way time presentation, it should be visible as an offset of the up-going wave.

Extending Measurements from the Borehole   513

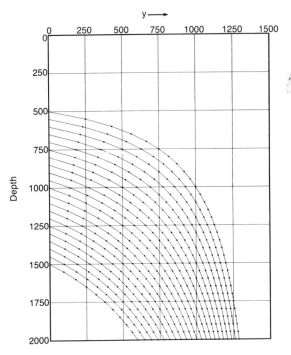

Figure 21-20. Offset coverage for a source located 3000' from vertical for a range of station depths between 500' and 1500'.

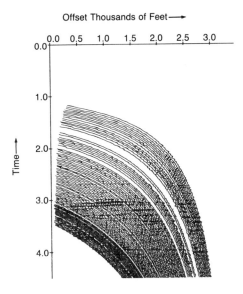

Figure 21-21. Example of an offset VSP data display in which each recorded waveform has been plotted along the line of minimum travel as computed in the previous two figures. From Dillon and Thomson.[7]

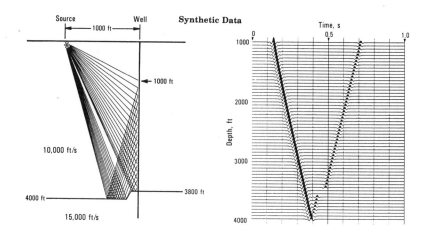

Figure 21-22. Use of synthetic seismograms to predict the effect of a pinchout on a hypothetical VSP experiment. Courtesy of Schlumberger.

In addition to the two possible types of offset VSP experiments, another alternative exists in deliberately deviated wells. In this case, there is the possibility of examining the reservoir laterally. One possibility for collecting the data, for easy imaging, is to position the surface source precisely (there is some relaxation on this as a result of the Fresnel criteria) above the tool location for any given depth. With some planning and by proper positioning, it would be possible to obtain measurements along equally spaced distances from the well, as indicated in the upper part of Fig. 21-23. To illustrate the possibility of imaging from this type of experiment, consider a hypothetical reservoir, which contains three prominent reflectors; the uppermost is horizontal, while the two lower ones are dipping, and one does not cross the well bore. For this model, the anticipated set of wavetrains for first arrival and the subsequent up-going arrivals is shown in the bottom portion of the figure. In this presentation, the traces are spaced in scale with their true vertical depth.

A quick visualization can be made by isolating the up-going waves and putting them on a two-way transit time basis, as shown in the upper portion of Fig. 21-24. A quite realistic seismogram can be obtained from this display by flipping it about the dotted diagonal representing the line of first break times. This is shown in the lower portion of Fig. 21-24. In this case, each vertical trace represents a distance offset from the well at which the source was located, and the track of the deviated well bore can be sketched in. The reconstruction accurately reflects the model input, as can be seen by comparison with Fig. 21-23.

Extending Measurements from the Borehole   515

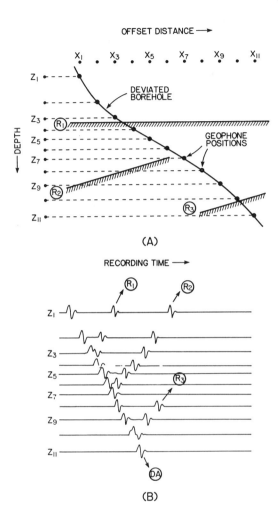

Figure 21–23. Use of a deviated well to obtain a VSP mapping. The source is to be positioned above the geophone location at each point. The location of VSP recording locations and corresponding shot points are indicated. In the lower portion of the figure, the expected waveform is sketched in the conventional VSP presentation at each vertical depth. From Hardage.[4]

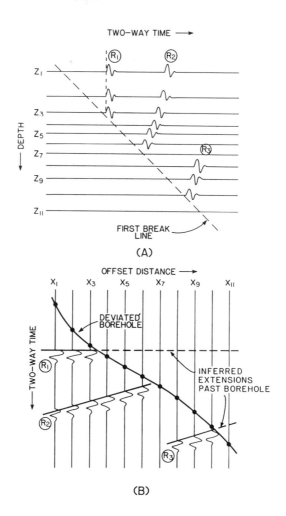

Figure 21–24. Generating an image of the subsurface in two steps. First the conventional VSP traces of Fig. 21–23 are aligned to emphasize upgoing waves. In the bottom figure, the traces are then suspended vertically according to their shot-point position, and the three reflectors appear. From Hardage.[4]

Appendix

ACOUSTIC IMPEDANCE AND REFLECTION COEFFICIENTS

The concept of acoustic impedance can be thought of in analogy with electrical resistance. In the case of resistance or resistivity, a voltage applied to a material produces a current. The characteristic of the material which relates the produced current to the voltage drop is the resistivity. For the acoustic analogy, an acoustic (or pressure) wave is applied to a material. A subsequent wave of local particle displacement or particle velocity is produced. What is the characteristic of the material which links these two phenomena?

Before performing the algebraic manipulation which will give the desired result, what can we obtain from physical reasoning or dimensional analysis? We wish to find an expression linking pressure P to local particle velocity $\dot{\xi}$, which is the time derivative of the local particle displacement coordinate ξ. The form of the equation sought is:

$$P = z\,\dot{\xi}.$$

What are the dimensions of z? Since the last term in the equation has dimensions of length (L) per unit time (T), and the first has force (ML/T^2) per unit area (L^2), we can write:

$$[P] = [z]\,[\dot{\xi}],$$

or

$$\left[\frac{MLT^{-2}}{L^2}\right] = [z]\left[\frac{L}{T}\right].$$

Thus the property of the formation sought has the dimensions of:

$$[z] = \left[\frac{\dfrac{ML}{L^2}\dfrac{1}{T^2}}{\dfrac{L}{T}}\right] = \left[\frac{M}{L^3}\right]\left[\frac{L}{T}\right],$$

which is suggestive of the product of density and velocity.

To obtain an expression for z in terms of the material properties, refer to Fig. 21–25 which shows a pressure pulse impinging on an elemental volume of surface area δA. The equation of motion for this volume can be written from a consideration of the pressure drop along the dimension z as:

$$\frac{\partial P}{\partial z}\,\delta A\,\Delta z = \rho\,\delta A \Delta z \frac{\partial \dot{\xi}}{\partial t}.$$

Since this wave equation has solutions which are functions of (z - vt), we can assume the following forms:

$$P = P_0 e^{i(z - vt)}$$

and

$$\dot{\xi} = \dot{\xi}_0 e^{i(z - vt)}.$$

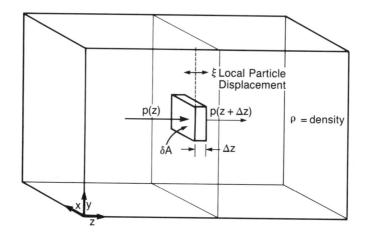

Figure 21–25. An elemental volume in an elastic medium which undergoes displacement due to the passage of a pressure wave.

Performing the required differentiation yields, when replaced in the equation of motion:

$$iP_0 = i\rho v \dot{\xi}_0.$$

Thus the acoustic impedance is seen to be the product of velocity and density, ρv.

This information can be used to obtain an expression for the reflection coefficient in the case of normal incidence. For this we need to consider the boundary conditions. Fig. 21–26 illustrates an incident and reflected pressure wave and a transmitted wave. Since the pressure must balance at the interface (to avoid infinite accelerations), we can immediately write this condition as:

$$P_i + P_r = P_t.$$

We must also have continuity of particle displacement or velocity, so:

$$\dot{\xi}_i - \dot{\xi}_r = \dot{\xi}_t,$$

where we have noted that the sign of the reflected particle velocity $\dot{\xi}_r$ is negative.

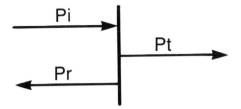

Figure 21-26. The boundary conditions for pressure at an interface. The transmitted pressure must be equal to the sum of the incident and reflected pressure.

Using the acoustic impedance relationship, the pressure continuity equation may be rewritten as:

$$\frac{\dot{\xi}_i}{\rho_1 v_1} + \frac{\dot{\xi}_r}{\rho_1 v_1} = \frac{\dot{\xi}_t}{\rho_2 v_2}$$

$$= \frac{\dot{\xi}_i - \dot{\xi}_r}{\rho_2 v_2}.$$

This expression may easily be solved for the reflection coefficient R, which is defined as the ratio of particle velocity $\dfrac{\dot{\xi}_r}{\dot{\xi}_i}$ and is found to be:

$$R = \frac{\rho_2 v_2 - \rho_1 v_1}{\rho_1 v_1 + \rho_2 v_2}.$$

References

1. Runge, R. J., Worthington, A. E., and Lucas, D. R., "Ultra-Long Spaced Electric Log (ULSEL)," Paper H, Trans. SPWLA Annual Logging Symposium, 1969.

2. Kretzschmar, J. L., Kibbe, K. L., and Witterholt, E. J., "Tomographic Reconstruction Techniques for Reservoir Monitoring," Paper SPE 10990, SPE Annual Technical Conference, 1982.

3. Somerstein, S. F. et al., "Radio-frequency Geotomography for Remotely Probing the Interiors of Operating Mini- and Commercial-sized Oil-shale Retorts," *Geophysics*, Vol. 49, No. 1, 1984.

4. Hardage, B. A., "Part A: Principles," in *Seismic Exploration*, Vol. 14A: *Vertical Seismic Profiling*, Geophysical Press, London and Amsterdam, 1983.

5. Balch, A. H., and Lee, M. W., *Vertical Seismic Profiling: Techniques, Applications, and Case Histories*, IHRDC, Boston, 1984.

6. Dobrin M. B., *Introduction to Geophysical Prospecting*, McGraw-Hill, New York, 1976.

7. Dillon, P. B., and Thomson, R. C., "Image Reconstruction for Offset Source VSP Surveys," 45th Annual EAEG Meeting, Oslo, 1983.

INDEX

A

Absorption cross section
  mass normalized, 284
  thermal neutron macroscopic, 234, 239, 247, 250, 258, 279, 281, 282, 285, 457
Acoustic amplitude, 492
Acoustic attenuation, 493
Acoustic impedance, 201, 494, 499, 517-518
Acoustic logging, 339-412
Acoustic velocity, 359, 360
  compressional velocity, 350-353, 360, 365, 370-371, 375, 391, 400, 402
  shear velocity, 352, 365, 371, 375, 381, 394, 400
Acoustic waves, 9, 342, 352, 359, 370, 489, 494
Activation, 162, 441, 458, 461-463
Aluminum, 173, 307, 440, 463
  activation, 458
Anhydrite, 31, 225, 297, 417, 428
Annihilation, 171, 176
Archie, G. E., 61
Archie relation, 57, 63, 64 439, 471, 478, 479, 484

B

Barite, 19, 188, 222, 225
Barn, 166, 213, 282
Bed thickness, 52, 87, 119, 384
  shoulder bed, 75, 82, 83, 85, 90, 122
Beryllium, 233, 243
Borehole diameter, 24, 82, 388
Borehole environment, 13-18, 201, 490
  borehole rugosity, 26, 398
  compensation, 388-389

Boron, 239-240, 284, 290, 456, 457, 460
Bulk density, 8, 21, 161, 170, 172, 197, 201, 207-211, 228, 234, 258-261, 418, 420, 431
  and acoustic properties, 366, 369, 385
  and gamma ray attenuation, 201-204
  and porosity, 202, 222

C

Calibration, 59, 181, 186, 187, 209, 251, 252, 253, 322, 498
Caliper, 24, 29, 211, 385
Capacitance, 142
Capture cross section,
  thermal neutron, 239, 281, 284, 286, 291, 304, 446, 447, 457
Capture gamma rays, 162, 285, 287, 288, 297, 441, 458, 463
Carbon, 176, 185, 231, 281, 294, 296, 297
C/O, 296
Carbonates, 34, 267, 291, 337, 384, 399, 445
Casing, 281, 342, 409, 458
Cation exchange, 139-140, 185, 442-443, 446, 471, 481
Cement, 7 18
  cased hole, 290, 342, 381, 409, 457-458
  rock fabric, 220, 356, 417, 445, 459
Cementation exponent, 64, 71, 72, 152, 471, 473-474
Chlorine, 49, 239, 281, 284, 290, 297, 442, 457
Clay mineral, 8, 138, 139, 181, 183, 185, 189, 190, 192, 194,

521

222, 254, 439-466, 481
Clays, 5, 8, 42, 124, 190, 239, 254, 284, 390, 439-466, 478-485
  dispersed, 444-445, 448
  laminated, 355, 448
  structural, 444
Compensation, 30, 206, 211, 250, 278, 385, 388-389
Compton scattering, 167-169, 170, 172-177, 201-204, 213, 269
Conductivity, 1, 8, 38, 42, 43, 59, 64, 66, 99, 106, 108, 110-114, 118, 127, 139, 146-149, 152-153, 439, 441, 478-486, 491
  electrical conductivity, 1, 8, 486
  electrolytic, 38, 42, 43, 45, 481
Core analysis, 72, 190, 191, 406,
Core measurements, 59-61, 72, 462
Cross plot, 149, 192-195, 258, 418-432, 448-456, 461, 472
  neutron-density cross plot, 258, 419, 428, 430, 431, 449, 456 457-463
Cross section, 162-172, 202, 213, 220, 227-240, 243-247, 250, 258, 279, 281-286, 457
  macroscopic cross section, 170, 234, 281, 282
  microscopic cross section, 166

D

Decay constant, 163, 288
Decay rate, 288, 300, 409
Decay time, 175, 286-288, 299-304, 318, 333
Density,
  ρ_b, 21, 166, 170-173, 187, 201-211, 219, 222-225, 234, 262, 418-419, 425, 427, 430-432, 452
  $\Delta\rho$, 30, 206, 209, 210, 211, 385
  ρ_e, 202, 204, 220, 221, 430, 431, 452
  electron density, 201, 202, 204, 220, 221, 430
  matrix density, 201, 222, 427, 428, 452
  particle density, 39, 49, 349
Density log, 26, 30, 191, 201-222, 421-422, 475, 489 501
Depth of investigation, 25, 94, 111-114, 117, 124-125, 147, 197, 243, 264, 265, 269-277, 304, 321, 382-383, 394, 489-492, 501
Dielectric constant, 124, 142, 145, 146, 147, 148, 151-154
Dielectric log, 145
Dielectrics, 142
Diffusion, 46-49, 239, 245-252, 287-300, 327
Dipmeter, 126, 127, 128, 498

E

Effective stress, 370, 398, 401
Elastic collision, 230
Elastic constants, 345, 348, 350, 366
Elastic media, 339, 342, 344, 348
Electric current, 100
Electric field, 38-40, 47, 48, 65, 66, 68, 106, 107, 114, 116, 117, 143, 144, 171, 382
Electrochemical potential, 46
Electrode device, 25, 71, 72, 73, 90, 92, 99, 118, 123-125
Electrofacies, 433, 434
Electromagnetic wave, 114, 491
Electromotive force, 104
Electron density, 201, 202, 204, 220, 221, 430
Environmental correction, 121, 278, 322
Evaporites, 183

F

Faraday, 104, 106
Fast neutron, 228, 230, 234, 243, 247, 294
Filtrate, 17, 20, 22, 36, 49, 51, 95, 118, 128-133, 243, 333
  invasion, 17-25, 36, 92-98 119-121

Index 523

mud filtrate, 20, 22, 36, 49, 51, 95, 118, 128-133, 159, 222, 260-264, 333
Finite elements, 81, 90
Fluid identification, 400
Fluid properties 368
Fluid saturation,
 gas saturation, 261, 263, 264, 279, 365, 402
 water saturation, 6, 9, 16, 21, 22, 57-69, 123-134, 151, 159, 201, 259, 281, 284, 290-291, 296, 365, 379, 439, 445, 457, 472-476, 484-486
Flushed zone, 17, 18, 20, 22, 123, 134
Focused electrode device, 25, 73, 84, 87, 90, 99, 111, 112, 125
 spherical focusing, 87, 88, 125
Formation factor, 61, 62, 64, 129, 133, 140, 142, 478, 485
Fractures, 2, 6, 9, 128, 190, 342, 381, 397, 403, 404, 405, 406

G

GR, 21, 24, 29, 32, 181, 187, 188, 189, 291, 442, 448, 449, 463
Gamma ray detector, 173, 174, 181, 187, 298, 458
Gamma ray index, 188, 189, 190
Gamma ray interaction, 167, 171, 173, 175, 178
 Compton scattering, 167-174, 176-177, 201-204, 213, 269
 pair production, 167, 171, 176, 177, 205
 photoelectric absorption, 167-177, 201-202, 212-225 430, 432
Gamma ray log, 167-172, 176-181, 187-188, 190, 204, 213, 297, 448
 spectral gamma ray log, 187, 188, 190, 192, 460
Gamma rays, 161-178, 201-222
Gamma ray source, 205, 206, 227, 269

Gamma ray transport, 201 201, 244, 245, 269, 342
Gas, 6, 30, 45, 174, 240, 243, 259-263, 279, 302, 365, 368, 370, 401, 402, 421-422, 450
Gas detection, 422, 449
Gas effect, 30, 261, 422
Gauss, 143, 310, 313, 320, 338
Geiger counter, 186
Geochemistry, 2, 185, 417, 486
Geometric factor, 91, 92, 99, 100, 107, 109-118, 125, 197, 264-273,383, 491
Grain density, 8, 190, 431, 448, 455, 460

H

Half-life, 163, 180, 205, 458
He, 162, 233, 234, 240
Hooke, 344, 348
Huygen's principle, 352
Hydrocarbons, 1-9, 15-18, 30, 33, 37, 64, 132, 123, 146, 339, 484
Hydrostatic pressure, 19, 402, 405

I

Illite, 184, 185, 194, 254, 280, 440, 441, 445, 455, 456, 457, 458, 460, 463
Impedance, 201, 494, 495, 497, 499, 517, 518
Induced polarization, 483
Induction, 25, 99-120, 134, 272, 305, 313, 337, 491
Induction log, 25, 29, 100, 105, 119, 121, 134, 149, 304, 305
Inelastic scattering, 231
Invasion, 19, 25, 36, 92, 134
Ions, 46, 47, 49, 139, 140, 315, 327, 442, 482, 483
Iron, 18, 222, 239, 278, 315, 445, 446, 451, 454-460
 Fe, 441, 454, 462, 463

K

Kaolinite, 194, 254, 440-445, 455-460
Kozeny, 331, 332

L

Laplace, 79, 80, 81
Laterolog, 73, 84-96 118-122, 125-126, 134
Lifetime
  neutron, 284
  neutron die-away, 281, 284
Lithology, 168, 188, 201, 212, 217, 258, 297, 400, 417-435, 495
Lithology identification, 297, 417, 433
Logging speed, 3

M

Macroscopic cross section, 170, 234, 281, 282
Magnetic field, 100-107, 305-316, 321, 326, 327, 334, 382
  nuclear magnetism, 9, 304, 305, 306, 307, 308, 323, 331, 332, 333, 337
Magnetic moment, 305-313, 316, 319
Magnetic permeability, 153
Magnetic polarization, 306, 324
Magnetic resonance, 305, 337
Magnetism, 310, 312, 337
Magnetite, 225, 321
Magnetometer, 127, 305, 306, 307
Maxwell, 38, 116, 144, 152
Mean free path, 166, 197, 234, 269-273, 286, 490, 492
Membrane potential, 46, 49
Microelectrode, 124, 125, 126, 127
Microlaterolog, 125
Microresistivity, 94, 119, 125, 126, 159
Microscopic cross section, 166,
Monte Carlo, 197, 198, 215, 245, 264

Mudcake, 19, 20, 26, 30, 35, 36, 125, 206-211, 222, 249, 280, 333, 407
Mud filtrate, 20, 22, 36, 49, 51, 95, 118, 128-133, 133, 243, 333

N

Natural gamma ray, 269, 461, 462
Natural radioactivity, 32, 182, 463
Neutron capture, 231, 239, 281-285, 287, 293, 297, 457
Neutron decay time, 302
Neutron-density cross plot, 258, 419, 428, 430, 431, 449, 456
Neutron detector, 240, 250, 284
Neutron die-away, 281, 284
Neutron diffusion, 272, 287
Neutron energy, 162, 229, 230, 232, 240, 247, 293
  epithermal neutron, 228, 230, 240, 243, 247, 249, 261, 264, 271, 272, 284
  fast neutron, 228, 230, 234, 243, 247, 294
  thermal neutron, 228, 239-243, 247-251, 272, 279, 281, 284-287
Neutron induced
  activation 162, 441, 458, 461, 462, 463
  gamma ray 175, 227, 270, 281, 293
Neutron interaction, 228, 231, 232, 234, 240, 243, 293
  elastic scattering, 227, 230, 231, 232, 234
  inelastic scattering, 231
  radiative capture, 231, 232
  thermal capture, 281, 282
Neutron log, 231, 243, 253, 258, 267, 291, 439, 449, 454
Neutron logging, 231, 243-266
Neutrons, 8, 227-240, 243-266, 281-299
  diffusion length, 236, 238, 247, 250, 252, 272

diffusion of, 239, 245, 247, 250, 252, 272, 287-300
slowing-down length, 236, 238, 247-254, 257-263, 272, 279, 285, 422, 454, 455
Neutron source, 227, 233, 245, 253, 281
Normal device, 73, 78, 85, 88, 125, 491
short normal, 73, 75, 76-81, 94
Nuclear magnetic logging, 9, 304-335
Nuclear magnetic moment, 305
Nuclear magnetic resonance, 305, 337
Nuclide, 240

O

Ohm's law, 38, 39, 40, 41
Oil saturation, 16, 22, 71, 132, 284, 296, 307
Overburden, 201, 356, 360, 364, 369, 404
Overpressure, 6, 19, 381, 402, 442, 495, 509
Oxygen, 281, 294, 296, 440, 453

P

Pair production, 167, 171, 176, 177,
P_e, 168, 213, 218-223, 423, 430-432, 451, 452-454
205
Permeability, 9, 62, 307, 331, 332, 337, 370, 406, 408, 439, 445
Phase velocity, 155
Photoelectric cross section, 168, 213, 220
Photoelectric effect, 167, 168, 213
Photomultiplier, 174, 178
Photons, 162, 177
Poisson distribution, 164, 180
Pore space, 5, 6, 8, 16, 17, 19, 42, 43, 139, 243, 328, 439, 444, 454
Pore throat, 445
Porosity, 5-9, 16-22, 26, 27, 30-36, 53, 62, 64, 70, 71, 72, 94, 123, 128-134, 150-151, 159, 180, 201, 211-212, 222, 243-266 260-264, 267-271, 288, 290-291, 296, 307, 331, 333, 337, 339, 355, 359, 360, 365, 370, 381-400 417-435, 439, 441, 445, 449-459, 460, 472, 474, 478, 483-485
density porosity, 31, 53, 260-263, 420, 455-456, 460
neutron porosity, 21, 30, 31, 231-240, 243-268 385, 441, 446, 449, 454-466
secondary porosity, 399, 426
Porosity log, 133, 244, 253, 254, 355, 425, 472
Porosity unit, 26, 234, 439, 455
Potassium, 182-185, 188-195, 307, 442, 446, 455-463
Potential, 46, 49, 65-68, 76, 81, 85-89, 102, 181, 491
Pressure, 6-9, 19, 280, 345, 349-352, 355, 356, 360-365, 368-370, 375, 379, 382, 402-407, 517-518
Pressure wave, 371, 375, 407, 492, 518
Pulse height, 174
Pulsed neutron, 227, 231, 281-300, 462
Pyrite, 225

R

Radioactive decay, 163, 164, 180,
Radioactivity, 32, 163, 181, 182, 183, 185, 186, 188, 190, 463 285
Radius of investigation,
see Depth of investigation
Residual oil, 71, 132, 307, 333
Residual water, 130, 337
Resistance, 17, 37, 38, 39, 40, 41, 42, 55, 71, 76, 77, 517
Resistivity, 17-33, 37-52, 57-69, 73-96, 117-119, 123-154, 279, 402, 471-479 517

R_t, 17-22, 30, 63-66, 68-72, 76, 92, 118, 123, 130, 134, 160, 472, 474
R_{xo}, 18, 30, 92, 123, 128, 130, 132, 134, 152
Resistivity index, 484, 486
Resistivity log, 25, 126, 128, 192, 291, 472, 476
  shoulder, 20, 75, 82, 83, 85, 90, 119
Rugosity, 26, 398

S

Salinity, 19, 46, 56, 146, 280, 290
Saturation, see Fluid saturation
Scattering, 162, 167-176, 202, 205, 215-217, 227, 230, 244, 245, 247
Secondary porosity, 399, 426
Semblance, 73, 300
Shale, 9, 21, 24-29, 49, 51, 138, 181, 184-192, 254, 260, 261, 267, 331, 390, 402, 439-466, 478-479, 480-485
Shaly sand, 451, 486
  dual water model, 130, 337, 483, 486
Shear modulus, 346-348, 352, 365, 368-369, 375, 379, 403-404
Shear stress, 347
Shear velocity, 352, 365, 371, 375, 379, 381, 394, 400
Σ, 38, 47, 66-67, 80, 106-108, 116-118, 143, 148-154, 164-172, 202, 228, 234, 239, 244, 245, 247, 279-291 299, 300-304, 345-348, 457, 459, 460
Silicon, 297, 440, 451, 453, 463
Skin depth, 117, 147, 153, 490-492
Skin effect, 100, 114, 117, 121, 152
Sodium chloride (NaCl), 19, 42, 43, 46, 50, 56, 71, 201, 279, 284, 302, 303, 304, 379, 483
Sonic log, 18, 21, 26, 341, 374, 381-411, 418, 472, 492, 498-499
SP. 18, 21, 24, 27, 29, 32, 49, 50, 51, 119, 472
Spectroscopy, 161, 176, 181, 190, 281, 293, 297, 441, 461-462
Spontaneous potential, 8, 18, 21, 36, 37, 38, 46, 49, 58
Statistics, 162, 163, 164, 165, 297, 490
Stoneley, 370, 374, 375, 381, 396, 406
Strain, 344, 346, 347, 348, 350, 352, 360
Stress, 342, 344, 345, 347, 348, 360, 364, 370, 390, 398, 401, 404, 405
Sulfur, 297, 307
Surface area, 6, 44, 443-447, 483
Synthetic seismogram, 499, 501

T

Tadpole plot, 127
Televiewer, 408
Tension, 343, 344, 404, 489, 500, 510
Thermal decay time, see Lifetime
Thermal neutron, 228, 230, 236, 239, 240, 243, 247, 249, 250, 251, 258, 261, 264, 271, 272, 281-287, 297, 299, 302, 418, 454-455
Thorium, 182, 183, 184, 185, 190, 192, 194, 195, 463
Time-average, 341, 355, 359, 360, 364, 370, 379, 397
Time constant, 286-288, 300, 314, 316-318, 321-324
Time-lapse, 291
Tortuosity, 64, 331
Transducers, 382, 389, 408, 409
Transition zone, 20
Transport,
  gamma ray, 201
  neutron, 244-245
Tube wave, 375, 406

U

U, macroscopic photoelectric cross section, 220-222, 431-433. 468
Uranium, 182, 183, 184, 185, 190, 191, 463

V

Variable density, 209
Variable intensity (VDL), 391, 406,
Velocity,
  compressional velocity, 350-353, 360, 365, 370, 371, 375, 379, 391, 400, 402
  shear velocity, 352, 365, 371, 375, 379, 381, 394, 400
Velocity log, 341
Velocity survey, 340, 341
Viscosity, 6, 44, 45, 46, 322, 327, 333, 370, 408
Volume fraction, 8, 21, 35, 134, 188, 220, 221, 322, 331, 379, 430, 439, 446, 455, 480

W

Water,
  bound, 483
Water resistivity, 64, 94, 118, 151, 152
Water saturation, 6, 9, 57, 59, 62, 64, 71, 130, 281, 284, 290, 291, 365, 472-476
Wavelength, 349, 408, 492, 493,
Wave velocity, 148, 151, 365, 370, 375, 391, 403, 406 497, 502, 503, 510
Waxman-Smits, 140, 483, 486
Weight fraction, 454
Wyllie, 341, 355-359, 370, 397

X

X-ray, 441, 463

Y

Young's modulus, 344, 345, 349, 350, 352

CREDITS

The author thanks the many publishers and firms that allowed the use of tables and figures. Credit is also due to Schlumberger for permitting the use of illustrations from their numerous publications.

Adams, JS and Weaver, CE. Bull AAPG, 42, 1958; with permission American Association of Petroleum Geologists; Fig. 9.4

Almon, WR. Paper WW, SPWLA Symposium, 1979; Fig. 19.4

Almon, WR and Davies, DK. Clays and the Resource Geologist, Longstaffe, FJ. ed. Mineralogical Assn. of Canada, Toronto, 1981; Table 19.1

Anderson, B. Paper II, SPWLA 27th Ann Logging Symposium, 1986; Fig. 6.12a

Archie, GE. Trans. AIME 146, 1942; copyright 1942 SPE-AIME; Figs. 4.6, 4.7, 4.8

Arnold, DM and Smith, HD Jr. Paper W, SPWLA 22nd Ann Logging Symposium, 1981; Figs. 12.1, 12.5

Asquith, GB and Gibson, CR. Basic Well Log Analysis for Geologists. AAPG, Tulsa, 1982; with permission of American Association of Petroleum Geologists; Figs. 3.10, 12.14, 12.15

Ausburn JR. Paper F, Trans. SPWLA, 1977; Figs. 17.12, 17.24

Balch, AH and Lee, MW. Vertical Seismic Profiling: Techniques, Applications, and Case Histories. IHRDC, Boston, 1984; Figs. 21.5, 21.10, 21.14

Berner, RA. Principles of Chemical Sedimentology. McGraw-Hill, New York, 1971; Fig. 20.9

Bertozzi, W, et al. Geophysics, 46(10), 1981; Figs. 10.11, 10.13, 10.14, 10.15, 10.16, Table 10.1

Bloch, F. The Physical Review, 70, 1946; Fig. 14.6

Brindley, GW. Mineralogical Assoc. of Canada, Toronto, 1981; Fig. 19.1

Brown, RJS. Nature, 189; copyright 1961 Macmillan Magazines Limited; Fig. 14.18

Chemali, R, et al. Paper UU, Trans SPWLA Ann Symposium, 1983; Figs. 5.19, 5.20

Chemali, R, et al. Ninth S.A.I.D. Colloquium, Paper 13, 1984; Fig. 17.2

Clavier, CL, et al. JPT, 23, 1971; Tables 13.3, 13.4

Clavier, C, et al. Society of Petroleum Engineering Journal, April 1984; Figs. 7.14, 7.15, 7.16

Davis, JC Jr. Advanced Physical Chemistry Molecules, Structure and Spectra, Ronald Press, New York, 1965; copyright 1965 John Wiley & Sons, Inc; Table 14.1

Delfiner, PC, et al. Paper SPE 13290, SPE 59th Ann Technical Conference, 1984; copyright 1984 SPE-AIME; Figs. 18.15, 18.16

Dewan, JT. Essentials of Modern Open-hole Log Interpretation. PennWell Publ. Co., Tulsa, 1983; Figs. 2.3, 2.5, 3.8, 18.4, 18.5, 18.6, 18.7

Dillon, PB and Thomson, RC. 45th Annual EAEG Meeting, Oslo, 1983; EAEG, Blackwell Scientific Publ., Ltd., Oxford, England; Figs. 21.17, 21.21

Dobrin, MB. Introduction to Geophysical Prospecting. McGraw-Hill, New York, 1976; Table 21.1

Doll, HG. Pet. Trans. AIME 186, 1949; copyright 1949 SPE-AIME; Fig. 6.6, 6.11

Doll, HG, et al. Electrical Logging, in Petroleum Production Handbook, SPE, 1962; copyright 1962 SPE-AIME; Fig. 5.10

Domenico, SN. The Log Analyst, 18, 38–46 1977.

Dresser Atlas, Dresser Industries, 1983; Figs. 6.9, 6.10, 6.12, 7.8, 9.7

Edmundson, H and Raymer, LL. SPWLA 20th Ann Logging Symposium, 1979; Figs. 12.3, 12.4, 12.6

Ekstrom, MP, et al. Paper BB, SPWLA 27th Ann Logging Symposium, 1986; Fig. 7.5a

Ellis, DV. First Break, 4(3), 1986; Figs. 19.9, 19.10

Ellis, DV. SPE Petroleum Production Handbook, Bradley, H. ed. SPE, Dallas, 1987; copyright 1987 SPE-AIME; Figs. 8.1, 8.2, 9.6, 9.8, 9.9, 9.10, 11.1, 11.5, 11.7, 11.8, 11.11, 12.7, 12.9, 12.11, 12.12, 12.16, 12.17, 12.18

Ellis, DV, et al. Paper SPE 12048 SPE Ann Technical Conference and Exhibition, 1983; copyright 1983 SPE-AIME; Figs. 10.3, 10.8, 10.12

Evans, RD. The Atomic Nucleus. McGraw-Hill, New York, 1967; Figs. 8.7, 8.8

Feynman, RP, et al. Feynman Lectures on Physics, Vol. 1. Addison-Wesley, Reading, Mass., 1965; Figs. 11.9, 11.10

Feynman, RP, et al. Feynman Lectures on Physics, Vol. 2. Addison-Wesley, Reading, Mass., 1965; Figs. 3.6, 6.1, 6.3, 7.17, 14.1, 14.2, 14.4, 15.5, 15.6, 15.7, 15.8, 15.9

Fukishima, E and Roeder, S. Experimental Pulse NMR: A Nuts and Bolts Approach. Addison-Wesley, Reading, Mass., 1965; Fig. 14.8

Gardner, GHF and Harris, MH. Paper H, Trans. SPWLA 9, 1968; Fig. 16.8

Gardner, GHF, et al. Geophysics, 39, 770–80, 1974; Figs. 16.4, 16.5

Garrels, RM and MacKenzie, FT. Evolution of Sedimentary Rocks. W.W. Norton, New York, 1971; Fig. 9.3

Goetz, JF, et al. APEA Journal, 19, 1979; Figs. 17.10, 17.17

Hardage, BA. Vertical Seismic Profiling, Part A: Principles, 2 enl ed. Geophysical Press, Amsterdam, 1985; Figs. 21.8, 21.12, 21.13, 21.16, 21.23, 21.24

Havira, RM. Trans SPWLA 23rd Ann Logging Symposium, 1982; Fig. 17.31

Hearst, JR and Nelson, P. Well Logging for Physical Properties. McGraw-Hill, New York, 1985; Fig. 3.11

Herron, M. INDC(NDS)-184/GM, IAEA, Vienna, 1987; Figs. 19.15, 19.16, 19.17, 19.18

Hertzog, RC. Paper SPE 7430, 53rd Ann Technical Conference, 1978; copyright 1978 SPE-AIME; Figs. 13.8, 13.9, 13.10, 13.11, 13.12

Hilchie, DW. Applied Openhole Log Interpretation. DW Hilchie, Golden, Colorado, 1978; Figs. 20.3, 20.4

Hodsen, GW, et al. The Log Analyst, 17(1), 1976; Fig. 9.11

Hottman, CE and Johnson, RK. JPT June 1965; copyright 1965 SPE-AIME; Fig. 17.25

Jackson, JA. The Log Analyst, Sept–Oct, 1984; Figs. 14.26, 14.27

Kenyon, WE, et al. Paper SPE 15643, SPE Ann Technical Conference, 1986; copyright 1986 SPE-AIME; Fig. 14.14

Leslie, HD, and Mons, F. Paper GG, SPWLA 23rd Ann Logging Symposium, 1982; Fig. 17.23

Leverette, MC. Trans. AIME 132, 1938; copyright 1938 SPE-AIME; Figs. 4.3, 4.4, 4.5

Lynch, EJ. Formation Evaluation. Harper & Row, New York, 1962; Fig. 7.13, 17.7

Martin, M, et al. Geophysics, 3, 1938; Figs, 4.1, 4.2

Misk, A, et al. SAID, Third Annual Logging Symposium, June 1976; Fig. 17.11

Morris, CF, et al. Paper SPE 13285, 59th Ann Technical Conference, Houston, 1984; copyright 1984 SPE-AIME; Figs. 17.18, 17.19, 17.20

Neuman, CH and Brown, RJS. Journal of Petroleum Technology, 34 1982 copyright 1982 SPE-AIME; Figs. 14.23, 14.24, 14.25

Patchett, JG. Paper U, SPWLA Symposium, 1975; Fig. 19.2

Pople, JA, et al. High-resolution Nuclear Magnetic Resonance. McGraw-Hill, New York, 1959; Figs. 14.16, 14.17, 14.19

Poupon A, et al. JPT, July 1970; copyright 1970 SPE-AIME; Fig. 19.3

Ramo S, et al. Fields and Waves in Communication Electronics. John Wiley, New York, 1965; copyright 1965 John Wiley & Sons, Inc.; Fig. 7.19

Ransom, RC. The Log Analyst 18, 47–62, May–June 1977; Fig. 17.6

Ruhovets, N and Fertl, WH. Paris SPWLA Symposium, 1981; Fig. 9.12

Rydin, RA. Nuclear Reactor Theory and Design. University Publications, Blacksburg, 1977; Figs. 11.2, 11.3, 11.4

Scholle, PA, et al. AAPG Memoir 33, Tulsa, 1983; with permission of American Association of Petroleum Geologists; Fig. 2.19

Schultz, WE, et al. Paper CC, CWLS-SPLA Symposium, 1983; Figs. 13.3, 13.4

Schweitzer, JS and Manente, RA. American Institute of Physics Conf. Proc., vol. 125, 1985; Fig. 13.13

Seevers, DO. Paper L. SPWLA Trans, 1967; Fig. 14.20

Serra, O. Fundamentals of Well-Log Interpretation. Elsevier, Amsterdam, 1984; Figs. 5.14, 5.16, 7.1, 7.2, 7.4, 9.5, Tables 1.1, 9.1

Sherman, H and Locke, S. Paper Q, SPWLA 16th Ann Logging Symposium, 1975; Figs. 12.19, 12.20, 12.26

Staal, JJ and Robinson, JD. Paper SPE 6821 SPE of AIME, 1977; copyright 1977 SPE-AIME; Fig. 17.27

Thomas, DH. The Log Analyst, Jan–Feb 1977; Fig. 17.16

Timur, A. Journal of Petroleum Technology, 21, 1969; copyright 1969 SPE-AIME; Fig. 14.22

Timur, A. SPE Petroleum Production Handbook, Bradley, H. ed. SPE, Dallas, 1987; copyright 1987 SPE-AIME; Figs. 16.1, 17.3, 17.5, 17.13, 17.15

Tittman, J. Geophysical Well Logging. Academic Press, Orlando, 1986; Figs. 4.11, 17.1, Table 3U

Toksoz, MN, et al. Geophysics, 41, 621–45, 1976; Figs. 16.7, 16.10

Wahl, JS. Geophysics, 48(11), 1983; Fig. 9.15

Wahl, JS, et al. Paper presented at 39th SPE Ann Meeting, 1964; copyright 1964 SPE-AIME; Fig. 10.7

White, JE. Underground Sound: Application of Seismic Waves. Elsevier, Amsterdam, 1983; Table 15.1

Williams, DM, et al. Trans SPWLA 25th Ann Logging Symposium, 1984; Fig. 17.28

Winsauer, WO and McCardell, WM. Petroleum Trans. AIME, 198, 1953; copyright 1953 SPE-AIME; Fig. 20.8

Worthington, PF. The Log Analyst, Jan–Feb, 1985; Figs. 20.6, 20.7, 20.10, Table 20.1

Wyllie, HRJ, et al. Geophysics 21, 41–70, 1956; Table 16.1

Wyllie, MRJ, et al. Geophysics 23, 1958; copyright 1958 SPE-AIME; Figs. 15.13, 16.2, 16.3, 16.6

Yariv, S and Cross, H. Geochemistry of Colloid Systems for Earth Scientists. Springer-Verlag, Berlin, 1979; Fig. 20.9, Tables 16.1, 19.2